Predrag B. Petrović

The Riemann Zeta Function

Also of Interest

Algebra and Number Theory.
A Selection of Highlights
Fine, Moldenhauer, Schürenberg, Rosenberger, Wienke, 2023
ISBN 978-3-11-078998-0, e-ISBN (PDF) 978-3-11-079028-3, e-ISBN (EPUB) 978-3-11-079039-9

Topics in Complex Analysis.
Romik, 2023
ISBN 978-3-11-079678-0, e-ISBN (PDF) 978-3-11-079681-0, e-ISBN (EPUB) 978-3-11-079688-9

Number Theory.
Multiplicative and Additive with Factorization and Primality Testing
Schumer, 2025
ISBN 978-3-11-157867-5, e-ISBN (PDF) 978-3-11-157928-3, e-ISBN (EPUB) 978-3-11-157954-2

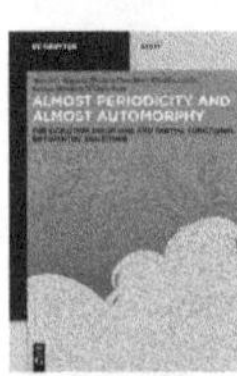

Almost Periodicity and Almost Automorphy.
For Evolution Equations and Partial Functional Differential Equations
Afoukal, Es-sebbar, Ezzinbi, N'Guérékata, 2022
ISBN 978-3-11-168361-4, e-ISBN (PDF) 978-3-11-168471-0, e-ISBN (EPUB) 978-3-11-168514-4

Casimir Force, Casimir Operators and the Riemann Hypothesis. Mathematics for Innovation in Industry and Science
Van Dijk, Wakayama (Eds.), 2010
ISBN 978-3-11-022612-6, e-ISBN (PDF) 978-3-11-022613-3

Predrag B. Petrović

The Riemann Zeta Function

Integrals, Multiple Sums, Mordell's Sums, Ramanujan's Formula

DE GRUYTER

Mathematics Subject Classification 2020
Primary: 11M06, 11M26, 11M36; Secondary: 11M41, 11N05, 11M50

Author
Predrag B. Petrović
University of Kragujevac
Department of Electronics
Faculty of Technical Sciences Čačak
Svetog Save 65
32000 Čačak
Serbia
predrag.petrovic@ftn.kg.ac.rs
https://orcid.org/0000-0002-8849-4230

ISBN 978-3-11-223326-9
e-ISBN (PDF) 978-3-11-223327-6
e-ISBN (EPUB) 978-3-11-223328-3

Library of Congress Control Number: 2026931063

Bibliographic information published by the Deutsche Nationalbibliothek
The Deutsche Nationalbibliothek lists this publication in the Deutsche Nationalbibliografie; detailed bibliographic data are available on the Internet at http://dnb.dnb.de.

Cover image: Jan Homann / https://de.wikipedia.org/wiki/Datei:Complex_zeta.jpg
Typesetting: VTeX UAB, Lithuania

www.degruyterbrill.com
Questions about General Product Safety Regulation:
productsafety@degruyterbrill.com

Preface

Highly intricate mathematical concepts often emerge from the pursuit of answers to questions that are, in essence, elementary. A prominent example is the *Riemann* zeta function. Its systematic study originated from investigations into the distribution of prime numbers [55, 224]. Prime numbers themselves are simple and familiar objects; however, their distribution reveals a striking level of complexity. Although identifying the next prime may appear straightforward at first glance, in practice it frequently proves to be far from trivial. For generations, mathematicians have sought a general law governing the distribution of prime numbers of arbitrary magnitude. This endeavor ultimately led *Bernhard Riemann* to apply the theory of complex functions as a framework for describing and analyzing the distribution of prime numbers.

The *Riemann* zeta function occupies a central role in numerous areas of complex analysis, particularly in number theory, including the study of irrational numbers and the distribution of prime numbers [59, 70, 127, 128, 220, 223, 224]. It also serves as a fundamental tool in signal analysis and in applied fields such as technology and cryptography, machine learning and artificial intelligence (AI). Historically [40, 150, 191, 192, 218, 231], significant attention has been devoted to closed-form evaluations of the *Riemann* zeta function at positive integers, since these special values determine essential properties of the associated mathematical and physical systems. In condensed matter physics, for example, the *Sommerfeld* expansion, used to compute particle numbers and the internal electron energy, involves the *Riemann* zeta function at even integers [82, 129]. Conversely, the spin–spin correlation function of isotropic spin-$\frac{1}{2}$ systems in the *Heisenberg* model [221] is expressed in terms of $\ln 2$ and the *Riemann* zeta function at odd integers [202, 205].

The *Riemann* zeta function underpins key theoretical aspects of modern machine learning and AI by providing analytic tools for understanding high-dimensional distributions, spectral properties, and kernel methods. Its connections to prime number theory and complex analysis inform randomness generation and hashing algorithms, which are critical in cryptographically secure training and optimization. Moreover, zeta-function-based series and functional relations appear in regularization schemes, loss landscapes, and neural network generalization bounds, enabling rigorous mathematical analysis of convergence, stability, and expressivity in large-scale AI models.

The computation of the *Riemann* zeta function and related series is of significant importance in computational mathematics [38, 51, 133, 140, 166, 214, 222, 226, 230], often relying on advanced software tools such as *Mathematica*. Classical approaches are based on the *Euler–Maclaurin* and *Riemann–Siegel* formulas [77, 181, 183, 185], yet new algorithms and techniques continue to be developed [211, 225, 227, 229]. In practice, most numerical methods are restricted to specific domains. Consequently, for the *Riemann* zeta function at odd integer arguments, specialized methods are required to relate these values to those at even integers. The monograph presents and develops such methods, offering a systematic framework and guiding future research toward new results in this domain.

https://doi.org/10.1515/9783112233276-202

This monograph is the culmination of decades of research by Professor Milorad Stevanović on the *Riemann* ζ function and its applications to the evaluation of diverse sums and integrals, which have relevance across various fields of science and technology [86, 198, 201, 202]. The complexity of this domain within complex analysis motivated the author to rigorously define and prove the fundamental properties of the function. The significance of this task is underscored by over two centuries of contributions from numerous researchers. Accordingly, the first three chapters are dedicated to these foundational aspects. The author has made a concerted effort to present a clear, innovative, and systematic exposition of the principal features of the *Riemann* ζ function, with all theorem proofs entirely original. These chapters establish a solid theoretical basis for the subsequent sections, where the focus shifts to the evaluation of multiple sums. The first two chapters survey the latest developments regarding the ζ function, while the next three chapters present the original contributions of Professor Stevanović and the author in determining explicit values of integrals and sums involving the *Riemann* zeta function.

The first problem addressed in detail concerns the computation of the coefficients $F_{(p,q)}(-1)$, $G_{(p,q)}(1)$, $G_{(p,q)}(-1)$, and $H_{(p,q)}(-1)$. The significance of this problem, first noted by *Euler* and *Goldbach* [117, 123, 161, 162, 176, 195], lies in its connection to the values of basic multiple sums of order 2 at the points −1 and 1, with some sums additionally evaluated at $z = i$. The derivation of functional equations in the region $|z| \leq 1$, particularly on the boundary $|z| = 1$, relies on decomposition techniques such as partial fractions and summation over specially defined subsets of indices m and n. This approach, along with the methods used to obtain the results, is entirely original. The functions $F_{(p,q)}(x)$, $G_{(p,q)}(x)$, and $H_{(p,q)}(x)$ for $|x| \leq 1$ are constructed to establish functional relations among them, especially at $x = -1$ and $x = 1$. The monograph demonstrates that these results are independent of *Nielsen*'s findings and are more general. Additionally, special cases of these coefficients, previously absent from the literature, are presented.

For instance, solutions for the coefficients when $p + q$ is odd are provided in Chapter 4, alongside certain even cases. Reciprocal relations among the coefficients are established, and all 16 values of $\omega_j(p, q)$ are calculated: 10 for $p + q = 2k + 1$ and 6 for $p + q = 2k$. Formulas are derived for sums of the form

$$\sum_{n=1}^{\infty} \frac{(-1)^n}{n^r}\left(1 + \frac{1}{2} + \cdots + \frac{1}{n} - \log n\right), \quad r = 1, 2k,$$

and

$$\sum_{n=1}^{\infty} \frac{1}{n^r}\left(1 + \frac{1}{2} + \cdots + \frac{1}{n} - \log n\right),$$

generalizing one of *Hardy*'s formulas from the 1920s [75, 76, 124, 142]. This analysis also yields novel combinatorial identities. Open problems remain for cases complementary to those solved.

Chapter 5 develops the integration of $f(z) = \frac{\log^2(1-z)}{z}$ along appropriate contours to derive generating relations for multiple summations, which are amenable to successive differentiation and integration [10, 80, 110, 138]. Summation formulas involving $\zeta(3)$ and the Catalan constant G are obtained, and, using the previously determined values for $G_{(2k,1)}(-1)$ (Chapter 4), the relation

$$\sum_{m=1}^{\infty} \frac{\cos ma}{m^{2k}} \sum_{n=1}^{\infty} \frac{1}{n}, \quad 0 \le a \le \pi$$

is established, with explicit evaluation for $a = \pi/3$. It is shown that triple sums can be reduced to double sums via combinatorial considerations. Some formulas are listed without proof, where the possibilities for generalization are clear. Contour integration combined with combinatorial techniques facilitates evaluation of triple sums of the form

$$\sum_{m=1}^{\infty} \sum_{n=1}^{\infty} \sum_{p=1}^{\infty} \frac{(-1)^{\lambda m+\mu n+\nu p}}{mnp(m+n+p)}, \quad \lambda, \mu, \nu \in \{0, 1\},$$

covering all four combinations of the parameters. These sums are directly related to the integrals

$$\int_0^1 \frac{\log^3(1-x)}{x}\, dx, \quad \int_0^1 \frac{\log^3(1+x)}{x}\, dx,$$

whose values are determined using results from Chapter 4. In addition, formulas are obtained for sums of the form

$$\sum_{m=1}^{\infty} \frac{\lambda^m}{m^2} \sum_{n=1}^{\infty} \frac{u^n}{n} \sum_{p=1}^{\infty} \frac{\nu^p}{p}, \quad \lambda, u, \nu \in \{-1, 1\},$$

using combinatorial identities and prior results. Chapter 5 thus extends the evaluation of multiple sums systematically.

$$\sum_{m=1}^{\infty} \sum_{n=1}^{\infty} \sum_{p=1}^{\infty} \frac{1}{mnp(m+n+p)^3}, \quad \sum_{m=1}^{\infty} \frac{1}{m^a} \sum_{n=1}^{\infty} \frac{1}{n^b} \sum_{p=1}^{\infty} \frac{1}{p}$$

for $(a, b) = (3, 1)$ and $(a, b) = (2, 2)$ have also been calculated. The problems that remain open are the following:

(1) To derive a formula for the sum

$$\sum_{m=1}^{\infty} \frac{\lambda^m}{m^a} \sum_{n=1}^{\infty} \frac{\mu^n}{n^b} \sum_{p=1}^{\infty} \frac{\nu^p}{p^c}, \quad \lambda, \mu, \nu \in \{-1, 1\},$$

in the cases where $a + b + c = 2k$ and $a + b + c = 2k + 1$.

(2) To derive the formula for the sum

$$\sum_{m=1}^{\infty}\sum_{n=1}^{\infty}\sum_{p=1}^{\infty}\frac{(-1)^{\lambda m+\mu n+\nu p}}{m^a n^b p^c (m+n+p)^d}, \quad \lambda,\mu,\nu\in\{0,1\},$$

in the cases where $a+b+c+d=2k$ and $a+b+c+d=2k+1$.

In the course of this analysis, a fundamental problem in the theory of multiple sums is addressed, namely how to express sums of multiplicity s in terms of sums of lower multiplicity $r\leqslant s-1$. The monograph derives several nontrivial partial results, although a complete solution to this problem is not obtained. All resulting expressions involve special values of the *Riemann* ζ function, highlighting its central role in the structural analysis of multiple sums.

Moreover, certain formulas contain constants of the form $\beta_r(-1)$ and $G_{(2,2)}(-1)$. A natural expectation is that, if $\beta_{2r}(-1)$ does not admit a representation in terms of previously known quantities, such a phenomenon should already arise among sums of multiplicity 2. Nevertheless, several identities expressed in terms of $G_{(2,2)}(-1)$ provide evidence against this assumption. This leads naturally to the problem of determining a minimal generating set of constants for sums of a given multiplicity.

Chapter 5 discusses the *Mordell* sums of the form

$$S_n=\sum_{r=1}^{\infty}\sum_{s=1}^{\infty}\frac{1}{r^n s^n (r+s)^n}, \quad n\in\mathbb{N}.$$

The formula for S_{2k} was obtained by *Subbarao* and *Sitaramachandrarao* in 1985 [147, 148]. The formula for S_{2k+1} has been generally unknown so far, although for S_1, the formula was familiar even to *Euler*. Based on some of the basic results cited in the first chapters, the authors have come up with the formulas for S_3 and S_5. Based on the results and formulas derived in Chapter 4, the formula for each S_{2k+1} has been obtained, and the formula for S_n has also been provided. In relation to the result of *L. Tornheim*: "S_{2k+1} is a polynomial of $\zeta_2(1),\zeta_3(1),\dots,\zeta_{6k+3}(1)$ with rational coefficients", it can be deduced from the derived relations for S_{2k+1} that S_{2k+1} is a second-degree polynomial in relation to $\zeta_2(1),\zeta_3(1),\dots,\zeta_{6k+3}(1)$ with integers. In this chapter, the sum

$$W(m,n,p)=\sum_{r=1}^{\infty}\sum_{s=1}^{\infty}\frac{(-1)^{\lambda r+\mu s}}{r^m s^n (r+s)^p}, \quad \lambda,\mu\in\{0,1\},\ m+n+p=2k+1,$$

has also been calculated, giving a generalization of the *Mordell* sums and sums similar to them. To obtain the above results, the methods of calculating finite combinatorial sums and their double summation have been used. In all the cases, complementary sums appear, which have enabled us to obtain elegant analytical expressions.

Chapter 6 is devoted to the double sums of the form

$$\sum_{m=1}^{\infty} \frac{\lambda^m}{r^a} \sum_{n=1}^{m} \frac{\mu^n}{s^b}, \quad a+b=4,\ \lambda,\mu \in \{-1,1\},\ r \in \{m, 2m-1\},\ s \in \{n, 2n-1\}.$$

In total, 40 sums arise, all of which are structurally analogous to those studied in Chapter 4. Their evaluation is motivated by the fact that they constitute the first nontrivial instance of the case $p+q = 2k$. Although closed forms are not obtained for all of these sums, explicit formulas are derived for a substantial subset, and nontrivial relations among several of them are established. In particular, the constants $\beta_4(-1)$ and $G_{(2,2)}(-1)$ appear systematically in the resulting expressions, naturally leading to the question of whether a direct functional relation exists between $\beta_4(-1)$ and $G_{(2,2)}(-1)$. The formulas obtained in this chapter are derived by multiplying carefully chosen power-series expansions and by combining the associated double sums with suitable product identities.

Chapter 6 further develops a collection of functional relations valid in the interval $-1 \leqslant x \leqslant 1$, as well as in relevant subregions. The derivation of these relations is initiated from the *Spencer* identity [17], which provides an explicit representation of $\zeta_3(\frac{x}{x-1})$ for $0 \leqslant x \leqslant \frac{1}{2}$. As illustrative applications, explicit expressions for $G_{(2,1)}(x)$ and $G_{(1,2)}(x)$ are obtained in the same region. Moreover, corresponding decompositions of these functions are established for the larger intervals $0 \leqslant x \leqslant 1$ and $0 \leqslant x < 1$. This analysis is followed by explicit formulas for integrals of the form

$$\int_0^x \frac{\log^k(1-t)}{t}\,dt, \quad \int_0^x \frac{\log^k(1+t)}{t}\,dt, \quad k=1,2,3,$$

in various regions, with decompositions of the functions $G_{(2,2)}(x)$ and $G_{(3,1)}(x)$ in the region $0 \leqslant x \leqslant \frac{1}{2}$. For $\zeta_k(\frac{1}{2})$, the formulas for $k = 1,2,3$ are known. The relations between $\zeta_4(\frac{1}{2})$ and $G_{(2,2)}(-1)$ are presented. This provides the basis for the claim that $G_{(2,2)}(-1)$ can be expressed in terms of $\log^4 2$, $\zeta_2(1)\log^2 2$, $\zeta_3(1)\log 2$, and $\zeta_4(1)$. This is supported by the formulas for some integrals in which $G_{(2,2)}(-1)$ appears. The values for the expression $G_{(p,q)}(\frac{1}{2})$ have been calculated when $p+q=4$. For the functions given by triple sums

$$G_{(a,b,c)}(x) = \sum_{m=1}^{\infty} \frac{x^m}{m^a} \sum_{n=1}^{m} \frac{1}{n^b} \sum_{p=1}^{n} \frac{1}{p^c}, \quad a+b+c=4,$$

the formulas in the region $-1 \leqslant x \leqslant \frac{1}{2}$ have been obtained, as well as the following formulas:

$$G_{(2,1,1)}(x) = -G_{(1,3)}\left(\frac{x}{x-1}\right), \quad G_{(1,2,1)}(x) = -G_{(2,2)}\left(\frac{x}{x-1}\right),$$
$$G_{(1,1,2)}(x) = -G_{(3,1)}\left(\frac{x}{x-1}\right), \quad -1 \leqslant x \leqslant \frac{1}{2}.$$

The following formula is true in the same region:

$$G_{(1,1,\ldots,1)_r}(x) = -\zeta_r\left(\frac{x}{x-1}\right), \quad r \geqslant 2.$$

Furthermore, in Chapter 6, for the functions

$$F_r(x) = \sum_{k_1=1}^{\infty} \frac{x^{k_1}}{k_1} \sum_{k_2=1}^{k_1} \frac{x^{k_2}}{k_2} \cdots \sum_{k_r=1}^{k_{r-1}} \frac{x^{k_r}}{k_r}, \quad r \geqslant 2,$$

$$F_0(x) = 1, \quad F_1(x) = \zeta_1(x), \quad -1 \leqslant x < 1,$$

one functional and, as a consequence, one recurrent relation have been obtained. The values of $F_r(x)$ have been specified for $r = 0, 1, \ldots, 6$. The problem remains to determine the formula for $\zeta_k(\frac{1}{2})$, $k \geqslant 4$ (if we do not think that the derived formula is true for $k = 4$). The results reported in this chapter have been obtained by combining the combinatorial summation and multiple summation methods and using various functional relations for the functions given either by multiple sums or by integrals. The method used for proving the functional relation for $F_r(x)$ could be named the "entry-exit" method for multiple sums.

Motivated by the results obtained by *H. M. Srivastava* in his 1988 study concerning several classes of summation formulas involving series with the *Riemann* zeta function—originally investigated by *Euler* and *Goldbach* [58, 111, 194, 199]—Chapter 7 is devoted to a systematic analysis of series of this type. For each formula presented in the introduction to the chapter, a corresponding generalization is derived that holds for all natural numbers n. These generalizations are expressed in terms of the following families of sums, for which explicit formulae are provided:

$$\sum_{k=1}^{\infty} \frac{\zeta(2k)-1}{k+n}, \quad n \geqslant 1, \qquad \sum_{k=1}^{\infty} \frac{\zeta(2k)}{k(2k+n)}, \quad n \geqslant 1,$$

$$\sum_{k=2}^{\infty} (-1)^k \frac{\zeta(k)-1}{k+n}, \quad n \geqslant 1, \qquad \sum_{k=1}^{\infty} \frac{\zeta(2k)-1}{2k+2n+1)}, \quad n \geqslant 0,$$

$$\sum_{k=1}^{\infty} \frac{\zeta(2k+1)-1}{k+n}, \quad n \geqslant 1, \qquad \sum_{k=2}^{\infty} (-1)^k \frac{\zeta(k)}{2^k(k+n)}, \quad n \geqslant 1.$$

The problem that remains open refers to the calculation of the sum

$$\sum_{k=2}^{\infty} (-1)^k \frac{\zeta(k)-1}{(k+n)^r}, \quad r \geqslant 2, \ r, n \in \mathbb{N}.$$

To derive the stated results, the author employs a method based on integrating functions of the form $P(x) \log \Gamma(x)$, where $P(x)$ is a polynomial. For the logarithm of the gamma function, the corresponding *Kummer* series expansion valid in the region

$0 < x < 1$ is used [40, 145, 159]. A unifying feature of all problems treated in the monograph is that the resulting multiple sums and integrals admit closed representations in terms of the *Riemann* zeta function. Although the results presented constitute only a small subset of the identities that can, in principle, be derived, they nonetheless provide compelling evidence of the fundamental role played by the zeta function in the theory of multiple series and integral summation.

Throughout the monograph, the following notation is adopted. For positive quantities A and B, the relation $B = O(A)$ (equivalently $B \ll A$) means that there exists an absolute constant $c > 0$ such that $|B| \leq cA$. Moreover, the logarithm log denotes the natural logarithm. When sums are taken over the nontrivial zeros of the *Riemann* ζ-function, the zeros are ordered by increasing modulus of their imaginary parts; ties are resolved arbitrarily.

Finally, it must be noted with regret that Professor Stevanović passed away in 2010. His death confronted the author, who was his close collaborator and colleague, with significant and at times difficult challenges in adequately presenting, interpreting, and extending the results that Professor Stevanović achieved during his exceptionally productive scientific career. The author can only hope that his dedication and effort have resulted in a work of high quality, and that the monograph in its present form will reach and be appreciated by the professional scientific community.

This text probably would not have seen the light of day if it were not for the wife of Professor Stevanović, Vera, who prepared a huge piece of material that had been in the form of handwritten notes for further revision and computer processing. Mladen Janjić, a late professor's student, performed the fracture and prepared the monograph for printing, which is why we would like to express our sincere gratitude to him.

Furthermore, the authors would like to thank Dr. Lena Tica for proofreading and editing the English version of the manuscript.

Contents

Notation

Due to the nature of this text, absolute consistency in notation could not be attained, although standard notation is used whenever possible. Notation used commonly through the text is explained there, whereas specific notation introduced in the proof of a theorem or lemma is given at the proper place in the body of the text.

- $k, l, m, n, \ldots$: Natural numbers (positive integers).
- p: A generic prime number.
- $\mathbb{N}, \mathbb{Z}, \mathbb{R}, \mathbb{C}$: The sets of natural numbers, integers, real and complex numbers, respectively.
- $A, B, C, C_1, \ldots$: Absolute positive constants (not necessarily the same at each occurrence).
- ε: An arbitrarily small positive number, not necessarily the same at each occurrence.
- s, z, w: Complex variables ($\mathfrak{Re}\, s$ and $\mathfrak{Im}\, s$ denote the real and imaginary parts of s, respectively; common notation is $\sigma = \mathfrak{Re}\, s$ and $t = \mathfrak{Im}\, s$).
- t, x, y: Real variables.
- $\mathrm{res}_{s=s_0} F(s)$: The residue of $F(s)$ at the point $s = s_0$.
- $\zeta(s)$: The *Riemann* zeta-function defined as $\zeta(s) = \sum_{n=1}^{\infty} n^{-s}$ for $\mathfrak{Re}\, s > 1$ and otherwise by analytic continuation.
- $\Gamma(s) = \int_0^{\infty} x^{s-1} e^{-x}\, \mathrm{d}x$ for $\mathfrak{Re}\, s > 0$; otherwise, by analytic continuation by $s\Gamma(s) = \Gamma(s+1)$. This is the *Euler* gamma function.
- γ: *Euler's* constant $\gamma = -\Gamma'(1) = 0.5772157\ldots$.
- $\chi(s)$: The function defined by $\zeta(s) = \chi(s)\zeta(1-s)$, so that by the functional equation for $\zeta(s)$ we have $\chi(s) = \frac{(2\pi)^s}{2\Gamma(s)\cos(\frac{\pi s}{2})}$.
- $\theta(t) = \mathfrak{Im}\{\log \Gamma(\frac{1}{4} + \frac{1}{2} it)\} - \frac{1}{2} t \log \pi$ for real t.
- $\rho = \beta + i\gamma$: A complex zero of $\zeta(s)$; $\beta = \mathfrak{Re}\, \rho$, $\gamma = \mathfrak{Im}\, \rho$.
- $N(T)$: The number of zeros $\rho = \beta + i\gamma$ of $\zeta(s)$, counted with multiplicities, for which $0 < \gamma \le T$.
- $N(\sigma, T)$: The number of zeros ρ of $\zeta(s)$ for which $\beta \ge \sigma$, $|\gamma| \le T$.
- $S(T) = \frac{1}{\pi} \arg \zeta(\frac{1}{2} + iT)$.
- $\mu(\sigma) = \limsup_{t\to\infty} \frac{\log|\zeta(\sigma+it)|}{\log t}$ for real σ. The *Mobius* function is defined as $\mu(n) = (-1)^k$ if $n = p_1 \cdots p_k$ (the p_j's being different primes) and zero otherwise, and $\mu(1) = 1$.
- $\exp(z) = e^z$.
- $e(z) = e^{2\pi i z}$.
- $\log x = \log_e x \equiv \ln x$.
- $[x]$: The greatest integer not exceeding a real number x.
- $\{x\} = x - [x]$, the fractional part of x.
- $\sum_{n\le x} f(n)$: A sum taken over all natural numbers n not exceeding x; the empty sum by definition is equal to zero.
- $\sum_{d|n}$: A sum taken over all positive divisors of n.
- $\Lambda_k(n)$: The generalized *von Mangoldt* function defined as $\Lambda_k(n) = \sum_{d|n} \mu(d)(\log \frac{n}{d})^k$; $\Lambda_1(n) = \Lambda(n)$, the ordinary *von Mangoldt* function.
- $\prod_j$: The product taken over all possible values of the index j; the empty product by definition is equal to unity.
- $\psi(x) = x - [x] - \frac{1}{2}$, or $\psi(z) = \frac{\Gamma'(z)}{\Gamma(z)}$.
- $\psi_k(x) = \sum_{n\le x} \Lambda_k(n)$.
- $\pi(x) = \sum_{p\le x} 1$, the number of primes not exceeding x.
- $\theta(x) = \sum_{p\le x} \log p$.
- $M(x) = \sum_{n\le x} \mu(x)$.
- $r(n)$: The number of ways n can be written as a sum of two integer squares.

https://doi.org/10.1515/9783112233276-204

- $d_k(n)$: The number of ways n can be written as a product of $k \geq 2$ fixed factors; $d_2(n) = d(n)$ is the number of divisors of n.
- $\Delta(x) = \sum_{n\leq x} d(n) - x(\log x + 2\gamma - 1)$, the error term in the *Dirichlet* divisor problem.
- $\varphi(n)$: *Euler*'s function defined as $\varphi(n) = n\prod_{p|n}(1-\frac{1}{p})$, where the product is over all prime divisors of n.
- $\omega(n)$: The number of distinct prime divisors of n.
- $\Omega(n)$: The number of all prime divisors of n.
- $f(x) \sim g(x)$: Means that $\lim_{x\to x_0}\frac{f(x)}{g(x)} = 1$, with x_0 not necessarily finite.
- $f(x) = O(g(x))$: Means that $|f(x)| \leq Cg(x)$ for $x \geq x_0$ and some constant $C > 0$. Here $f(x)$ is a complex function of the real variable x, and $g(x)$ is a positive function of $x \geq x_0$. $f(x) = O_{\alpha,\beta,\ldots}(g(x))$ means that the implied constant depends on $\alpha, \beta, \ldots$.
- $f(x) \ll g(x)$: Means the same as $f(x) = O(g(x))$. Likewise, $f(x) \ll_{\alpha,\beta,\ldots} g(x)$ means that the implied constant depends on $\alpha, \beta, \ldots$.
- $f(x) \gg g(x)$: Means the same as $g(x) = O(f(x))$.
- (a, b): Means the interval $a < x < b$.
- $[a, b]$: Means the interval $a \leq x \leq b$.
- $f(x) = o(g(x))$ as $x \to x_0$: Means that $\lim_{x\to x_0}\frac{f(x)}{g(x)} = 0$, with x_0 possibly infinite.
- $f(x) = \Omega(g(x))$: Means that $f(x) = o(g(x))$ does not hold as $x \to \infty$.
- $\int_{(c)} G(s)\,ds = \lim_{T\to\infty}\int_{c-iT}^{c+iT} G(s)\,ds$.

1 Definition and properties of the function $\zeta(s)$

1.1 Definition of the function $\zeta(s)$ and its relation to arithmetic functions

The series

$$\zeta(s) = \sum_{n=1}^{\infty} \frac{1}{n^s}, \quad s = \sigma + it \ (\sigma > 1), \tag{1.1}$$

absolutely and uniformly converges in any finite region of the s plane in which $\sigma \geqslant 1 + \delta > 1\ (\delta > 0)$. According to the *Weierstrass* theorem on uniformly convergent series of analytic functions, the series (1.1) specifies an analytic function for $\sigma > 1$.

Definition 1.1. The analytic function defined by relation (1.1) is called the *Riemann* zeta function.

Theorem 1.1. *For $\sigma > 1$, we have*

$$\zeta(s) = \prod_p \left(1 - \frac{1}{p^s}\right)^{-1}, \tag{1.2}$$

where the product runs over all prime numbers p.

Proof. By multiplying finitely many absolutely convergent series, we obtain

$$\prod_{p \le P} \left(1 - \frac{1}{p^s}\right)^{-1} = \prod_{p \le P} \left(1 + \frac{1}{p^s} + \frac{1}{p^{2s}} + \cdots\right) = 1 + \frac{1}{n_1^s} + \frac{1}{n_2^s} + \cdots, \tag{1.3}$$

where $n_1, n_2, \ldots$ are natural numbers whose prime factors do not exceed P. Since all natural numbers up to P have such a form, it follows that

$$\begin{aligned} \left|\zeta(s) - \prod_{p \le P} \left(1 - \frac{1}{p^s}\right)^{-1}\right| &= \left|\sum_{n=1}^{\infty} \frac{1}{n^s} - \left(1 + \frac{1}{p^s} + \frac{1}{p^{2s}} + \cdots\right)\right| \\ &\leqslant \left(\frac{1}{(P+1)^\sigma} + \frac{1}{(P+1)^\sigma} + \cdots\right) \longrightarrow 0 \quad (P \longrightarrow \infty). \end{aligned} \tag{1.4}$$

□

The equivalence of (1.1) and (1.2) is one of the most significant features of the function $\zeta(s)$.

https://doi.org/10.1515/9783112233276-001

Definition 1.2. The *Möbius* function $\mu(n)$ is defined for all natural numbers as follows:

$$\mu(1) = 1, \quad \mu(1) = (-1)^k \tag{1.5}$$

if n is the product of different k prime numbers, and $\mu(n) = 0$ if n contains a square of some prime number. For $\mu(n)$, the following formulas are valid (see [103], p. 26):

$$\sum_{d|n} \mu(d) = \begin{cases} 0 & (a > 1), \\ 1 & (a = 1), \end{cases} \tag{1.6}$$

$$\sum_{d|n} \frac{\mu(d)}{d^s} = \begin{cases} (1 - \frac{1}{p_1^s})(1 - \frac{1}{p_2^s}) \cdots (1 - \frac{1}{p_k^s}) & (a > 1), \\ 1 & (a = 1), \end{cases} \tag{1.7}$$

where $a = p_1^{\alpha_1} p_2^{\alpha_2} \cdots p_k^{\alpha_k}$ (canonical development of a number a).

Lemma 1.1. *For $\sigma > 1$, we have*

$$\frac{1}{\zeta(s)} = \sum_{n=1}^{\infty} \frac{\mu(n)}{n^s}. \tag{1.8}$$

Proof. For $\sigma > 1$,

$$\zeta(s) \neq 0. \tag{1.9}$$

Indeed, for $\sigma > 1$,

$$\left(1 - \frac{1}{2^s}\right)\left(1 - \frac{1}{3^s}\right) \cdots \left(1 - \frac{1}{p^s}\right)\zeta(s) = 1 + \frac{1}{m_1^s} + \frac{1}{m_2^s} + \cdots, \tag{1.10}$$

where $m_1, m_2, \ldots$ are natural numbers whose all prime factors exceed P. Therefore

$$\left|\left(1 - \frac{1}{2^s}\right)\left(1 - \frac{1}{3^s}\right) \cdots \left(1 - \frac{1}{p^s}\right)\zeta(s)\right| \geq 1 - \frac{1}{(P+1)^s} - \frac{1}{(P+2)^s} - \cdots > 0 \tag{1.11}$$

for sufficiently large P. It follows that $|\zeta(s)| > 0$, therefore $\zeta(s) \neq 0$, so there exists $\frac{1}{\zeta(s)}$. If $N > 2$, then from (1.7) it follows that

$$\prod_{p \leqslant N}\left(1 - \frac{1}{p^s}\right) = \sum 0 < n \leq N \frac{\mu(n)}{n^s} + \Sigma' \frac{\mu(n)}{n^s}, \tag{1.12}$$

where in the second sum on the right, there are only n greater than N. Since the series $\sum_{n=1}^{\infty} \frac{1}{n^s}$ absolutely converges for $\zeta > 1$, $\Sigma' \frac{\mu(n)}{n^s} \longrightarrow 0$ $(N \longrightarrow \infty)$, and relation (1.8) follows. □

Definition 1.3. The *von Mangoldt* function $\Lambda(n)$ is defined for all natural numbers as follows:

$$\Lambda(n) = \log p \tag{1.13}$$

for $n = p^m$, where p is a prime number, and m is a natural number; $\Lambda(n) = 0$ for the remaining n.

For $\Lambda(n)$, we have

$$\sum_{d|n} \Lambda(d) = \log n \tag{1.14}$$

(see [112], p. 79).

Lemma 1.2. *For $\sigma > 1$, we have*

$$\frac{\zeta'(s)}{\zeta(s)} = -\sum_{\infty}^{n=2} \frac{\Lambda(n)}{n^s}. \tag{1.15}$$

Proof. Logarithmizing equation (1.2), we obtain

$$\log \zeta(s) = -\sum_p \log(1 - p^{-s}). \tag{1.16}$$

Since

$$\begin{aligned} \left|-\log(1-z\right| = \left|\log(1-z)\right| &= \left|z + \frac{z^2}{2} + \frac{z^3}{3} + \cdots\right| \leqslant |z| + \frac{|z|^2}{2} + \frac{|z|^3}{3} + \cdots \\ &= -\log(1-|z|) = \left|-\log(1-|z|)\right|, \quad (|z| < 1), \end{aligned} \tag{1.17}$$

we have

$$\left|\log(1-p^{-s})\right| \leqslant -\log(1-p^{-\sigma}) = p^{-\sigma} + \frac{(p^{-\sigma})^2}{2} + \frac{(p^{-\sigma})^3}{3} + \cdots < \frac{p^{-\sigma}}{1-p^{-\sigma}} < \frac{2}{p^\sigma}. \tag{1.18}$$

Therefore, according to the *Weierstrass* criterion, the series $-\sum_p \log(1-p^{-s})$ converges uniformly for $\sigma > 1$, and we have

$$\log \zeta(s) = -\sum_p \log(1-p^{-s}) = \sum_{m,p} \frac{1}{mp^{ms}}. \tag{1.19}$$

Differentiating (due to the uniform convergence) the series term-by-term, we obtain

$$-\frac{\zeta'(s)}{\zeta(s)} = \sum_p \frac{p^{-s}\log p}{1-p^{-s}} = \sum_{m,p} \frac{\log p}{p^{ms}} = \sum_p \log p \sum_{m=1}^{\infty} \frac{1}{p^{ms}} = \sum_{n=2}^{\infty} \frac{\Lambda(n)}{n^s}. \tag{1.20}$$

The series $\sum_p \frac{p^{-s}\log p}{1-p^{-s}}$ also converges uniformly, since

$$\left|\frac{p^{-s}\log p}{1-p^{-s}}\right| \leqslant \frac{p^{-\sigma}}{1-p^{-\sigma}}\log p = \frac{\log p}{p^{\sigma}-1} < 2\frac{\log p}{p^{\sigma}}, \tag{1.21}$$

where $\log n < n^{\frac{\sigma}{2}}$ for sufficiently large n. For $\sigma \geq 1+\delta > 1$, we have

$$\frac{\log n}{n^{\sigma}} < n^{-(\sigma-\frac{\delta}{2})} \leqslant n^{-(1+\frac{\delta}{2})}. \tag{1.22}$$

□

Definition 1.4. The *Euler* indicator $\varphi(n)$ is the number of integers among $0, 1, 2, \ldots, n-1$ that are relatively prime to n. For $n = p_1^{\alpha_1}p_2^{\alpha_2}\cdots p_k^{ak}$ presented in canonical development,

$$\varphi(n) = n\prod_{i=1}^{k}\left(1-\frac{1}{p_i}\right), \tag{1.23}$$

$$\sum_{a|n}\varphi(d) = 1 \tag{1.24}$$

(see [103], pp. 27–28).

Lemma 1.3. *For $\sigma > 2$, we have*

$$\sum_{n=1}^{\infty}\frac{\zeta(s-1)}{\zeta(s)}. \tag{1.25}$$

Proof.

$$\begin{aligned}
\frac{\zeta(s-1)}{\zeta(s)} &= \prod_p\left(1-\frac{1}{p^s}\right)\cdot\prod_p\left(1-\frac{1}{p^{s-1}}\right)^{-1} = \prod_p\frac{1-\frac{1}{p^s}}{1-\frac{1}{p^{s-1}}}\\
&= \prod_p\left(1-\frac{1}{p^s}\right)\left(1+\frac{p}{p^s}+\frac{p^2}{p^{2s}}+\cdots\right)\\
&= \prod_p\left\{1+\left(1-\frac{1}{p}\right)\left(\frac{p}{p^s}+\frac{p^2}{p^{2s}}+\cdots\right)\right\}\\
&= \prod_p\left(1+\frac{\varphi(p)}{p^s}+\frac{\varphi(p^2)}{p^{2s}}+\cdots\right)\sum_{n=1}^{\infty}\frac{\varphi(n)}{n^s}.
\end{aligned} \tag{1.26}$$

□

We now provide some more formulas that relate $\zeta(s)$ and other arithmetic functions (see [135, 103, 50]).

$$\zeta^2(s) = \sum_{d=1}^{\infty}\frac{d(n)}{n^s} \quad (\sigma > 1), \tag{1.27}$$

where $d(n)$ is the number of divisors of a number n (including 1 and n).

$$\zeta^k(s) = \sum_{n=1}^{a} \frac{d_k(n)}{n^s} \quad (\sigma > 1), \quad (k = 2, 3, 4, \dots). \tag{1.28}$$

In relation (1.28), $d_k(n)$ is the number of ways in which n can be expressed as the product of k factors, where the expressions with the same factors in different order are considered to be different. In addition, we have the following equation:

$$\frac{\zeta^2(s)}{\zeta(2s)} = \sum_{n=1}^{\infty} \frac{2^{\nu(n)}}{n^s} \quad (\sigma > 1), \tag{1.29}$$

where $\nu(n)$ is the number of different prime factors of a number n. Further, we have

$$\frac{\zeta(2s)}{\zeta(s)} = \sum_{n=1}^{\infty} \frac{\lambda(n)}{n^s} \quad (\sigma > 1), \tag{1.30}$$

where $\lambda(n) = (-1)^r$ if n has r prime factors, where a prime factor with multiplicity k is counted k times.

Definition 1.5. $\pi(x)$ is the number of prime numbers not greater than a positive real number x.

Theorem 1.2. *If* $\sigma > 1$, *then*

$$\log \zeta(s) = s \int_2^{\infty} \frac{\pi(x)}{x(x^s - 1)} \partial x. \tag{1.31}$$

Proof. If $\sigma > 1$, then $\zeta(s) \neq 0$ (see (1.9)), and if $s > 1$, then $\zeta(s) > 1$. Therefore we can define the branch of the function $\log \zeta(s)$, real and positive for $s > 1$ and one-liner for $\sigma > 1$. Then

$$\begin{aligned}
\log \zeta(s) &= -\sum_p \log(1 - p^{-s}) = -\sum_p (\pi(p) - \pi(p-1)) \log(1 - p^{-s}) \\
&= -\lim_{K \to \infty} \sum_{n=2}^{K} (\pi(n) - \pi(n-1)) \log(1 - n^{-s}) \\
&= -\lim_{K \to \infty} \Bigg\{ \sum_{n=2}^{K} \pi(n)[\log(1 - n^{-s}) - \log(1 - (n+1)^{-s})] \\
&\quad - \pi(1) \log(1 - 2^{-s}) + \pi(k) \log(1 - (K+1)^{-s}) \Bigg\} \\
&= \lim_{K \to \infty} \sum_{n=2}^{K} \pi(n)[\log(1 - (n+1)^{-s}) - \log(1 - n^{-s})] - \lim_{K \to \infty} \pi(K) \log(1 - (K+1)^{-s}).
\end{aligned} \tag{1.32}$$

Since

$$\left|\pi(K)\log(1-(K+1)^{-s}\right| \leqslant K\left|\log(1-(K+1)^{-s})\right| < \frac{2K}{|(K+1)^s|} \longrightarrow 0$$
$$(K\longrightarrow\infty),\quad (\sigma>1), \tag{1.33}$$

we obtain

$$\log\zeta(s) = \sum_{n=2}^{\infty}\pi(n)\int_n^{n+1}\left\{\frac{\partial}{\partial x}\log(1-x^{-s})\right\}\partial x = \sum_{n=2}^{\infty}\pi(n)\int_n^{n+1}\frac{s}{x(x^s-1)}\partial x$$
$$= s\sum_{n=2}^{\infty}\int_n^{n+1}\frac{\pi(x)}{x(x^s-1)}\partial x = s\int_2^{\infty}\frac{\pi(x)}{x(x^s-1)}\partial x. \tag{1.34}$$

□

1.1.1 The asymptotic law of the distribution of prime numbers

Theorem 1.3 (The asymptotic law of the distribution of prime numbers).

$$\lim_{x\longrightarrow\infty}\frac{\pi(x)}{\frac{x}{\log x}} = 1. \tag{1.35}$$

The theorem was proved in 1896 by *J. Hadamard* and *Ch. de la Vallée Poussin*, who developed it simultaneously and independently of one another [18, 19]. We prove this theorem following the proof given in [174].

Lemma 1.4 (*Abel* transformation). *Let a_n $(n = 1, 2, \ldots)$ be a series of complex numbers, let $A(x) = \sum_{n\leqslant x} a_n$, $x > 1$, and let $g(t)$ $(1 \leqslant t < \infty)$ be a continuously differentiable complex-valued function. Then*

$$\sum_{n\leqslant x} a_n g(n) = A(x)g(x) - \int_1^x A(t)g'(t)\,\partial t, \tag{1.36}$$

and if $\lim_{x\longrightarrow\infty} A(x)g(x) = 0$, *then*

$$\sum_{n=1}^{\infty} a_n g(n) = -\int_1^{\infty} A(t)g'(t)\,\partial t. \tag{1.37}$$

(For the prof see [174], pp. 38–39.)

If in Lemma 1.4, $a_n = 1$ $(n = 1, 2, \ldots)$ and $g(x) = (x+N)^{-s}$, then $A(x) = [x]$, and for $\sigma > 1$, $\lim_{x\longrightarrow\infty} A(x)g(x) = 0$, since

$$\frac{[x]}{|(x+N)^s|} = \frac{[x]}{(x+N)^\sigma} \leq \frac{x}{(x+N)^\sigma} < \frac{1}{x^{\sigma-1}} \longrightarrow 0 \quad (x \longrightarrow \infty). \tag{1.38}$$

It follows that

$$\begin{aligned}\sum_{n=N+1}^{\infty} \frac{1}{n^s} &= s\int_1^\infty \frac{[x]}{(x+N)^{s+1}}\,\partial x = s\int_0^\infty \frac{[x]}{(x+N)^{s+1}}\,\partial x = s\int_N^\infty \frac{[x-N]}{x^{s+1}}\,\partial x \\ &= s\int_N^\infty \frac{x-N-\{x\}}{x^{s+1}}\,\partial x = \frac{N^{1-s}}{s-1} - s\int_N^\infty \frac{\{x\}}{x^{s+1}}\,\partial x,\end{aligned} \tag{1.39}$$

that is,

$$\zeta(s) = \sum_{n=1}^{N} \frac{1}{n^s} + \frac{N^{1-s}}{s-1} - s\int_N^\infty \frac{\{x\}}{x^{s+1}}\,\partial x. \tag{1.40}$$

In particular, for $N = 1$ we get

$$\zeta(s) = 1 + \frac{1}{s-1} - s\int_1^\infty \frac{\{x\}}{x^{s+1}}\,\partial x \tag{1.41}$$

or

$$\zeta(s) = 1 + \frac{1}{s-1} - s\sum_{n=1}^{\infty}\int_n^{n+1} \frac{x-n}{x^{s+1}}\,\partial x. \tag{1.42}$$

Since

$$\left|\int_n^{n*1} \frac{x-n}{x^{s+1}}\,\partial x\right| \leqslant \frac{1}{n^{\sigma+1}}, \tag{1.43}$$

the series on the right-hand side of equation (1.42) uniformly converges in the region $\sigma \geqslant \delta > 0$. Each of the integrals is an analytic function in the same region. Therefore the right-hand side of the equation is an analytic function in the region $\sigma \geqslant \delta$, $s \neq 1$. Due to the arbitrariness of δ, $\sigma > 0$, $s \neq 1$, at the point $s = 1$, the function $\zeta(s)$ has a prime pole. Hence we have the following:

Theorem 1.4. *The right-hand side of relation* (1.42) *is an analytic extension of the function* $\zeta(s)$ *defined by* (1.1) *in the region* $\sigma > 0$. *In this region,* $\zeta(s)$ *has a unique singularity* $s = 1$, *which is a pole of order 1. Because of the uniqueness of the analytic extension, it follows that in the region* $\sigma > 0$, *relation* (1.40) *holds for each* N.

Lemma 1.5. *In the region $|t| \geq 3, 1 \leq \sigma \leq 2$ $(s = \sigma + it)$, we have the following inequalities:*

$$|\zeta(s)| \leqslant \frac{29}{6} \log |t|, \tag{1.44}$$

$$|\zeta'(s)| \leqslant \frac{61}{9} \log^2 |t|. \tag{1.45}$$

Proof. If $N = [|t|]$ in (1.40), then

$$\left|\sum_{n=1}^{N} \frac{1}{n^s}\right| \leqslant \sum_{n=1}^{N} \frac{1}{n^\sigma} \leqslant 1 + \int_1^N \frac{\partial x}{x} = 1 + \log N \leqslant 2 \log |t|. \tag{1.46}$$

Since

$$|s-1|^2 = (\sigma - 1)^2 + t^2 \geq t^2, \tag{1.47}$$

that is, $|s-1| \geq |t| \geq 3$, and

$$\left|\frac{N^{1-s}}{s-1}\right| \leqslant \frac{1}{3} N^{1-\sigma} \leqslant \frac{1}{3}, \tag{1.48}$$

as well as $N \leqslant |t| < N+1$, it follows that

$$\left|s \int_N^\infty \frac{\{x\}}{x^{s+1}} \partial x\right| \leqslant (\sigma + |t|) \int_N^\infty \frac{\partial x}{x^2} \leqslant \frac{|t|+2}{N} \leqslant \frac{|t|+2}{|t|-1} \leqslant \frac{5}{2}. \tag{1.49}$$

From (1.40) it follows that:

$$|\zeta(s)| \leqslant 2 \log |t| + \frac{1}{3} + \frac{5}{2} \leqslant \frac{29}{6} \log |t|. \tag{1.50}$$

Differentiating relation (1.40), we obtain

$$\zeta'(s) = -\sum_{n=1}^{N} \frac{\log n}{n^s} - \frac{N^{1-s}}{(s-1)^2} - \frac{N^{1-s} \log N}{s-1} - \int_N^\infty \frac{\{x\}}{x^{s+1}} \partial x + s \int_N^\infty \frac{\{x\} \log x}{x^{s+1}} \partial x. \tag{1.51}$$

Since the series on the right-hand side of the relation

$$\int_N^\infty \frac{\{x\}}{x^{s+1}} \partial x = \sum_{n=N}^\infty \int_n^{n+1} \frac{x-n}{x^{s+1}} \partial x \tag{1.52}$$

uniformly converges, we can differentiate it term-by-term, that is,

$$\frac{d}{ds}\left(\sum_{n=N}^\infty \int_n^{n+1} \frac{x-n}{x^{s+1}} \partial x\right) = -\sum_{n=N}^\infty \int_n^{n+1} \frac{x-n}{x^{s+1}} \log x \, \partial x = -\int_N^\infty \frac{\{x\} \log x}{x^{s+1}} \partial x. \tag{1.53}$$

The following estimates are evident:

$$\left|\frac{N^{1-s}}{(s-1)^2}\right| \le \frac{1}{9}, \quad \left|\frac{N^{1-s}\log N}{s-1}\right| \le \frac{1}{3}\log N \le \frac{1}{3}\log|t|, \tag{1.54}$$

$$\left|\int_N^\infty \frac{\{x\}}{x^{s+1}}\,\mathfrak{d}x\right| \le \int_N^\infty \frac{\mathfrak{d}x}{x^2} = \frac{1}{N} \le \frac{1}{|t|-1} \le \frac{1}{2}, \tag{1.55}$$

$$\begin{aligned}\left|s\int_N^\infty \frac{\{x\}\log x}{x^{s+1}}\,\mathfrak{d}x\right| &\le (|t|+2)\int_N^\infty \frac{\log x}{x^2}\,\mathfrak{d}x = (|t|+2)\left(\frac{\log N}{N}+\frac{1}{N}\right)\\ &\le \frac{|t|+2}{|t|-1}(\log|t|+1) < \frac{5}{2}(\log|t|+1).\end{aligned} \tag{1.56}$$

Since $\frac{\log x}{x}$ monotonically decreases for $x \ge \mathfrak{e}$, we have

$$\begin{aligned}\left|\sum_{n=2}^N \frac{\log n}{n^s}\right| &\le \sum_{n=2}^N \frac{\log n}{n} \le \frac{\log 2}{2} + \frac{\log 3}{3} + \int_3^N \frac{\log x}{x}\,\mathfrak{d}x\\ &= \frac{\log 2}{2} + \frac{\log 3}{3} + \frac{\log^2 N}{2} - \frac{\log^2 3}{2}\\ &< \frac{5\log 3}{6} + \frac{\log^2|t|}{2} - \frac{\log^2 3}{2} < \frac{\log^2 3}{3} + \frac{\log^2|t|}{2}\\ &\le \frac{\log^2|t|}{3} + \frac{\log^2|t|}{2} < \frac{5}{6}\log^2|t|.\end{aligned} \tag{1.57}$$

From (1.51) it follows that

$$|\zeta'(s)| < \frac{1}{9} + \frac{1}{3}\log|t| + \frac{1}{2} + \frac{5}{2}(\log|t|+1) + \frac{5}{6}\log^2|t| < \frac{61}{9}\log^2|t|. \tag{1.58}$$

□

Lemma 1.6. *For* $0 < r < 1$ *and* $\varphi \in \mathbb{R}$, *we have*

$$\left|(1-r)^3(1-r\mathfrak{e}^{\mathrm{i}\varphi})^4(1-r\mathfrak{e}^{2\mathrm{i}\varphi})\right|^{-1} \ge 1. \tag{1.59}$$

Proof. For every z with $|z| < 1$, we have

$$-\log(1-z) = \sum_{n=1}^\infty \frac{z^n}{n}. \tag{1.60}$$

Since $\mathfrak{Re}(\log t) = \log|t|$, denoting the left-hand side of relation (1.59) by $F(r,\varphi)$, we have

$$\begin{aligned}\log F(r,\varphi) &= -3\log(1-r) - 4\log|1-re^{i\varphi}| - \log|1-re^{2i\varphi}| \\ &= -3\log(1-r) - 4\,\mathfrak{Re}\log(1-re^{i\varphi}) - \mathfrak{Re}\log(1-re^{2i\varphi}) \\ &= \mathfrak{Re}[-3\log(1-r) - 4\log(1-re^{i\varphi}) - \log(1-re^{2i\varphi})] \\ &= \mathfrak{Re}\left\{\sum_{n=1}^{\infty}\frac{r^n}{n}(3+4e^{in\varphi}+e^{2in\varphi})\right\} \\ &= \sum_{n=1}^{\infty}\frac{r^n}{n}(3+4\cos n\varphi+\cos 2n\varphi) \\ &= 2\sum_{n=1}^{\infty}\frac{r^n}{n}(1+\cos n\varphi)^2 \ge 0 \quad\Longrightarrow\quad F(r,\varphi)\ge 1. \end{aligned} \tag{1.61}$$

□

Theorem 1.5 (Hadamard, de la Vallée Poussin). *If $t \neq 0$, then $\zeta(1+it) \neq 0$.*

Proof. Let $\sigma > 1$. From the *Euler* relation (1.2) it follows that

$$|\zeta^3(\sigma)\zeta^4(\sigma+it)\zeta(\sigma+2it)| = \prod_p |(1-p^{-\sigma})^3(1-p^{-\sigma-it})^4(1-p^{-\sigma-2it})|^{-1}. \tag{1.62}$$

If in Lemma 1.6, $r = p^{-5}$ and $\varphi = -t\log p$, i. e., $e^{i\varphi} = p^{-it}$, then we get

$$|\zeta^3(\sigma)\zeta^4(\sigma+it)\zeta(\sigma+2it)| \ge 1, \tag{1.63}$$

that is,

$$|(\sigma-1)\zeta(\sigma)|^3\left|\frac{\zeta(\sigma+it)^4}{\sigma-1}\right||\zeta(\sigma+2it)| \ge \frac{1}{\sigma-1} \quad (\sigma>1). \tag{1.64}$$

From (1.41) it follows that for $s = \sigma > 1$,

$$\zeta(\sigma) = 1 + \frac{1}{\sigma-1} - \sigma\int_1^{\infty}\frac{\{x\}}{x^{\sigma+1}}\,\partial x, \tag{1.65}$$

that is,

$$(\sigma-1)\zeta(\sigma) = \sigma - \sigma(\sigma-1)\int_1^{\infty}\frac{\{x\}}{x^{\sigma+1}}\,\partial x. \tag{1.66}$$

Then

$$\lim_{\sigma\to 1}(\sigma-1)\zeta(\sigma) = 1. \tag{1.67}$$

Further, for $\sigma > 1$, we have

$$|\zeta(s)| = \left|\sum_{n=1}^{\infty} \frac{1}{n^s}\right| \leqslant \sum_{n=1}^{\infty} \frac{1}{n^\sigma} \leqslant 1 + \int_1^{\infty} \frac{\partial x}{x^\sigma} = \frac{\sigma}{\sigma - 1}. \tag{1.68}$$

From this equation and from (1.67) it follows that

$$|(\sigma - 1)\zeta(\sigma)| \leqslant 2 \quad (1 \leqslant \sigma \leqslant 2). \tag{1.69}$$

Based on (1.64), we have

$$2^3 \left|\frac{\zeta(\sigma + it)}{\sigma - 1}\right|^4 |\zeta(\sigma + it)| \geq \frac{1}{\sigma - 1}. \tag{1.70}$$

Contrary to the statement, suppose that $\zeta(1 + it_0) = 0$ for some $t_0 \neq 0$. Since $\zeta(s)$ is an analytic function at the point $s = 1 + it_0$, from

$$\lim_{\sigma \to 1} \frac{\zeta(\sigma + it_0) - \zeta(\sigma + it_0)}{\sigma - 1} = \zeta'(\sigma + it_0) \tag{1.71}$$

it follows that

$$\left|\frac{\zeta(\sigma + it_0)}{\sigma - 1}\right| \leqslant K \quad (\sigma \to 1). \tag{1.72}$$

Since the function $\zeta(\sigma + it_0)$ is continuous and therefore bounded for $1 \leqslant \sigma \leqslant 2$, taking the limit as $\sigma \downarrow 1$ in (1.72), we get a contradiction. □

Lemma 1.7. *In the region* $\{s = \sigma + it \mid |t| \geq 3, 1 \leqslant \sigma \leqslant 2\}$ *with* $c = \frac{29 \cdot 2^7 \cdot (61)^4}{3^9}$, *we have*

$$\left|\frac{\zeta'(s)}{\zeta(s)}\right| \leqslant c \log^9 |t|. \tag{1.73}$$

Proof. For σ, t such that

$$\sigma_1 = 1 + \frac{1}{c \log^9 |t|} \leqslant \sigma \leqslant 2, \quad |t| \geq 3, \tag{1.74}$$

by (1.69) it follows that

$$|\zeta(s)| \leqslant \frac{2}{\sigma - 1} \leqslant 2c \log^9 |t|, \tag{1.75}$$

and from Lemma 1.5 we get

$$|\zeta(\sigma + 2it)| \leqslant \frac{29}{6} \log 2|t| < \frac{29}{3} \log |t|. \tag{1.76}$$

By (1.63), (1.75), and (1.76) we have

$$\begin{aligned}|\zeta(\sigma+it)| &\geq |\zeta(\sigma)|^{-\frac{3}{4}}|\zeta(\sigma+2it)|^{-\frac{1}{4}} \geq (2c\log^9|t|)^{-\frac{3}{4}}\cdot\left(\frac{29}{3}\log|t|\right)^{-\frac{1}{4}}\\ &= \left(\frac{3c}{8\cdot 29}\right)0^{\frac{1}{4}}\cdot\frac{1}{c}\log^{-7}|t|.\end{aligned} \tag{1.77}$$

For σ in the region $1 \leqslant \sigma \leqslant \sigma_1$, from Lemma 1.5 it follows that

$$|\zeta(\sigma+it)-\zeta(\sigma_1+it)| = \left|\int_{\sigma}^{\sigma_1}\zeta'(u+it)\,\partial u\right| \leqslant \frac{61}{9}(\sigma_1-\sigma)\log^2|t| \leqslant \frac{61}{9c}\log^{-7}|t|. \tag{1.78}$$

From (1.78) and (1.77), with σ_1 instead of σ, it follows that

$$\begin{aligned}|\zeta(\sigma+it)| &\geq |\zeta(\sigma_1+it)| - |\zeta(\sigma_1+it)-\zeta(\sigma+it)|\\ &\geq \left[\left(\frac{3c}{8\cdot 9}\right)^{\frac{1}{4}} - \frac{61}{9}\right]\frac{1}{c}\log^{-7}|t| \quad (1 \leqslant \sigma \leqslant \sigma_1),\end{aligned} \tag{1.79}$$

This, together with (1.78), gives

$$|\zeta(s)| \geqslant \left[\left(\frac{3c}{8\cdot 29}\right)^{\frac{1}{4}} - \frac{61}{9}\right]\frac{1}{c}\log^{-7}|t| \quad (|t| \geq 3,\ 1 \leqslant \sigma \leqslant 2). \tag{1.80}$$

From Lemma 1.5 and (1.80) it follows that

$$\left|\frac{\zeta'(s)}{\zeta(s)}\right| \leqslant \frac{\frac{61}{9}\log^2|t|}{[(\frac{3c}{8\cdot 29})^{\frac{1}{4}} - \frac{61}{9}]\frac{1}{c}\log^{-7}|t|} = \frac{c\log^9|t|}{\frac{61}{9}(\frac{3c}{8\cdot 29})^{\frac{1}{4}} - 1} = c\log^9|t|, \tag{1.81}$$

and the lemma is proved. □

To prove the asymptotic law of the distribution of prime numbers, it suffices to prove that

$$\lim_{x\to\infty}\frac{\psi(x)}{x} = 1, \tag{1.82}$$

where

$$\psi(x) = \sum_{p\leqslant x}\left[\frac{\log x}{\log p}\right]\log p \quad (x > 0) \tag{1.83}$$

(see [112, 113], pp. 90–91).

Instead of $\psi(x)$, it is more convenient to test the function

$$\omega(x) = \int_1^x \frac{\psi(x)}{t}\,\partial t. \tag{1.84}$$

Taking into consideration this function, to prove the asymptotic law of the distribution of the prime numbers, it suffices to prove that

$$\psi(x) = x + o(x) \quad (x \to \infty) \tag{1.85}$$

(see [174], p. 46).

Lemma 1.8. *For $a, b > 0$, we have*

$$\frac{1}{2\pi \mathrm{i}} \int_{a-\mathrm{i}\infty}^{a+\mathrm{i}\infty} \frac{b^s}{s^2}\,\partial s = \begin{cases} \log b, & b \geq 1, \\ 0, & 0 < b < 1. \end{cases} \tag{1.86}$$

Proof. Let

$$I_2 = \frac{1}{2\pi \mathrm{i}} \lim_{R\to\infty} \int_{-R}^{R} \frac{b^{a+\mathrm{i}t}}{(a+\mathrm{i}t)^2}\,\mathrm{i}\,\partial t = -\frac{b^a}{2\pi} \lim_{R\to\infty} \int_{-R}^{R} \frac{b^{\mathrm{i}t}}{(t-a\mathrm{i})^2}\,\partial t. \tag{1.87}$$

For $b \geq 1$, we will integrate the following function along the semicircle of radius R with center 0 (the coordinate origin) above the x-axis and part of the x-axis from $-R$ to R:

$$f(z) = \frac{b^{\mathrm{i}z}}{(z-a\mathrm{i})^2}. \tag{1.88}$$

It should be taken into account that

$$\operatorname*{res}_{z=a\mathrm{i}} f(z) = \mathrm{i}b^{-a} \cdot \log b,$$

$$|I_R| = \left| \int_0^{\pi} \frac{b^{\mathrm{i}Re^{\varphi\mathrm{i}}}}{(Re^{\varphi\mathrm{i}} - a\mathrm{i})^2}\,\mathrm{i}Re^{\varphi\mathrm{i}}\,\partial\varphi \right| \leq \int_0^{\pi} \frac{Rb^{-R\sin\varphi}}{(R-a)^2}\,\partial\varphi \leqslant \frac{8\pi}{\log b}\,\frac{1-b^{-R}}{R^2} \to 0 \quad (R\to\infty). \tag{1.89}$$

In the estimation of integrals we applied the inequality $\sin\varphi \geq \frac{2}{\pi}\varphi$ $(0 \leq \varphi \leq \frac{\pi}{2})$. If $0 < b < 1$, then we integrate the same function along the lower semicircle of radius R with center 0 and the part of the x-axis from $-R$ to R. Then

$$|I_R| \leq \frac{8\pi}{\log\frac{1}{b}} \cdot \frac{1-(\frac{1}{b})^{-R}}{R^2} \to 0 \quad (R\to\infty). \tag{1.90}$$

□

Remark 1.1. The formula used in Lemma 1.8 is a particular case of the following formula:

$$\frac{1}{2\pi}\int_{a-i\infty}^{a+i\infty}\frac{b^s}{s^n}\,\partial s=\begin{cases}\frac{(\log b)^{n-1}}{(n-1)!}, & b\geq 1,\\ 0, & 0<b<1,\end{cases}\tag{1.91}$$

where $a>0$ and $n\in\mathbb{N}$.

Lemma 1.9. *If for $a>1$ and $x\geq 2$, we define*

$$\omega(x)=\frac{1}{2\pi i}\int_{a-i\infty}^{a+i\infty}\left(-\frac{\zeta'(s)}{\zeta(s)}\right)\frac{x^s}{s^2}\,\partial s.\tag{1.92}$$

Then the integral defined by (1.92) *converges absolutely.*

Proof. From Lemma 1.2 it follows that for $\sigma=a$,

$$\left|\frac{\zeta'(s)}{\zeta(s)}\right|\leq\sum_{n+1}^{\infty}\frac{\Lambda(n)}{n^a}\leq\sum_{n+1}^{\infty}\frac{\log n}{n^a}.\tag{1.93}$$

It directly follows that integral (1.92) converges absolutely. Taking $a_n=\Lambda(n)$ and $g(t)=\log\frac{x}{t}$ in Lemma 1.4, it follows that

$$A(t)=\sum_{n\nleq t}\Lambda(n)=\sum_{p^k\leq t}\log p=\psi(t).\tag{1.94}$$

By Lemma 1.4 it follows that

$$\sum_{n\leq x}\Lambda(n)\log\frac{x}{n}=\psi(x)\log 1+\int_1^x\frac{\psi(t)}{t}\,\partial t=\omega(x).\tag{1.95}$$

From Lemma 1.8 it follows that

$$\frac{1}{2\pi i}\int_{a-i\infty}^{a+i\infty}\left(\frac{x}{n}\right)^s\frac{1}{s^2}\,\partial s=\begin{cases}\log\frac{x}{n}, & n\leq x,\\ 0, & n>x,\end{cases}\tag{1.96}$$

whereas from Lemma 1.2 it follows that

$$-\frac{\zeta'(s)}{\zeta(s)}\cdot\frac{x^s}{s^2}=\sum_{n=1}^{\infty}\Lambda(n)\left(\frac{x}{n}\right)^s\cdot\frac{1}{s^2}.\tag{1.97}$$

If we integrate the series in (1.97) term-by-term along the line $\sigma=a$, then (1.92) follows from (1.95) and (1.96). It remains to examine the possibility of integrating term-by-term the series in (1.97). For $\sigma=a$, we have

$$\left|\frac{\Lambda(n)}{s^2}\cdot\left(\frac{x}{n}\right)^s\right| \le \frac{\log n}{n^a}\cdot\frac{x^a}{a^2} \le K\frac{\log n}{n^a}. \tag{1.98}$$

By the *Weierstrass* criterion it follows that the series (1.97) uniformly converges along s on the line $\sigma = a$. Integrating, for each $u > 0$, the series (1.97) term-by-term on the segment with endpoints $a - iu$ and $a + iu$, we obtain

$$\frac{1}{2\pi i}\int_{a-iu}^{a+iu}\left(-\frac{\zeta'(s)}{\zeta(s)}\right)\frac{x^s}{s^2}\,\partial s = \sum_{n=1}^{\infty}\frac{1}{2\pi i}\int_{a-iu}^{a+iu}\frac{\Lambda(n)}{s^2}\cdot\left(\frac{x}{n}\right)^s\partial s. \tag{1.99}$$

Since

$$\left|\frac{1}{2\pi i}\int_{a-iu}^{a+iu}\frac{\Lambda(n)}{s^2}\left(\frac{x}{n}\right)^s\partial s\right| \le \frac{\log n}{2\pi}\left(\frac{x}{n}\right)^a\int_{-\infty}^{+\infty}\frac{\partial t}{a^2+t^2} = \frac{\log n}{2a}\left(\frac{x}{n}\right)^a, \tag{1.100}$$

it follows that the series (1.100) converges uniformly for $u > 0$. Therefore it is possible to move to the limes in (1.100) as $u \to +\infty$, and by (1.95) and (1.96) we obtain expression (1.92). □

Lemma 1.10. *Let $\zeta(s)$ be not equal to zero on a closed rectangle $\eta \le \sigma \le 1$, $-T \le t \le T$ $(0 < \eta < 1, T > 0)$. Then $\omega(x) = x(1 + R(x))$, where*

$$R(x) = \frac{1}{2\pi i}\int_{\Gamma(T,\eta)}\left(-\frac{\zeta'(s)}{\zeta(s)}\right)\frac{x^{s-1}}{s^2}\,\partial s, \tag{1.101}$$

and $\Gamma(T,\eta)$ is the line defined by Figure 1.1.

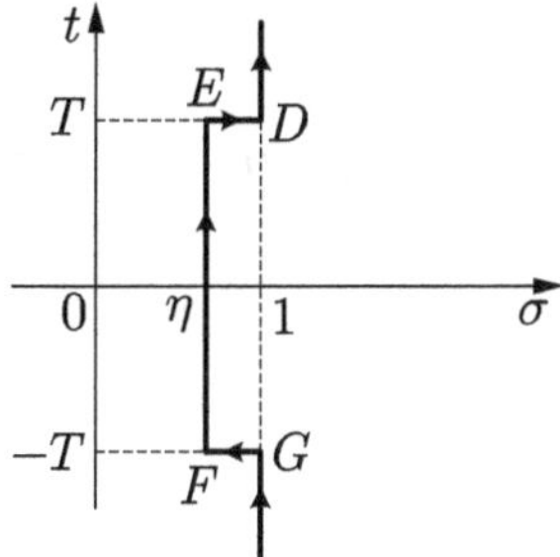

Figure 1.1: Integration contour of the function defined by (1.101).

Proof. For $U > T$, let $\Gamma(U,T,\eta)$ be the contour of the polygon P shown in Figure 1.2.

The existence of the rectangle $DEFG$ follows from the facts that $\zeta(s)$ is an analytic function for $\sigma > 0$ (except for $s = 1$) and that $\zeta(1+it) \ne 0$ for $t \ne 0$, as it is already proved in Theorem 1.3. We have

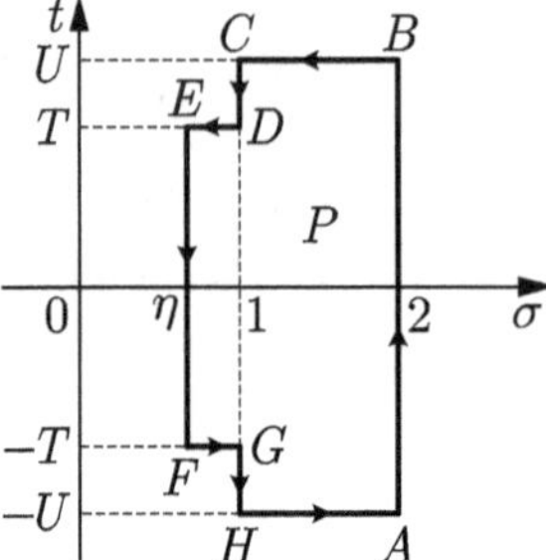

Figure 1.2: Contour that defines the polygon P.

$$\zeta(s) = \frac{1}{s-1} + f(s) \tag{1.102}$$

for $f(s)$, which is defined as an analytic function for $\sigma > 0$. It follows that

$$\frac{\zeta'(s)}{\zeta(s)} = -\frac{1-(s-1)^2 f'(s)}{1+(s-1)f(s)} \cdot \frac{1}{s-1}, \tag{1.103}$$

and since $\zeta(s)$ is different from zero on the rectangle $DEFG$, from Lemma 1.1 (relation (1.9)) it follows that $(-\frac{\zeta'(s)}{\zeta(s)})\frac{x^s}{s^2}$ on the polygon P has a unique singularity, a pole of order 1 at $s = 1$. The remainder at that point is x. It follows that

$$\frac{1}{2\pi i} \int\limits_{\Gamma(U,T,\eta)} \left(-\frac{\zeta'(s)}{\zeta(s)}\right)\frac{x^s}{s^2}\, \partial s = x. \tag{1.104}$$

Using Lemma 1.7, the value of the integral along the line BC can be estimated as

$$\left|\frac{1}{2\pi i}\int\limits_{2+iU}^{1+iU}\left(-\frac{\zeta'(s)}{\zeta(s)}\right)\frac{x^s}{s^2}\,\partial s\right| = \left|\frac{1}{2\pi i}\int\limits_{1}^{2}\frac{\zeta'(\sigma+iU)}{\zeta(\sigma+iU)}\cdot\frac{x^{\sigma+iU}}{(\sigma+iU)^2}\,\partial\sigma\right|$$
$$\leq \frac{1}{2\pi} c\log^9 U \cdot \int\limits_1^2 \frac{x^\sigma}{\sigma^2+U^2}\,\partial\sigma < cx^2\frac{\log^9 U}{U^2} \to 0 \quad (U \to +\infty). \tag{1.105}$$

By analogy the integral along the line AH converges to 0 as $U \to +\infty$. Switching to (1.104) in the case where $U \to +\infty$, Lemma 1.10 follows from Lemma 1.9. □

If we prove that $R(x) = o(1)$, then from Lemma 1.10 it follows that $\omega(x) \sim x$ $(x \to \infty)$, which would prove relation (1.85), equivalent to proving the asymptotic law of prime number distribution.

Theorem 1.6. *We have the following estimate:*

$$R(x) = o(1). \tag{1.106}$$

Proof. Let $\varepsilon > 0$ be arbitrary. For $|t| \geq 3$, the following estimate follows from Lemma 1.7:

$$\left|\frac{\zeta'(1+\mathrm{i}t)}{\zeta(1+\mathrm{i}t)} \cdot \frac{x^{\mathrm{i}t}}{(1+\mathrm{i}t)^2}\right| \leq \frac{c\log^9|t|}{1+t^2}. \tag{1.107}$$

Therefore we can choose $T = T(\varepsilon) > 3$ (which does not depend on x) such that the following inequalities are valid:

$$\begin{aligned} J_1 &= \left|\frac{1}{2\pi\mathrm{i}} \int\limits_{1+\mathrm{i}T}^{1+\mathrm{i}\infty} \left(-\frac{\zeta'(s)}{\zeta(s)}\right)\frac{x^{s-1}}{s^2}\,\partial s\right| = \left|\frac{1}{2\pi}\int\limits_{T}^{\infty}\left(-\frac{\zeta'(1+\mathrm{i}t)}{\zeta(1+\mathrm{i}t)}\right)\frac{x^{\mathrm{i}t}}{(1+\mathrm{i}t)}\,\partial t\right| \\ &\leq \frac{1}{2\pi}\int\limits_{T}^{\infty} c\frac{\log^9|t|}{1+t^2}\,\partial t < \frac{\varepsilon}{5}, \end{aligned} \tag{1.108}$$

$$\begin{aligned} J_2 &= \left|\frac{1}{2\pi\mathrm{i}} \int\limits_{1-\mathrm{i}\infty}^{1-\mathrm{i}t} \left(-\frac{\zeta'(s)}{\zeta(s)}\right)\frac{x^{s-1}}{s^2}\,\partial s\right| = \left|\frac{1}{2\pi}\int\limits_{T}^{\infty}\left(-\frac{\zeta'(1-\mathrm{i}t)}{\zeta(1-\mathrm{i}t)}\right)\frac{x^{-\mathrm{i}t}}{(1-\mathrm{i}t)^2}\,\partial t\right| \\ &\leq \frac{1}{2\pi}\int\limits_{T}^{\infty} c\frac{\log^9|t|}{1+t^2}\,\partial t < \frac{\varepsilon}{5}. \end{aligned} \tag{1.109}$$

Let us choose $\eta = \eta(t)$, $0 < \eta < 1$, so that the $\zeta(s)$ is not equal to zero in the rectangle $\eta \leq \sigma \leq 1$, $-T \leq t \leq T$. In addition, the function $|\frac{\zeta'(s)}{\zeta(s)} \cdot \frac{1}{s^2}|$ is continuous on the curve *DEFG* (Figure 1.2). Therefore there exists a constant $M = M(T, \eta)$ such that at every point on that curve, the following condition is satisfied:

$$\left|\frac{\zeta'(s)}{\zeta(s)} \cdot \frac{1}{s^2}\right| \leq M. \tag{1.110}$$

Therefore we have the following inequalities:

$$\begin{aligned} J_3 &= \left|\frac{1}{2\pi\mathrm{i}} \int\limits_{1+\mathrm{i}T}^{\eta+\mathrm{i}T} \left(-\frac{\zeta'(s)}{\zeta(s)}\right)\frac{x^{s-1}}{s^2}\,\partial s\right| = \left|\frac{1}{2\pi\mathrm{i}}\int\limits_{1}^{\eta}\left(-\frac{\zeta'(\sigma+\mathrm{i}t)}{\zeta(\sigma+\mathrm{i}t)}\right)\frac{x^{\sigma-1+\mathrm{i}t}}{(\sigma+\mathrm{i}T)^2}\,\partial\sigma\right| \\ &\leq \frac{1}{2\pi}M\int\limits_{\eta}^{1} x^{\sigma-1}\,\partial\sigma = \frac{M}{2\pi}\left(1-\frac{1}{x^{1-\eta}}\right)\frac{1}{\log x} < \frac{M}{2\pi\log x}, \end{aligned} \tag{1.111}$$

$$\begin{aligned} J_4 &= \left|\frac{1}{2\pi\mathrm{i}} \int\limits_{\eta-\mathrm{i}T}^{1-\mathrm{i}T} \left(-\frac{\zeta'(s)}{\zeta(s)}\right)\frac{x^{s-1}}{s^2}\,\partial s\right| = \left|\frac{1}{2\pi\mathrm{i}}\int\limits_{\eta}^{1}\left(-\frac{\zeta'(\sigma-\mathrm{i}t)}{\zeta(\sigma-\mathrm{i}t)}\right)\frac{x^{\sigma-1-\mathrm{i}t}}{(\sigma-\mathrm{i}T)^2}\,\partial\sigma\right| \\ &\leq \frac{1}{2\pi}M\int\limits_{\eta}^{1} x^{\sigma-1}\,\partial\sigma < \frac{M}{2\pi\log x}, \end{aligned} \tag{1.112}$$

$$J_5 = \left| \frac{1}{2\pi i} \int_{\eta - iT}^{\eta + iT} \left(-\frac{\zeta'(s)}{\zeta(s)} \right) \frac{x^{s-1}}{s^2} \, \partial s \right| = \left| \frac{1}{2\pi} \int_{-T}^{T} \left(-\frac{\zeta'(\eta + it)}{\zeta(\eta + it)} \right) \frac{x^{\eta - 1 + it}}{(\eta + it)^2} \, \partial t \right|$$
$$\leq \frac{1}{2\pi} M x^{\eta - 1} \int_{-T}^{T} \partial t = \frac{T}{\pi} M x^{\eta - 1}. \tag{1.113}$$

From these inequalities it follows that there exists $x_0 = x_0(M, T, \eta)$ such that for all $x > x_0$, we have the inequalities

$$J_3 < \frac{\varepsilon}{5}, \quad J_4 < \frac{\varepsilon}{5}, \quad J_5 < \frac{\varepsilon}{5}. \tag{1.114}$$

From (1.108), (1.109), and (1.114) it follows that for $x > x_0$,

$$|R(x)| \leqslant J_1 + J_2 + J_3 + J_4 + J_5 < \varepsilon, \tag{1.115}$$

that is,

$$R(x) = o(1). \tag{1.116}$$

□

Lemma 1.11. *We have the following equation:*

$$-\frac{\zeta'(s)}{\zeta(s)} = s \int_{1}^{\infty} \frac{\psi(x)}{x^{s+1}} \, \partial x \quad (\sigma > 1). \tag{1.117}$$

Proof. Taking $a_n = \Lambda(n)$ and $g(x) = x^{-s}$ in Lemma 1.4, we can write

$$A(x) = \sum_{n \leq x} a_n = \Psi(x). \tag{1.118}$$

Since

$$\Psi(x) = \sum_{p \leq x} \left[\frac{\log x}{\log p} \right] \log p \leq \sum_{p \leq x} \frac{\log x}{\log p} \log p = \pi(x) \log x < x \log x, \tag{1.119}$$

we have

$$|A(x) g(x)| < \frac{\log x}{x^{\sigma - 1}}, \tag{1.120}$$

and since

$$\lim_{x \to \infty} \frac{\log x}{x^{\sigma - 1}} = 0 \quad (\sigma > 1), \tag{1.121}$$

we have

$$\lim_{x\to\infty} A(x)g(x) = 0. \tag{1.122}$$

From Lemma 1.4 it follows that

$$\sum_{n=1}^{\infty} \frac{\Lambda(n)}{n^s} = s\int_1^{\infty} \frac{\psi(x)}{x^{s+1}}\,\partial x \quad (\sigma > 1). \tag{1.123}$$

This, together with formula

$$\sum_{n=1}^{\infty} \frac{\Lambda(n)}{n^s} = -\frac{\zeta'(s)}{\zeta(s)} \quad (\sigma > 1) \tag{1.124}$$

and Lemma 1.2, gives equation (1.117). □

Theorem 1.7. *From*

$$\lim_{x\to\infty} \frac{\Psi(x)}{x} = 1 \tag{1.125}$$

it follows that

$$\zeta(1+it) \neq 0 \quad (\forall t). \tag{1.126}$$

Proof. From (1.17) it follows that

$$\int_1^{\infty} \frac{\psi(x)-x}{x^{s+1}}\,\partial x = -\frac{\zeta'(s)}{s\zeta(s)} + \frac{1}{1-s} \equiv \Phi(s). \tag{1.127}$$

According to Theorem 1.4, the function $\Phi(s)$ is regular for $\sigma > 0$, except at the points where $\zeta(s) = 0$ (the point s_0 such that $\zeta(s_0) = 0$ is a pole of order 1 for $\Phi(s)$) and at the point $s = 1$. From $\lim_{x\to\infty} \frac{\Psi(x)}{x} = 1$ it follows that $\Psi(x) = x + o(x)$, i. e., for $\varepsilon > 0$, we have $|\Psi(x) - x| < \varepsilon x$ for $x \geq x_0(\varepsilon) > 1$. Then for $\sigma > 1$, we have

$$|\Phi(s)| < \int_1^{x_0} \frac{|\psi(x)| - x}{x^2}\,\partial x + \int_{x_0}^{\infty} \frac{|\psi(x)| - x}{x^{\sigma+1}}\,\partial x, \tag{1.128}$$

$$\int_{x_0}^{\infty} \frac{|\psi(x)| - x}{x^{\sigma+1}}\,\partial x < \varepsilon\int_{x_0}^{\infty} \frac{\partial x) - x}{x^{\sigma}} < \varepsilon\int_{x_0}^{\infty} \frac{\partial x)}{x^{\sigma}} = \frac{\varepsilon}{\sigma-1}. \tag{1.129}$$

It follows that

$$|\Phi(s)| < K + \frac{\varepsilon}{\sigma-1} \quad (\sigma > 1), \tag{1.130}$$

where $K = K(x_0) = K(\varepsilon)$, and thus

$$(\sigma - 1)|\Phi(s)| < K(\sigma - 1) + \varepsilon \quad (\sigma > 1). \tag{1.131}$$

If $\sigma \downarrow 1$, then assuming that $\sigma < \frac{\varepsilon}{K} + 1$, we have $(\sigma - 1)|\Phi(s)| < 2\varepsilon$, i. e., for fixed t, $\lim_{\sigma\downarrow 1}(\sigma - 1)\Phi(\sigma + it) = 0$. If $1 + it$ $(t \neq 0)$ is a zero of the function $\zeta(s)$, then it follows that

$$\begin{aligned}\lim_{\sigma\downarrow 1}(\sigma - 1)\Phi(\sigma + it) &= \lim_{\sigma\downarrow 1}(\sigma - 1)\left\{-\frac{\zeta'(\sigma + it)}{(\sigma + it)\zeta(\sigma + it)} + \frac{1}{1 - \sigma - it}\right\} \\ &= \lim_{\sigma\downarrow 1}\left\{-\frac{\zeta'(\sigma + it)}{\frac{\zeta(\sigma+it)-\zeta(\sigma+it)}{\sigma-1}} \cdot \frac{1}{\sigma + it} + \frac{\sigma - 1}{1 - \sigma - it}\right\} = -\frac{1}{1 + it} \neq 0, \end{aligned} \tag{1.132}$$

which is impossible. Therefore it is clear that $\zeta(1 + it) \neq 0$ $(t \neq 0)$. □

Remark 1.2. From Theorems 1.6 and 1.7 it follows that

$$\lim_{x\to\infty} \frac{\psi(x)}{x} = 1 \quad \text{and} \quad \zeta(1 + it) \neq 0 \quad (\forall t) \tag{1.133}$$

are equivalent statements.

1.2 Analytic extensions of the function $\zeta(s)$

1.2.1 Extensions using *Dirichlet* series

One way in which we can obtain an analytic extension of the function $\zeta(s)$ defined by relation (1.1) for $\sigma > 1$ is given by relations (1.40) and (1.41). The second way is based on the application of *Dirichlet* series.

Definition 1.6. Let $\{\lambda_n\}_1^\infty$ be a strictly increasing series of positive numbers that tends to infinity, and let $\{a_n\}_1^\infty$ be a series of complex numbers. Series of the form

$$\sum_{n=1}^{\infty} a_n e^{-\lambda_n s} \tag{1.134}$$

are called *Dirichlet* series.

Lemma 1.12. (a) *If a series*

$$\sum_{n=1}^{\infty} a_n e^{-\lambda_n s} \quad (s_0 = \sigma_0 + it_0) \tag{1.135}$$

converges, then the series (1.134) *uniformly converges in the region*

$$|\arg(s - s_0)| \leq \gamma < \frac{\pi}{2}. \tag{1.136}$$

(b) *If a series*

$$\sum_{n=1}^{\infty} a_n e^{-\lambda_n s} \quad (s_0 = \sigma_0 + it_0) \tag{1.137}$$

converges, then the series (1.134) *converges for every* $s = \sigma + it$, $\sigma > \sigma_0$, *and uniformly converges on any segment of the semiplane* $\sigma > \sigma_0$.

(c) *If a series*

$$\sum_{n=1}^{\infty} a_n e^{-\lambda_n s} \quad (s_0 = \sigma_0 + it_0) \tag{1.138}$$

absolutely converges, then the series (1.134) *absolutely converges for* $s = \sigma + it$, $\sigma > \sigma_0$.

Proof. (a) Let $s = s_0 + s'$, $s = \sigma' + it'$, $\sigma' > \sigma$, $|\arg s'| < \gamma$. We introduce the following notations:

$$\sum_{n\le m} a_n e^{-\lambda_n s_0} = A_m, \quad (s_0) = A_m, \tag{1.139}$$

$$\sum_{n=1}^{\infty} a_n e^{-\lambda_n s_0} = S = \lim_{m\to\infty} A_m. \tag{1.140}$$

For $\sigma' > 0$, we have

$$|\tan(\arg s')| = \frac{|t'|}{\sigma'} \le \tan\gamma = M < \infty, \tag{1.141}$$

and for $q > p \ge 2$,

$$\begin{aligned}
\sum_{p\le n\le q} a_n e^{-\lambda_n s} &= \sum_{p\le n\le q} a_n e^{-\lambda_n s_0} e^{-\lambda_n s'} = \sum_{p\le n\le q} (A_n - A_{n-1}) e^{-\lambda_n s'} \\
&= \sum_{p\le n\le q} [(A_n - S) - (A_{n-1} - S)] e^{-\lambda_n s'} \\
&= (A_q - S)e^{-\lambda_q s'} - (A_{p-1} - S)e^{-\lambda_p s'} + \sum_{p\le n\le q-1} (A_n - S)(e^{-\lambda_n s'} - e^{-\lambda_{n+1} s'}).
\end{aligned} \tag{1.142}$$

Let $\varepsilon > 0$ be arbitrary, and let p be chosen so that $|A_n - S| < \varepsilon$ for $n \ge p-1$ (which is fulfilled because S exists). Then

$$\begin{aligned}
\left|\sum_{p\le n\le q-1} a_n e^{-\lambda_n s}\right| &\le |A_q - S| e^{-\lambda_q \sigma'} + |A_{p-1} - S| e^{-\lambda_p \sigma'} + \sum_{p\le n\le q-1} |A_n - S| \left|e^{-\lambda_n s'} - e^{-\lambda_{n+1} s'}\right| \\
&< \varepsilon \sum_{p\le n\le q-1} \left|e^{-\lambda_n s'} - e^{-\lambda_{n+1} s'}\right| + 2\varepsilon.
\end{aligned} \tag{1.143}$$

Since

$$|e^{-\lambda_n s'} - e^{-\lambda_{n+1} s'}| = |s'| \left| \int_{\lambda_n}^{\lambda_{n+1}} e^{-\mu s'} \, \partial u \right| \leq |s'| \int_{\lambda_n}^{\lambda_{n+1}} e^{-\mu \sigma'} \, \partial u$$
$$= \frac{|s'|}{\sigma'} (e^{-\lambda_n \sigma'} - e^{-\lambda_{n+1} \sigma'}) \leq (M+1)(e^{-\lambda_n \sigma'} - e^{-\lambda_{n+1} \sigma'}), \tag{1.144}$$

it follows that

$$\left| \sum_{p \leqslant n \leqslant q} a_n e^{-\lambda_n s} \right| < \varepsilon(M+1) \sum_{p \leqslant n \leqslant q-1} (e^{-\lambda_n \sigma'} - e^{-\lambda_{n+1} \sigma'}) + 2\varepsilon$$
$$= \varepsilon(M+1)(e^{-\lambda_p \sigma'} - e^{-\lambda_q \sigma'}) + 2\varepsilon = \varepsilon(M+3). \tag{1.145}$$

Because of the arbitrariness of the numbers p and q, this proves proposition (a) of Lemma 1.12.

(b) Follows directly from (a).

(c) If a series $\sum_{n=1}^{\infty} a_n e^{-\lambda_n s_0}$ absolutely converges, then the series $\sum_{n=1}^{\infty} |a_n| e^{-\lambda_n \sigma_0}$ converges, and hence, according to (a), the series $\sum_{n=1}^{\infty} |a_n| e^{-\lambda_n \sigma}$ converges for $s_0 = \sigma_0$, $s = \sigma > \sigma_0$, i. e., the series $\sum_{n=1}^{\infty} a_n e^{-\lambda_n s}$ absolutely converges for $\sigma > \sigma_0$.

The lemma is proved. □

We will now present some formulas by which analytic extensions of the function $\zeta(s)$ are expressed.

Proposition 1.1. *For $\sigma > 0$, we have*

$$\sum_{n=1}^{\infty} \frac{(-1)^{n-1}}{n^s} = (1 - 2^{1-s})\zeta(s). \tag{1.146}$$

Proof. Based on (1.1), for $\sigma > 1$, formula (1.146) can be easily checked. In Lemma 1.12, we take $\lambda_n = \log n$ and $a_n = (-1)^{n-1}$. For $s_0 = \sigma_0 > 0$, the series (1.146) converges according to the *Leibnitz* criterion. By Lemma 1.12 it follows that the series converges for $\sigma > \sigma_0 > 0$. Because of the arbitrariness of the number $\sigma_0 > 0$, the series (1.146) converges for $\sigma > 0$. Based on the uniqueness of the analytic extension, (1.146) is valid for $\sigma > 0$. □

For $\Re(\mu) > 0$ and $\Re(s) = \sigma > 0$, we have

$$\int_0^{+\infty} \frac{x^{s-1}}{e^{\mu x} + 1} \, \partial x = \frac{1}{\mu^s} (1 - 2^{1-s}) \Gamma(s) \zeta(s) \tag{1.147}$$

and, in particular, for $\mu = 1$,

$$\int_0^{+\infty} \frac{x^{s-1}}{e^x + 1} \, \partial x = (1 - 2^{1-s}) \Gamma(s) \zeta(s) \quad (\sigma > 0). \tag{1.148}$$

Definition 1.7. The function defined as

$$\Gamma(s) = \int_0^{+\infty} x^{s-1} e^{-x} \, \partial x \quad (\sigma > 0) \tag{1.149}$$

and extended along the whole s-plane (apart from $s = 0, -1, -2, \ldots$) by the relation

$$\Gamma(s+1) = s\Gamma(s) \tag{1.150}$$

is called the gamma function.

Proof. Let

$$I = \int_0^{+\infty} \frac{x^{s-1}}{e^{\mu x} + 1} \partial x = |\mu x = u| = \frac{1}{\mu^s} \int_0^{\infty} \frac{u^{s-1}}{e^u + 1} \partial u. \tag{1.151}$$

We integrate the function $f(z) = \frac{z^{s-1}}{e^z+1}$ along the curve presented in Figure 1.3.

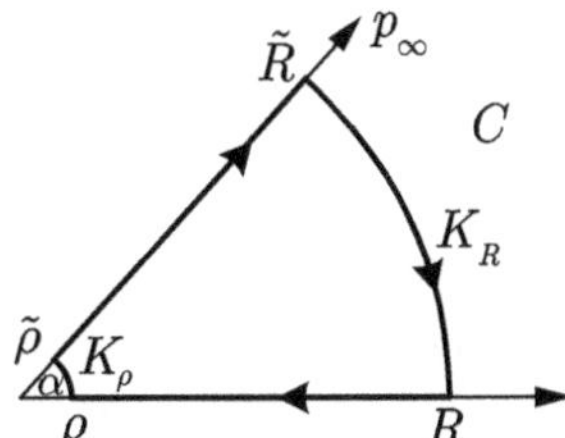

Figure 1.3: The curve along which the integration of the function $f(z)$ is performed.

Since for $0 \le \varphi \le \alpha$ ($\alpha = \arg \mu$), we have

$$\left|1 + e^{Re^{\varphi i}}\right|^2 \ge \left|1 + e^{Re^{\alpha i}}\right|^2 \ge \left(e^{R\cos\alpha} - 1\right)^2. \tag{1.152}$$

Then

$$|I_R| = \left| \int_\alpha^0 \frac{(Re^{\varphi i})^s}{e^{Re^{\varphi i}} + 1} i \, \partial\varphi \right| \le K \int_0^\alpha \frac{R^\sigma}{|e^{Re^{\varphi i}} + 1|} \partial\varphi \le K_1 \frac{R^\sigma}{e^{R\cos\alpha} - 1}. \tag{1.153}$$

However, since

$$\lim_{R\to\infty} \frac{R^\sigma}{e^{R\cos\alpha} - 1} = \cdots = L \lim_{R\to\infty} \frac{R^{\sigma-n}}{e^{R\cos\alpha}} = 0 \quad (n > \sigma > 0), \tag{1.154}$$

it follows that

$$\lim_{R\to\infty} I_R = 0 \quad (\sigma > 0). \tag{1.155}$$

In addition, for $\sigma > 0$, we can write

$$|I_\rho| \leqslant K \int_0^\alpha \frac{\rho\sigma}{|e^{\rho e^{\varphi i}} + 1|}\, \partial\varphi \to 0 \quad (\rho \to 0), \tag{1.156}$$

so it follows that

$$\lim_{\rho\to 0} I_\rho = 0 \quad (\sigma > 0). \tag{1.157}$$

Due to the regularity of $f(z)$ within the contour C (Figure 1.3), we can conclude that

$$\int_0^{p_\infty} \frac{u^{s-1}}{e^u + 1}\, \partial u = \int_0^\infty \frac{x^{s-1}}{e^x + 1}\, \partial x. \tag{1.158}$$

From (1.152) and (1.158) it follows that

$$I = \frac{1}{u^s} \int_0^\infty \frac{x^{s-1}}{e^x + 1}\, \partial x. \tag{1.159}$$

For $\sigma > 0$, relation (1.148) follows from the equation

$$\begin{aligned}\int_0^\infty \frac{x^{s-1}}{e^x+1}\,\partial x &= \int_0^\infty x^{s-1}\frac{e^{-x}}{1+e^{-x}}\,\partial x = \int_0^\infty \left(x^{s-1}\sum_{n=1}^\infty (-1)^{n-1} e^{-nx}\right)\partial x \\ &\stackrel{*}{=} \sum_{n=1}^\infty (-1)^{n-1}\int_0^\infty x^{s-1}e^{-nx}\,\partial x = |nx = v| = \sum_{n=1}^\infty (-1)^{n-1}\frac{\Gamma(s)}{n^s} \\ &= \Gamma(s)\sum_{n=1}^\infty \frac{(-1)^{n-1}}{n^s} \stackrel{(1.146)}{=} (1-2^{1-s})\Gamma(s)\zeta(s).\end{aligned} \tag{1.160}$$

The sign of equality $\stackrel{*}{=}$ is due to the fact that the series behind the sign $\stackrel{*}{=}$ absolutely converges for $\sigma > 1$. Indeed,

$$\sum_{n=1}^\infty \left|(-1)^{n-1}\int_0^\infty x^{s-1}e^{-nx}\,\partial x\right| \le \sum_{n=1}^\infty \int_0^\infty x^{\sigma-1}e^{-nx}\,\partial x = \sum_{n=1}^\infty \frac{\Gamma(\sigma)}{n^\sigma} = \Gamma(\sigma)\zeta(\sigma). \tag{1.161}$$

As the series defined by (1.146) converges for $\sigma > 0$, relation (1.148) is valid for $\sigma > 0$. Now (1.147) follows from (1.148) and (1.159). □

The following formulas are similar to those already given (the first one was derived by *Riemann* himself):

$$\int_0^\infty \frac{x^{s-1}}{e^x - 1}\,\partial x = \Gamma(s)\zeta(s) \quad (\sigma > 1), \tag{1.162}$$

$$\int_0^\infty \frac{x^{s-1}}{e^{\mu x} - 1}\,\partial x = \frac{1}{\mu^s}\Gamma(s)\zeta(s) \quad (\Re e(\mu) > 0,\ \sigma > 1), \tag{1.163}$$

$$\int_0^\infty \frac{x^{s-1}e^x}{e^{2x} - 1}\,\partial x = \frac{2^s - 1}{2^s}\Gamma(s)\zeta(s) \quad (\sigma > 1). \tag{1.164}$$

We will also need the following two formulas:

$$\int_0^{+\infty} e^{-x^2}\cos 2bx\,\partial x = \frac{\sqrt{\pi}}{2}e^{-b^2}, \tag{1.165}$$

$$\int_0^{+\infty} e^{-\lambda x^2}\cos(2a\lambda x)\,\partial x = \sqrt{\frac{\pi}{\lambda}}e^{-\lambda a^2} \quad (\lambda > 0,\ a \geq 0). \tag{1.166}$$

Proof. Integrating the function $f(s) = e^{-s^2}$ along the contour of the rectangle with vertices $(-R, O)$, (R, O), (R, b), $(-R, b)$ and using the relation $\int_0^{+\infty} e^{-x^2}\,\partial x = \sqrt{\pi}$, we obtain formula (1.165). After appropriate replacements, (1.166) follows from (1.165). □

Lemma 1.13. *Let for $\tau > 0$,*

$$\theta(\tau) = \sum_{n=-\infty}^{\infty} e^{-n^2\pi\tau}. \tag{1.167}$$

Then

$$\theta(\tau) = \frac{1}{\sqrt{\tau}}\theta\left(\frac{1}{\tau}\right). \tag{1.168}$$

Proof. We will integrate the function $f(s) = \frac{e^{-s^2\pi\tau}}{e^{2\pi si}-1}$ along the contour of the rectangle C_N with vertices $(-(N+\frac{1}{2}), -i)$, $(N+\frac{1}{2}, -i)$, $(N+\frac{1}{2}, i)$, $(-(N+\frac{1}{2}), i)$. The singularities of the function $f(s)$ within C_N are $s_k = k$ $(k = 0, \pm 1, \ldots, \pm N)$. Decomposing the function $e^{2\pi si} - 1$ in the series of degrees from $s - s_k$, we conclude that the points s_k are poles of order 1. In addition,

$$\operatorname*{res}_{s=s_k} f(s) = \frac{e^{-k^2\pi\tau}}{2\pi i}. \tag{1.169}$$

Further, we can write

$$|I_1| = \left|\int_{N+\frac{1}{2}-i}^{N+\frac{1}{2}+i} \frac{e^{-s^2\pi\tau}}{e^{2\pi si} - 1}\,\partial s\right| = \left|\int_{-1}^{1} \frac{e^{-\pi\tau[(N+\frac{1}{2})+\mu i]^2}}{e^{2\pi\tau[(N+\frac{1}{2})+\mu i]} - 1}\, i\,\partial u\right|$$

$$\le e^{-\pi\tau(N+\frac{1}{2})^2}\int_{-1}^{1}\frac{e^{\pi\tau u^2}|e^{-2\pi\tau(N+\frac{1}{2})ui}|}{|-e^{-2\pi u}-1|}\,\mathrm{d}u \le Ke^{-\pi\tau(N+\frac{1}{2})^2}\to 0 \quad (N\to\infty), \tag{1.170}$$

$$|I_2| = \left|\int_{-(N+\frac{1}{2})-i}^{-(N+\frac{1}{2})+i}\frac{e^{-s^2\pi\tau}}{e^{2\pi s i}-1}\,\mathrm{d}s\right| \le Ke^{-\pi\tau(N+\frac{1}{2})^2}\to 0 \quad (N\to\infty). \tag{1.171}$$

Taking the limits as $N\to\infty$, based on the above relations, according to *Cauchy*'s residue theorem, it follows that

$$\theta(\tau) = \int_{-\infty-i}^{+\infty-i}\frac{e^{-s^{2\pi\tau}}}{e^{2\pi s i}-1}\,\mathrm{d}s - \int_{-\infty+i}^{+\infty+i}\frac{e^{-s^2\pi\tau}}{e^{2\pi s i}-1}\,\mathrm{d}s. \tag{1.172}$$

For $s=\mu-i$, we have $|e^{-2\pi s i}| = e^{-2\pi} < 1$ and

$$\begin{aligned}P &= \int_{-\infty-i}^{+\infty-i}\frac{e^{-s^2\pi\tau}}{e^{2\pi s i}-1}\,\mathrm{d}s = \lim_{R\to\infty}\int_{-R-i}^{R-i}\frac{e^{-s^2\pi\tau}}{e^{s^2\pi\tau}-1}\,\mathrm{d}s = \lim_{R\to\infty}\int_{-R-i}^{R-i}\left(e^{-s^2\pi\tau}\sum_{n=1}^{\infty}e^{-2n\pi s i}\right)\mathrm{d}s\\ &= \int_{-\infty}^{+\infty}\left(e^{-\pi\tau(u-i)^2}\sum_{n=1}^{\infty}e^{-2n\pi i(u-i)}\right)\mathrm{d}u \overset{*}{=} \sum_{n=1}^{\infty}e^{-2n\pi}\cdot e^{\pi\tau}\int_{-\infty}^{+\infty}e^{\pi\tau u^2}\cdot e^{2\pi i u(\tau-n)}\,\mathrm{d}u.\end{aligned} \tag{1.173}$$

It follows that

$$\begin{aligned}\mathfrak{Re}(P) &= \sum_{n=1}^{\infty}e^{-2n\pi}\cdot e^{\pi\tau}\int_{-\infty}^{+\infty}e^{-\pi\tau u^2}\cos[2\pi(\tau-n)u]\,\mathrm{d}u \overset{(1.166)}{=} \left|\begin{matrix}\lambda=\pi i\\ \alpha=\frac{\tau-n}{\tau}\end{matrix}\right|\\ &= \sum_{n=1}^{\infty}e^{-2n\pi}\cdot e^{\pi\tau}\cdot\frac{1}{\sqrt{\tau}}\cdot e^{-\pi\frac{(\tau-n)^2}{\tau}} = \frac{1}{\sqrt{\tau}}\sum_{n=1}^{\infty}e^{-\frac{n^2\pi}{\tau}}.\end{aligned} \tag{1.174}$$

In addition, we have

$$\sum_{n=1}^{\infty}e^{-\frac{n^2\pi}{\tau}} < \sum_{n=1}^{\infty}(e^{-\frac{\pi}{\tau}}) = \frac{e^{-\frac{\pi}{\tau}}}{1-e^{-\frac{\pi}{\tau}}}. \tag{1.175}$$

Since this series absolutely converges, in the equality denoted by $\overset{*}{=}$ in (1.173), the order of series and integral can be interchanged. Since $|e^{2\pi s i}| = e^{-2\pi} < 1$ for $s=u+i$, it follows that

$$\begin{aligned}Q &= -\int_{-\infty+i}^{+\infty+i}\frac{e^{-s^2\pi\tau}}{e^{2\pi s i}-1}\,\mathrm{d}s = \int_{-\infty+i}^{+\infty+i}\left(e^{-s^2\pi\tau}\sum_{n=0}^{\infty}e^{2n\pi s i}\right)\mathrm{d}s\\ &= \int_{-\infty}^{+\infty}\left(e^{-\pi\tau(u+i)^2}\sum_{n=0}^{\infty}e^{2n\pi i(u+i)}\right)\mathrm{d}u \overset{*}{=} \sum_{n=0}^{\infty}e^{-2n\pi}\cdot e^{\pi\tau}\int_{-\infty}^{+\infty}e^{-\pi\tau u^2}e^{2\pi u(n-\tau)i}\,\mathrm{d}u.\end{aligned} \tag{1.176}$$

Therefore we also have the following equality:

$$\mathfrak{Re}(Q) = \sum_{n=0}^{\infty} e^{-2n\pi} \cdot e^{\pi\tau} \int_{-\infty}^{+\infty} e^{-\pi\tau u^2} \cos[2\pi(n-\tau)u]\, \partial u \overset{(1.165)}{=} \frac{1}{\sqrt{\tau}} \sum_{n=0}^{\infty} e^{-\frac{n^2\pi}{\tau}}. \tag{1.177}$$

Finally, we have

$$\theta(\tau) = \mathfrak{Re}(P+Q) = \frac{1}{\sqrt{\tau}} \sum_{n=0}^{\infty} e^{-\frac{n^2\pi}{\tau}} = \frac{1}{\sqrt{\tau}} \theta\left(\frac{1}{\tau}\right). \tag{1.178}$$

□

1.2.2 Functional equation for $\zeta(s)$

Theorem 1.8. *For $s \neq 0, 1$, we have the relation*

$$\pi^{-\frac{s}{2}} \Gamma\left(\frac{s}{2}\right) \zeta(s) = \pi^{-\frac{1-s}{2}} \Gamma\left(\frac{1-s}{2}\right) \zeta(1-s), \tag{1.179}$$

which is known as the functional equation for $\zeta(s)$, first discovered by Bernhard Riemann.

Proof. For $\sigma > 0$, we can conclude that

$$\int_{\infty}^{\infty} x^{\frac{s}{2}-1} e^{-n^2\pi x}\, \partial x = \frac{\Gamma(\frac{s}{2})}{n^s \pi^{\frac{s}{2}}}. \tag{1.180}$$

However, for $\sigma > 1$,

$$\frac{1}{\pi^{\frac{s}{2}}} \Gamma\left(\frac{s}{2}\right) \zeta(s) = \frac{1}{\pi^{\frac{s}{2}}} \Gamma\left(\frac{s}{2}\right) \sum_{n=1}^{\infty} \frac{1}{n^s} \overset{(1.180)}{=} \sum_{n=1}^{\infty} \int_0^{\infty} x^{\frac{s}{2}-1} e^{-n^2\pi x}\, \partial x \overset{*}{=} \int_0^{\infty} \left(x^{\frac{s}{2}-1} \sum_{n=1}^{\infty} e^{-n^2\pi x} \right) \partial x. \tag{1.181}$$

Since

$$\sum_{n=1}^{\infty} \left| \int_{\infty}^{\infty} x^{\frac{s}{2}-1} e^{-n^2\pi x}\, \partial x \right| \leq \sum_{n=1}^{\infty} \int_0^{\infty} x^{\frac{\sigma}{2}-1} e^{-n^2\pi x}\, \partial x = \sum_{n=1}^{\infty} \frac{\Gamma(\frac{\sigma}{2})}{n^\sigma \pi^{\frac{\sigma}{2}}} = \frac{1}{\pi^{\frac{\sigma}{2}}} \Gamma\left(\frac{\sigma}{2}\right) \zeta(\sigma) \quad (\sigma > 1), \tag{1.182}$$

and the series absolutely converges, the equality $\overset{*}{=}$ (relation (1.181)) is valid. Let

$$\varphi(x) = \sum_{n=1}^{\infty} e^{-n^2\pi x}. \tag{1.183}$$

From relation (1.181) it follows that

$$\pi^{-\frac{s}{2}}\Gamma\left(\frac{s}{2}\right)\zeta(s) = \int_0^\infty x^{\frac{s}{2}-1}\varphi(x)\,\partial x \quad (\sigma > 1). \tag{1.184}$$

Since

$$\theta(x) = 1 + 2\varphi(x), \tag{1.185}$$

from Lemma 1.13 it follows that

$$2\varphi(x) + 1 = \frac{1}{\sqrt{x}}\left(2\varphi\left(\frac{1}{x}\right) + 1\right). \tag{1.186}$$

Based on (1.184), we obtain

$$\begin{aligned}
\pi^{-\frac{s}{2}}\Gamma\left(\frac{s}{2}\right)\zeta(s) &= \int_0^1 x^{\frac{s}{2}-1}\varphi(x)\,\partial x + \int_1^\infty x^{\frac{s}{2}-1}\varphi(x)\,\partial x \\
&= \int_0^1 x^{\frac{s}{2}-1}\left\{\frac{1}{\sqrt{x}}\varphi\left(\frac{1}{x}\right) + \frac{1}{2\sqrt{x}} - \frac{1}{2}\right\}\partial x + + \int_1^\infty x^{\frac{s}{2}-1}\varphi(x)\,\partial x \\
&= \frac{1}{s-1} - \frac{1}{s} + \int_0^1 x^{\frac{s-3}{2}}\varphi\left(\frac{1}{x}\right)\partial x + \int_1^\infty x^{\frac{s}{2}-1}\varphi(x)\,\partial x.
\end{aligned} \tag{1.187}$$

Hence it follows that

$$\pi^{-\frac{s}{2}}\Gamma\left(\frac{s}{2}\right)\zeta(s) = \frac{1}{s-1} - \frac{1}{s} + \int_1^\infty (x^{-\frac{s+1}{2}} + x^{\frac{s-2}{2}})\varphi(x)\,\partial x \quad (\sigma > 1). \tag{1.188}$$

For real u, we have

$$\int_1^\infty x^u\varphi(x)\,\partial x \leqslant \int_1^\infty x^u \sum_{n=1}^\infty e^{-n\pi x}\,\partial x = \int_1^\infty x^u \frac{e^{-\pi x}}{1-e^{-\pi x}}\,\partial x \leq \frac{1}{1-e^{-\pi}}\int_1^\infty x^u e^{-\pi x}\,\partial x. \tag{1.189}$$

Since the last integral converges for every real u, the integrals:

$$\int_1^\infty x^{-\frac{s+1}{2}}\varphi(x)\,\partial x, \quad \int_1^\infty x^{\frac{s-2}{2}}\varphi(x)\,\partial x \tag{1.190}$$

converge for every s. Therefore the function on the right-hand side of relation (1.188) is an analytic function of s for every $s \neq 0, 1$. Therefore relation (1.188) enables an analytic extension with $\sigma > 1$ for the whole complex plane apart from $s = 0, 1$. For the function

$f(s)$, equal to the right-hand side of relation (1.188), $f(1-s) = f(s)$ is obviously true, which leads to the equality defined by (1.179).

The theorem is thus proved. □

From the relations valid for the gamma function some other forms of the functional equation can be obtained. Taking $z = \frac{1-s}{2}$ in the *Legendre* duplicate formula for the gamma function,

$$\Gamma(2z) = 2^{2z-1}\pi^{-\frac{1}{2}}\Gamma(z)\Gamma\left(z + \frac{1}{2}\right), \tag{1.191}$$

we get

$$2^{-s}\Gamma\left(\frac{1-s}{2}\right)\Gamma\left(1 - \frac{s}{2}\right) = \sqrt{\pi}\Gamma(1-s). \tag{1.192}$$

Taking $z = \frac{s}{2}$ in the formula

$$\Gamma(z)\Gamma(1-z) = \frac{\pi}{\sin \pi z}, \tag{1.193}$$

we have

$$\Gamma\left(\frac{s}{2}\right)\Gamma\left(1 - \frac{s}{2}\right) = \frac{\pi}{\sin\frac{\pi s}{2}}. \tag{1.194}$$

From (1.191) and (1.194) we get

$$2^{-s} \cdot \sqrt{\pi} \cdot \Gamma\left(1 - \frac{s}{2}\right) = \sin\frac{\pi s}{2}\Gamma\left(\frac{s}{2}\right)\Gamma(1-s), \tag{1.195}$$

which, together with (1.179), leads to a new form of the functional equation for $\zeta(s)$:

$$\zeta(s) = 2^s\pi^{s-1}\sin\frac{\pi s}{2}\Gamma(1-s)\zeta(1-s). \tag{1.196}$$

From (1.196) and (1.193) we arrive at another form of the functional equation for $\zeta(s)$, i. e., to the equation

$$\zeta(1-s) = 2^{1-s}\pi^{-s}\cos\frac{\pi s}{2}\Gamma(s)\zeta(s). \tag{1.197}$$

Closely related to the functional equation for $\zeta(s)$ is the functional equation presented below, which contains the so-called incomplete gamma function [34].

Definition 1.8. The incomplete gamma function is defined as

$$\Gamma(s,x) = \int_x^\infty u^{s-1}e^{-u}\,\partial u. \tag{1.198}$$

This function occurs as part of Theorem 1.9.

Theorem 1.9. *We have the following equation:*

$$\Gamma\left(\frac{s}{2}\right)\zeta(s) = \sum_{n=1}^{\infty} \frac{1}{n^s}\Gamma\left(\frac{s}{2} \cdot \pi\tau n^2\right) + \pi^{s-\frac{1}{2}} \sum_{n=1}^{\infty} \frac{1}{n^{1-s}}\Gamma\left(\frac{1-s}{2} \cdot \frac{\pi n^2}{\tau}\right)$$
$$- \frac{\pi^{\frac{1}{2}}(\pi\tau)^{\frac{s-1}{2}}}{1-s} - \frac{(\pi\tau)^{\frac{s}{2}}}{s} \quad \left(|\arg\tau| < \frac{\pi}{2}\right). \tag{1.199}$$

Proof. By formula (1.181) it follows that

$$\pi^{-\frac{s}{2}}\Gamma\left(\frac{s}{2}\right)\zeta(s) = \int_{X}^{\infty} \left(x^{\frac{s}{2}-1} \sum_{n=1}^{\infty} e^{-n^2\pi x}\right) \partial x \quad (\sigma > 1). \tag{1.200}$$

We will integrate the function

$$f(z) = z^{\frac{s}{2}-1} \sum_{n=1}^{\infty} e^{-n^2\pi z} \tag{1.201}$$

along the contour $C = C_1 U C_2 U C_3 U C_4$, where $C_1 : z = x(x \uparrow_{\rho}^{R})$; $C_2 : z = Re^{\varphi i}(\varphi \uparrow_0^{\arg\tau})$; $C_3 : z = x\tau(x \downarrow_{\rho}^{R})$; $C_4 : z = \rho e^{\varphi i}$ $(\varphi \downarrow_0^{\arg\tau})$. Also,

$$\Re e(\tau) > 0, \quad z^{\frac{s}{2}-1} = e^{(\frac{s}{2}-1)(\log|z|+\arg z)}. \tag{1.202}$$

For $\alpha = \arg\tau$,

$$|I_R| = \left|\int_0^{\alpha} (Re^{\varphi i})^{\frac{s}{2}-1}\left(\sum_{n=1}^{\infty} e^{-n^2\pi Re^{\varphi i}}\right) iRe^{\varphi i}\, \partial\varphi\right| \le \int_0^{\alpha} Re^{\frac{\sigma}{2}} e^{-\frac{\varphi t}{2}} \sum_{n=1}^{\infty} e^{-n^2\pi R\cos\varphi}\, \partial\varphi. \tag{1.203}$$

From the inequalities

$$e^{-n^2\pi R\cos\varphi} \le e^{-n^2\pi R\cos\alpha} \le (e^{-R\pi\cos\alpha})^n \tag{1.204}$$

it follows that

$$|I_R| \le \int_0^{\alpha} Re^{\frac{\sigma}{2}} e^{-\frac{\varphi t}{2}} \frac{e^{-\pi R\cos\alpha}}{1 - e^{-\pi R\cos\alpha}}\, \partial\varphi = K\frac{R^{\frac{\sigma}{2}}}{e^{\pi R\cos\alpha} - 1} \to 0 \quad (R \to \infty). \tag{1.205}$$

We can further conclude that

$$\sum_{n=1}^{\infty} e^{-n^2\pi x} = O\left(\frac{1}{\sqrt{x}}\right) \quad (0 < x < 1). \tag{1.206}$$

Since

$$\sum_{n=1}^{\infty} e^{-n^2\pi x} = \varphi(x) = \frac{1}{\sqrt{x}}\left(\frac{1}{2} + \varphi\left(\frac{1}{x}\right)\right) - \frac{1}{2} \quad (x > 0) \tag{1.207}$$

and

$$\varphi\left(\frac{1}{x}\right) < \sum_{n=1}^{\infty} e^{-n^2\pi\frac{1}{x}} = \frac{1}{e^{\frac{\pi}{x}-1}} < \frac{x}{\pi} < \frac{1}{\pi} \quad (0 < x < 1), \tag{1.208}$$

we have

$$\varphi(x) < \left(\frac{1}{2} + \frac{1}{\pi}\right)\frac{1}{\sqrt{x}} + \frac{1}{2} < \left(1 + \frac{1}{\pi}\right)\frac{1}{\sqrt{x}} \quad (0 < x < 1). \tag{1.209}$$

Hence the required evaluation for $\varphi(x)$ follows. Based on this evaluation, it is also possible to make the following estimate for I_ρ:

$$\begin{aligned} |I_\rho| &= \left|\int_0^{\alpha} (\rho e^{\varphi i})^{\frac{s}{2}-1}\left(\sum_{n=1}^{\infty} e^{-n^2\pi\rho e^{\varphi i}}\right) i\rho e^{\varphi i}\, \partial\varphi\right| \le \int_0^{\alpha} \rho^{\frac{\sigma}{2}} e^{-\frac{\varphi t}{2}} \sum_{n=1}^{\infty} e^{-n^2\pi\rho\cos\alpha}\, \partial\varphi \\ &\leqslant K\int_0^{\alpha} \rho^{\frac{\sigma}{2}} e^{-\frac{\varphi t}{2}} \rho^{-\frac{1}{2}}\, \partial\varphi = L\rho^{\frac{\sigma-1}{2}} \to 0 \quad (\rho \to 0). \end{aligned} \tag{1.210}$$

It follows that

$$\begin{aligned} \int_0^{+\infty} x^{\frac{s}{2}-1}\left(\sum_{n=1}^{\infty} e^{-n^2\pi x}\right)\partial x &= \int_0^{+\infty} (\tau x)^{\frac{s}{2}-1}\left(\sum_{n=1}^{\infty} e^{-n^2\pi\tau x}\right)\partial(\tau x)\tau^{\frac{s}{2}} \int_0^{+\infty} x^{\frac{s}{2}-1}\left(\sum_{n=1}^{\infty} e^{-n^2\pi\tau x}\right)\partial x \\ &= \tau^{\frac{s}{2}}\int_0^{+\infty} x^{\frac{s}{2}-1}\left(\sum_{n=1}^{\infty} e^{-n^2\pi\tau x}\right)\partial x + \tau^{\frac{s}{2}}\int_0^{+\infty} x^{-\frac{s}{2}-1}\left(\sum_{n=1}^{\infty} e^{-n^2\pi\frac{\tau}{x}}\right)\partial x. \end{aligned} \tag{1.211}$$

Since

$$2\varphi\left(\frac{1}{x}\right) + 1 = \sqrt{x}(2\varphi(x) + 1) \quad (x > 0), \tag{1.212}$$

by means of analytic extension we have

$$2\varphi\left(\frac{\tau}{x}\right) = \sqrt{\frac{x}{\tau}}\left(2\varphi\left(\frac{x}{\tau}\right) + 1\right) - 1 \quad \left(\mathfrak{Re}\left(\frac{\tau}{x}\right) > 0\right), \tag{1.213}$$

where because of $x > 0$, the last relation is valid for $\mathfrak{Re}(\tau) > 0$. Therefore it follows that

$$\int_0^{+\infty} x^{\frac{s}{2}-1}\left(\sum_{n=1}^{\infty} e^{-n^2\pi x}\right)\partial x$$
$$= \tau^{\frac{s}{2}} \int_0^{+\infty} x^{\frac{s}{2}-1}\left(\sum_{n=1}^{\infty} e^{-n^2\pi x\tau}\right)\partial x + \tau^{\frac{s-1}{2}} \int_0^{+\infty} x^{-\frac{s}{2}-\frac{1}{2}}\left(\sum_{n=1}^{\infty} e^{-n^2\pi\frac{x}{\tau}}\right)\partial x - \frac{\tau^{\frac{s}{2}}}{s} - \frac{\tau^{\frac{s-1}{2}}}{1-s}. \quad (1.214)$$

Since

$$\int_0^{+\infty} x^{\frac{s}{2}-1} e^{-n^2\pi x\tau}\,\partial x = \pi^{-\frac{s}{2}} n^{-s} \tau^{-\frac{s}{2}} \Gamma\left(\frac{s}{2}, n^2\pi\tau\right), \quad (1.215)$$

$$\int_0^{+\infty} x^{\frac{s}{2}-\frac{1}{2}} e^{-n^2\pi\frac{x}{\tau}}\,\partial x = \tau^{-\frac{s}{2}+\frac{1}{2}} \pi^{\frac{s}{2}-\frac{1}{2}} n^{s-1} \Gamma\left(\frac{1-s}{2}, \frac{n^2\pi}{\tau}\right), \quad (1.216)$$

we get

$$\pi^{-\frac{s}{2}}\Gamma\left(\frac{s}{2}\right)\zeta(s) = \pi^{-\frac{s}{2}} \sum_{n=1}^{\infty} \frac{\Gamma(\frac{s}{2}, n^2\pi\tau)}{n^s} + \pi^{\frac{s}{2}-\frac{1}{2}} \sum_{n=1}^{\infty} \frac{\Gamma(\frac{1-s}{2}, \frac{n^2\pi}{\tau})}{n^{1-s}} - \frac{\tau^{\frac{s}{2}}}{s} - \frac{\tau^{\frac{s}{2}-\frac{1}{2}}}{1-s}, \quad (1.217)$$

and hence (1.199) directly follows. □

In particular, for $\tau = 1$, we obtain relation (1.188). In 1920, *Hardy* and *Littlewood* [32] obtained the following approximate functional equation for $\zeta(s)$:

$$\zeta(s) = \sum_{n \leqslant x} \frac{1}{n^s} + \chi(s) \sum_{n \leqslant y} \frac{1}{n^{1-s}} + O(x^{-\sigma}) + O(|t|^{\frac{1}{2}-\sigma} y^{\sigma-1}), \quad (1.218)$$

where

$$\chi(s) = \frac{\pi^{s-\frac{1}{2}}\Gamma(\frac{1-s}{2})}{\Gamma(\frac{s}{2})}, \quad (1.219)$$

where $h > 0$ is a constant, $0 < \sigma < 1$, $x > h$, $2\pi xy = |t|$.

We will prove the accuracy of formulas (1.220)–(1.221), which serve to obtain an analytic extension of the function $\zeta(s)$ by decomposing the corresponding function in a *Fourier* series:

$$\int_0^{+\infty} x^{s-1} e^{-ix}\,\partial x = \Gamma(s) e^{\frac{-i\pi s}{2}} \quad (0 < \sigma < 1), \quad (1.220)$$

$$\int_0^{+\infty} x^{s-1} e^{ix}\,\partial x = \Gamma(s) e^{\frac{i\pi s}{2}} \quad (0 < \sigma < 1), \quad (1.221)$$

$$\int_0^{+\infty} x^{s-1} \sin x\,\partial x = \Gamma(s) \sin\frac{\pi s}{2} \quad (0 < \sigma < 1). \quad (1.222)$$

Proof. We will integrate the function $f(z) = z^{s-1}\mathfrak{e}^{-z}$ $(0 < \sigma < 1)$ along the contour shown in Figure 1.4, where for z^{s-1}, we take $\mathfrak{e}^{(s-1)[\log|z|+\mathrm{i}\arg z]}$ $(\rho \leq x \leq R)$.

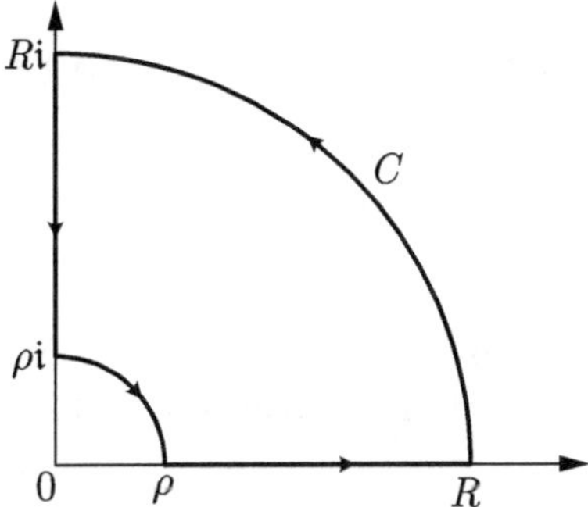

Figure 1.4: The contour of integration of the function $f(z)$.

According to *Cauchy*'s theorem,

$$\int_C f(z)\,\partial z = 0. \tag{1.223}$$

Since

$$|I_\rho| = \left| -\int_0^{\frac{\pi}{2}} \mathrm{i}\rho^s \mathfrak{e}^{us\mathrm{i}} \mathfrak{e}^{-\rho\mathfrak{e}^{u\mathrm{i}}}\,\partial u \right| \leqslant \int_0^{\frac{\pi}{2}} \rho^\sigma \mathfrak{e}^{-tu}\mathfrak{e}^{-\rho\cos u}\,\partial u \leqslant \left| \begin{matrix} \cos u \geqslant 1 - \frac{2}{\pi}u \\ \cos u \leq \frac{\pi}{2} \end{matrix} \right|$$

$$\leq \rho^\sigma \int_0^{\frac{\pi}{2}} \mathfrak{e}^{-tu-\rho(1-\frac{2}{\pi}u)}\,\partial u = \rho^\sigma \frac{\mathfrak{e}^{-\frac{t\pi}{2}} - \mathfrak{e}^{-\rho}}{\frac{2\rho}{\pi} - t} \longrightarrow 0 \quad (\rho \longrightarrow 0, \sigma > 0), \tag{1.224}$$

we have

$$\lim_{\rho \longrightarrow 0} I_\rho = 0 \quad (\sigma > 0). \tag{1.225}$$

In a similar way, we conclude that

$$|I_R| \leqslant R^{\sigma-1} \frac{\mathfrak{e}^{-\frac{t\pi}{2}} - \mathfrak{e}^{-R}}{\frac{2}{\pi} - \frac{t}{R}} \longrightarrow 0 \quad (R \longrightarrow \infty, \sigma < 1), \tag{1.226}$$

and therefore

$$\lim_{R \longrightarrow 0} I_R = 0 \quad (\sigma < 1). \tag{1.227}$$

Since $\mathrm{i}^s = \mathfrak{e}^{\frac{\pi s \mathrm{i}}{2}}$, passing to the limits as $R \longrightarrow \infty$ and $\rho \longrightarrow 0$, based on the definition of the gamma function, by *Cauchy*'s formula we obtain (1.220).

To prove formula (1.221), consider the same function $f(z)$ and the contour symmetric to the contour in Figure 1.4 with respect to the real axis. In this case, we have the following estimates:

$$|I_R| \leqslant \rho^{\sigma} \frac{e^{\frac{t\pi}{2}} - e^{-\rho}}{\frac{2\rho}{\pi} + t} \longrightarrow 0 \quad (\rho \longrightarrow 0, \sigma > 0), \tag{1.228}$$

$$|I_R| \leqslant R^{\sigma-1} \frac{e^{\frac{t\pi}{2}} - e^{-R}}{\frac{2}{\pi} + \frac{t}{R}} \longrightarrow 0 \quad (R \longrightarrow \infty, \sigma < 1). \tag{1.229}$$

Since $(-i)^s = e^{-\frac{\pi s i}{2}}$, passing to the limits as $R \longrightarrow \infty$ and $\rho \longrightarrow 0$, by the definition of the gamma function, (1.221) follows from *Cauchy*'s formula. Subtracting (1.220) from (1.221), we obtain (1.222). □

Further, from (1.41) it follows that

$$\zeta(s) = \frac{s}{s-1} - s \int_1^{+\infty} \frac{x - [x]}{x^{s+1}}\, \partial x \quad (\sigma > 0), \tag{1.230}$$

that is,

$$\zeta(s) = s \int_1^{\infty} \frac{[x] - x + \frac{1}{2}}{x^{s+1}}\, \partial x + \frac{1}{s-1} + \frac{1}{2} \quad (\sigma > 0), \tag{1.231}$$

because

$$\frac{s}{2} \int_1^{\infty} \frac{\partial x}{x^{s+1}} = \frac{1}{2} \quad (\sigma > 0). \tag{1.232}$$

For $0 < \sigma < 1$, we can derive a similar formula:

$$s \int_0^1 \frac{[x] - x}{x^{s+1}}\, \partial x = \frac{s}{s-1} \quad (\sigma < 1), \tag{1.233}$$

so from (1.230) it follows that

$$\zeta(s) = s \int_1^{\infty} \frac{[x] - x}{x^{s+1}}\, \partial x \quad (0 < \sigma < 1). \tag{1.234}$$

Let

$$f(x) = [x] - x + \frac{1}{2}, \quad f_1(x) = \int_1^x f(y)\, \partial y. \tag{1.235}$$

Since

$$\int_{n}^{n+1} f(y)\,\partial y = 0 \quad (\forall n \in \mathbb{N}), \tag{1.236}$$

the function $f_1(x)$ is bounded. Accordingly, it follows that

$$\int_{x_1}^{x_2} \frac{f(x)}{x^{s+1}}\,\partial x = \left[\frac{f_1(x)}{x^{s+1}}\right]_{x_1}^{x_2} + (s+1)\int_{x_1}^{x_2} \frac{f_1(x)}{x^{s+2}}\,\partial x \longrightarrow 0 \quad \begin{pmatrix} x_1 \longrightarrow 0 \\ x_2 \longrightarrow 0 \\ \sigma > -1 \end{pmatrix}. \tag{1.237}$$

By the last relation the integral defined by (1.231) converges for $\sigma > -1$, which determines its analytic extension when switching from $\sigma > 0$ to $\sigma > -1$. Since

$$s\int_{0}^{1} \frac{[x] - x + \frac{1}{2}}{x^{s+1}}\,\partial x = \frac{1}{s-1} + \frac{1}{2} \quad (\sigma < 0), \tag{1.238}$$

we have

$$\zeta(s) = s\int_{0}^{\infty} \frac{[x] - x + \frac{1}{2}}{x^{s+1}}\,\partial x \quad (-1 < \sigma < 0). \tag{1.239}$$

Decomposing the function $[x] - x + \frac{1}{2}$ in the *Fourier* series

$$[x] - x + \frac{1}{2} = \sum_{n=1}^{\infty} \frac{\sin 2n\pi x}{n\pi} \quad (x \notin Z), \tag{1.240}$$

by the last relation, from (1.239) we have

$$\begin{aligned} \zeta(s) &= s\int_{0}^{\infty} \frac{1}{x^{s+1}} \sum_{n=1}^{\infty} \frac{\sin 2n\pi x}{n\pi}\,\partial x \overset{*}{=} \frac{s}{\pi}\sum_{n=1}^{\infty}\frac{1}{n}\int_{0}^{\infty} \frac{\sin 2n\pi x}{x^{s+1}}\,\partial x \\ &= \frac{s}{\pi}\sum_{n=1}^{\infty}\frac{(2n\pi)^s}{n}\int_{0}^{\infty} x^{-s-1}\sin x\,\partial x \underset{-1<\sigma<0}{\overset{(1.220)}{=}} \frac{s}{\pi}(2\pi)^s\{-\Gamma(-s)\}\sin\frac{\pi s}{2}\zeta(1-s), \end{aligned} \tag{1.241}$$

that is,

$$\zeta(s) = -\frac{s}{\pi}(2\pi)^s\Gamma(-s)\sin\frac{\pi s}{2}\zeta(1-s). \tag{1.242}$$

The last relation gives an analytic extension of the function $\zeta(s)$ from the region $-1 < \sigma < 0$ to the whole s-plane, excluding the points $s = 0$ and $s = 1$. The equality $\overset{*}{=}$ (in relation (1.241)) is true if the integration is performed on any finite interval $[0, a]$, since in that case the series converges, except at appropriate points. It only remains to prove that

$$\lim_{x\longrightarrow\infty}\sum_{n=1}^{\infty}\frac{1}{n}\int_{x}^{\infty}\frac{\sin 2n\pi x}{x^{s+1}}\,\partial x = 0 \quad (-1<\sigma<0). \tag{1.243}$$

Since

$$\int_{x}^{\infty}\frac{\sin 2n\pi x}{x^{s+1}}\,\partial x = -\left[\frac{\cos 2n\pi x}{2n\pi x^{s+1}}\right]_{x}^{\infty} - \frac{s+1}{2n\pi}\int_{x}^{\infty}\frac{\cos 2n\pi x}{x^{s+2}}\,\partial x$$
$$= O\left(\frac{1}{nx^{\sigma+1}}\right) + O\left(\frac{1}{n}\int_{x}^{\infty}\frac{\partial x}{x^{\sigma+2}}\right) = O\left(\frac{1}{nx^{\sigma+1}}\right), \tag{1.244}$$

we get

$$\left|\sum_{n=1}^{\infty}\frac{1}{n}\int_{x}^{\infty}\frac{\sin 2n\pi x}{x^{s+1}}\,\partial x\right| \leq \sum_{n=1}^{\infty}\frac{1}{n}\cdot\frac{K}{nx^{s+1}} = \frac{K}{x^{s+1}} \longrightarrow 0 \quad (x\longrightarrow\infty,\ -1<\sigma<0). \tag{1.245}$$

Note that (1.196) follows from (1.242), since $-s\Gamma(-s) = \Gamma(1-s)$.

1.2.3 Some formulas for $\zeta(s)$

Definition 1.9. *Euler*'s γ constant [11] is defined as

$$\gamma = \lim_{n\longrightarrow\infty}\left(1+\frac{1}{2}+\cdots+\frac{1}{n}-\log n\right). \tag{1.246}$$

The following formula defines the relation between $\zeta(s)$ and γ:

$$\lim_{s\longrightarrow 1}\left\{\zeta(s)-\frac{1}{s-1}\right\} = \gamma. \tag{1.247}$$

Proof. We have

$$\lim_{s\longrightarrow 1}\left\{\zeta(s)-\frac{1}{s-1}\right\} \overset{(1.231)}{=} \lim_{s\longrightarrow 1}\left\{s\int_{1}^{\infty}\frac{[x]-x+\frac{1}{2}}{x^{s+1}}\,\partial x + \frac{1}{2}\right\}$$
$$= \left|\begin{array}{c}\text{since according to (1.42), the integral can be presented in}\\ \text{the form of a uniformly convergent series}\end{array}\right|$$
$$= \int_{1}^{\infty}\frac{[x]-x+\frac{1}{2}}{x^{s+1}}\,\partial x + \frac{1}{2} = \lim_{n\longrightarrow\infty}\left\{\sum_{m=1}^{n-1}\int_{m}^{m+1}\frac{\partial x}{x^{2}} - \log n + 1\right\}$$

$$= \lim_{n\to\infty}\left\{\sum_{m=1}^{n-1}\frac{1}{m+1} - \log n\right\} = \gamma. \tag{1.248}$$

□

The following relation is also true as $s \longrightarrow 1$:

$$\zeta(s) = \frac{1}{s-1} + \gamma + O(|s-1|), \tag{1.249}$$

whereas for γ, we can write

$$\gamma = \sum_{n=1}^{\infty}\frac{(-1)^n}{n}\left[\frac{\log n}{\log 2}\right], \tag{1.250}$$

(see [39, 131]).

Proof. Since there is a boundary value that determines the constant γ, it follows that

$$\gamma = \lim_{n\to\infty}\left(\sum_{k=1}^{2^n}\frac{1}{k} - \log 2^n\right) = \lim_{n\to\infty} A_n. \tag{1.251}$$

However, since

$$\begin{aligned} A_n &= \sum_{k=1}^{2^n}\frac{1}{k} - \log 2 - \log 2^{n-1} = \sum_{k=1}^{2^n}\frac{1}{k} + \sum_{k=1}^{\infty}\frac{(-1)^k}{k} - \log 2^{n-1} \\ &= 2\sum_{k=1}^{2^{n-1}}\frac{1}{2k} + \sum_{k=2^n+1}^{\infty}\frac{(-1)^k}{k} - \log 2^{n-1} = A_{n-1} + \sum_{k=2^n+1}^{\infty}\frac{(-1)^k}{k}, \end{aligned} \tag{1.252}$$

in the particular $n = 1$, we have

$$A_1 = 1 - \log 2^0 + \sum_{k=2^1+1}^{\infty}\frac{(-1)^k}{k} = 1 + \sum_{k=2^1+1}^{\infty}\frac{(-1)^k}{k}. \tag{1.253}$$

Hence it follows that

$$\begin{aligned} A_n &= 1 + \sum_{k=2^n+1}^{\infty}\frac{(-1)^k}{k} + \sum_{k=2^{n-1}+1}^{\infty}\frac{(-1)^k}{k} + \cdots + \sum_{k=2^1+1}^{\infty}\frac{(-1)^k}{k} \\ &= 1 + n\sum_{k=2^n+1}^{\infty}\frac{(-1)^k}{k} + \sum_{r=1}^{n-1} r \sum_{k=2^r+1}^{k=2^{r+1}}\frac{(-1)^k}{k}. \end{aligned} \tag{1.254}$$

Since

$$n\left|\sum_{k=2^n+1}^{\infty}\frac{(-1)^k}{k}\right| < n\left|\frac{(-1)^{2^n+1}}{2^n+1}\right| = \frac{n}{2^n+1} < \frac{1}{2^{\frac{n}{2}}}, \tag{1.255}$$

we have

$$\gamma = 1 + \sum_{n=1}^{\infty} n \sum_{k=2^n+1}^{\infty} \frac{(-1)^k}{k}. \tag{1.256}$$

For $k = 2^n + 1, 2^n + 2, 2^{n+1} - 1$, we have $[\frac{\log k}{\log 2}] = n$, whereas for $k = 2^{n+1}$, we have $[\frac{\log k}{\log 2}] = n + 1$, and so

$$\begin{aligned}
\gamma &= \sum_{\substack{k=3 \\ k\neq 2^{n+1}}}^{\infty} \frac{(-1)^k}{k}\left[\frac{\log k}{\log 2}\right] + \sum_{n=1}^{\infty} \frac{n}{2^{n+1}} + 1 \\
&= 1 - \sum_{n=1}^{\infty} \frac{1}{2^{n+1}} + \sum_{k=3}^{\infty} \frac{(-1)^k}{k}\left[\frac{\log k}{\log 2}\right] \\
&= \frac{1}{2} + \sum_{k=3}^{\infty} \frac{(-1)^k}{k}\left[\frac{\log k}{\log 2}\right] = \sum_{k=1}^{\infty} \frac{(-1)^k}{k}\left[\frac{\log k}{\log 2}\right].
\end{aligned} \tag{1.257}$$

□

Let us prove the following two relations:

$$\gamma = \frac{\log 2}{2} + \sum_{n=1}^{\infty} \frac{(-1)^n}{n} \cdot \frac{\log n}{\log 2}, \tag{1.258}$$

$$-\frac{\log 2}{2} = \sum_{n=1}^{\infty} \frac{(-1)^n}{n} \left\{\frac{\log n}{\log 2}\right\}. \tag{1.259}$$

Proof. The following equation is derived from (1.247):

$$\zeta(s) = \frac{1}{s-1} + \gamma + h(s), \tag{1.260}$$

where the function $h(s)$ is regular for every s, and $h(1) = 0$. Let

$$g(s) = (1 - 2^{1-s})\zeta(s), \quad \sigma > 0. \tag{1.261}$$

Then from (1.146) it follows that

$$g(s) = \sum_{n=1}^{\infty} \frac{(-1)^{n-1}}{n^s}, \quad \sigma > 0, \tag{1.262}$$

that is,

$$g(s) = (1 - 2^{1-s})\left(\frac{1}{s-1} + \gamma + h(s)\right). \tag{1.263}$$

Therefore

$$g'(s) = 2^{1-s}\log\left(\frac{1}{s-1} + \gamma + h(s)\right) + (1-2^{1-s})(h'(s) - \frac{1}{(s-1)^2}, \tag{1.264}$$

and so

$$\lim_{s\longrightarrow 1} g'(s) = \gamma \log 2 + \lim_{s\longrightarrow 1} \frac{2^{1-s}(s-1)\log 2 - (1-2^{1-s})}{(s-1)^2} = \gamma\log 2 - \frac{\log^2 2}{2}. \tag{1.265}$$

Based on Lemma 1.12, the series

$$\sum_{n=1}^{\infty} \frac{(-1)^{n-1}}{n^s} \tag{1.266}$$

converges uniformly on any compact in the region $\sigma > 0$. Therefore, for $\sigma \geq \delta > 0$, we can differentiate it term-by-term:

$$g'(s) = \sum_{n=1}^{\infty} \frac{(-1)^n \log n}{n^s}. \tag{1.267}$$

In particular,

$$g'(1) = \sum_{n=1}^{\infty} \frac{(-1)^n \log n}{n}, \tag{1.268}$$

from which relation (1.258) follows. Subtracting (1.250) from (1.258), we obtain relation (1.259). □

Furthermore,

$$\zeta(0) = -\frac{1}{2}, \quad \zeta(-2m) = 0 \quad (m = 1, 2, \ldots), \tag{1.269}$$

where $-2m$ are prime zeros of the function $\zeta(s)$.

Proof. From (1.188), by

$$\frac{s}{2}\Gamma\left(\frac{s}{2}\right) = \Gamma\left(1 + \frac{s}{2}\right) \tag{1.270}$$

we have

$$\pi^{-\frac{s}{2}}\Gamma\left(1+\frac{s}{2}\right)\zeta(s) = \frac{s}{2(s-1)} - \frac{1}{2} + \frac{s}{2}\int_1^{\infty}(x^{-\frac{s+1}{2}} + x^{\frac{s-2}{2}})\varphi(x)\,\partial x. \tag{1.271}$$

Passing to the limits as $s \longrightarrow 0$, since $\Gamma(1) = 1$, we have $\zeta(0) = -\frac{1}{2}$. Further, since the gamma function is not regular only at $s = 0, -1, -2, \ldots$, where these points are prime poles, from (1.197) we conclude that $s = -2m$ is a single zero of the function $\zeta(s)$. □

We will prove the following formula, which was first obtained by *Legendre*:

$$\int_0^\infty \frac{\sin ax}{e^{2\pi x}-1}\,\partial x = \frac{1}{4}\frac{e^a+1}{e^a-1} - \frac{1}{2a} \quad (a \neq 0). \tag{1.272}$$

Proof. Let us integrate the function

$$f(z) = \frac{e^{aiz}}{e^{2\pi z}-1} \tag{1.273}$$

along the contour shown in Figure 1.5.

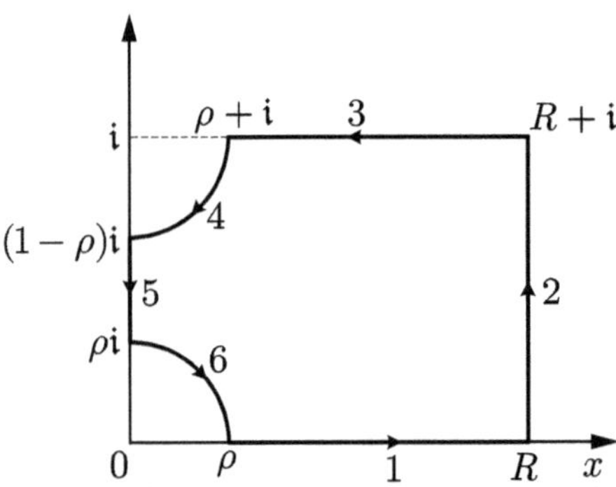

Figure 1.5: Contour of integration of the function defined by (1.273).

Then

$$\int_C f(z)\,\partial z = 0, \tag{1.274}$$

whereas

$$I_1 = \int_\rho^R \frac{e^{axi}}{e^{2\pi x}-1}\,\partial x, \quad I_3 = -e^{-a}I_1, \tag{1.275}$$

$$|I_2| = \left|\int_0^1 \frac{e^{ai(R+yi)}}{e^{2\pi(R+yi)}} i\,\partial y\right| \le \int_0^1 \frac{e^{-ay}}{e^{2R\pi}-1}\,\partial y \le \frac{K}{e^{2R\pi}-1} \longrightarrow 0 \quad (R\longrightarrow\infty), \tag{1.276}$$

$$I_5 = -i\int_\rho^{1-\rho} \frac{e^{-ay}}{e^{2\pi yi}-1}\,\partial y \le -\frac{1}{2}\int_\rho^{1-\rho} e^{-ay}\frac{\cos\pi y - i\sin\pi y}{\sin\pi y}\,\partial y. \tag{1.277}$$

From the last relation it follows that

$$I_m(I_5) = -\frac{1}{2a}\{e^{-a(1-\rho)} - e^{-a\rho}\}, \tag{1.278}$$

that is,

$$\lim_{\rho\longrightarrow 0} I_m(I_5) = \frac{1}{2a}(1 - e^{-a}). \tag{1.279}$$

By analogy we can conclude that

$$\lim_{\rho\longrightarrow 0} I_4 = \lim_{\rho\longrightarrow 0} \int_0^{-\frac{\pi}{2}} \frac{e^{ai(i+\rho e^{\varphi})}}{e^{2\pi(i+\rho e^{\varphi i})} - 1} i\rho e^{\varphi i}\, \partial\varphi = -\frac{i}{4}e^{-a}, \tag{1.280}$$

so that

$$\lim_{\rho\longrightarrow 0} I_m(I_4) = -\frac{e^{-a}}{4}. \tag{1.281}$$

Also,

$$\lim_{\rho\longrightarrow 0} I_6 = -\frac{i}{4}, \tag{1.282}$$

i. e.,

$$\lim_{\rho\longrightarrow 0} I_m(I_6) = -\frac{1}{4}. \tag{1.283}$$

It is clear that passing to the limits as $\rho \longrightarrow 0$ and $R \longrightarrow \infty$, we obtain (1.272). □

Definition 1.10. *Bernoulli's* numbers B_n are defined by the following relation:

$$\frac{z}{e^z - 1} = \sum_{n=0}^{\infty} \frac{B_n}{n!} z^n \quad (|z| < 2\pi). \tag{1.284}$$

For the numbers B_{2n}, we have

$$B_{2n} = (-1)^{n+1} \cdot 4n \int_0^{\infty} \frac{x^{2n-1}}{e^{2\pi x} - 1} \partial x \quad (n \in \mathbb{N}). \tag{1.285}$$

Proof. From (1.284) it follows that $B_0 = 1$ and $B_1 = -\frac{1}{2}$, and because of the parity of the function, it follows that

$$\frac{z}{e^z - 1} - 1 + \frac{z}{2} \implies B_{2n+1} = 0 \quad (n = 1, 2, \dots), \tag{1.286}$$

so (1.284) can be written as

$$\frac{z}{e^z - 1} = 1 - \frac{z}{2} + \sum_{n=0}^{\infty} \frac{B_{2n}}{(2n)!} z^{2n}. \tag{1.287}$$

From this it follows that

$$z \operatorname{ctan} z = iz\frac{e^{iz}+e^{-iz}}{e^{iz}-e^{-iz}} = iz\left(1+\frac{2}{e^{2iz}-1}\right) = iz + \frac{2iz}{e^{2iz}-1}$$
$$\overset{(1.287)}{=} 1+\sum_{k=0}^{\infty}(-1)^k\frac{2^{2k}B_{2k}z^{2k}}{(2k)!} \quad (|z|<\pi) \tag{1.288}$$

and

$$\int_0^{\infty}\frac{\sin px}{e^{\pi x}-1}\,\partial x = 2\int_0^{\infty}\frac{\sin 2px}{e^{2\pi x}-1}\,\partial x \overset{(1.272)}{=} \frac{1}{2}\frac{e^{2p}+1}{e^{2p}-1}-\frac{1}{2p}$$
$$= \frac{i}{2}\operatorname{ctan}(\pi) - \frac{1}{2p} = \sum_{k=0}^{\infty}\frac{2^{2k-1}B_{2k}p^{2k-1}}{(2k)!}. \tag{1.289}$$

Let

$$I_m = \int_0^{\infty}\frac{x^m\sin(px+m\cdot\frac{\pi}{2})}{e^{\pi x}-1}\,\partial x = \int_0^{\infty}\frac{\partial^m}{\partial p^m}\left(\frac{\sin px}{e^{\pi x}-1}\right)\partial x. \tag{1.290}$$

The integral I_m converges uniformly around the point $p = 0$. It follows that

$$\left|\frac{x^m\sin(px+m\frac{\pi}{2})}{e^{\pi x}-1}\right| \le \frac{x^m}{e^{\pi x}-1} \le \frac{K}{x^2}, \tag{1.291}$$

where

$$K = \frac{(m+2)!}{\pi^{m+2}}, \tag{1.292}$$

since

$$e^{\pi x}-1 \ge \frac{x^{m+2}\pi^{m+2}}{(m+2)!} \quad (x>0). \tag{1.293}$$

Since

$$\int_a^{\infty}\frac{K}{x^2}\,\partial x = \frac{K}{a} \quad (a>0) \tag{1.294}$$

and $\lim_{x\longrightarrow 0}\frac{x^m}{e^{\pi x}-1}$ is finite for $m \ge 1$, it follows that

$$\int_a^{\infty}\frac{x^m}{e^{\pi x}-1}\,\partial x \quad (a>0) \tag{1.295}$$

converges. Differentiating (1.289) $2n-1$ times and including $p = 0$, we get

$$\frac{2^{n-1}B_{2n}(2n-1)!}{(2n)!}=\frac{2^{n-1}B_{2n}(2n-1)}{(2n)}=\left(\frac{\partial^{2n-1}}{\partial p^{2n-1}}\left(\int\limits_a^\infty\frac{\sin px}{e^{\pi x}-1}\,\partial x\right)\right)_{p=0}$$

$$=\left(\int\limits_a^\infty\frac{\partial^{2n-1}}{\partial p^{2n-1}}\left(\frac{\sin px}{e^{\pi x}-1}\right)\partial x\right)_{p=0}=\left(\int\limits_a^\infty\frac{x^{2n-1}\sin(px+\frac{2n-1}{2}\pi)}{e^{\pi x}-1}\,\partial x\right)_{p=0}$$

$$=(-1)^{n+1}\int\limits_0^\infty\frac{x^{2n-1}}{e^{\pi x}-1}\,\partial x=(-1)^{n+1}\cdot 2^{2n}\int\limits_0^\infty\frac{x^{2n-1}}{e^{2\pi x}-1}\,\partial x, \tag{1.296}$$

from which (1.285) directly follows [151, 152, 179, 187]. □

Hermite proved the following relation for $\zeta(2m)$:

$$\zeta(2m)=(-1)^{m\ 1}\frac{2^{2m-1}\pi^{2m}}{(2m)!}B_{2m}\quad(m=1,2,\dots). \tag{1.297}$$

Proof. From (1.162) it follows that

$$\zeta(2m)=\frac{1}{\Gamma(2m)}\int\limits_0^\infty\frac{x^{2m-1}}{e^x-1}\,\partial x=\frac{(2\pi)^{2m}}{(2m-1)!}\int\limits_0^\infty\frac{x^{2m-1}}{e^{2\pi x}-1}\,\partial x$$

$$\overset{(1.285)}{=}(-1)^{m-1}\frac{2^{2m-1}\pi^{2m}}{(2m)!}B_{2m}. \tag{1.298}$$

□

For $\zeta(1-2m)$, we have the following formula:

$$\zeta(1-2m)=-\frac{B_{2m}}{2m}\quad(m=1,2,\dots). \tag{1.299}$$

Proof. Formula (1.299) follows from (1.197) and (1.297) for $s=2m$. □

1.2.4 Zeros of the function $\zeta(s)$

Definition 1.11. The function $\xi(s)$ is defined as

$$\xi(s)=\frac{1}{2}s(s-1)\pi^{-\frac{s}{2}}\Gamma\left(\frac{s}{2}\right)\zeta(s). \tag{1.300}$$

For the function $\xi(s)$, we have the following theorem.

Theorem 1.10. (a) *$\xi(s)$ is a whole-integer function such that $\xi(s)=\xi(1-s)$.*
(b) *$\xi(s)$ is a real function for $t=0$ and for $\sigma=\frac{1}{2}$.*
(c) $\xi(0)=\xi(1)=\frac{1}{2}$.

Proof. (a) It follows from (1.188) that $\xi(s)$ is an analytic function on any finite segment of the s-plane, i. e., it is a whole function. From the functional equation (1.179) it immediately follows that $\xi(s) = \xi(1-s)$.

(b) The function $\xi(s)$ takes real values for real s. Then by the *Riemann–Schwarz* symmetry principle it takes conjugated values at conjugated points. In particular, we have

$$\xi\left(\frac{1}{2}+it\right) = \overline{\xi\left(\frac{1}{2}-it\right)}. \tag{1.301}$$

However, since

$$\overline{\xi\left(\frac{1}{2}-it\right)} = \overline{\xi\left(1-\left(\frac{1}{2}-it\right)\right)} = \overline{\xi\left(\frac{1}{2}+it\right)}, \tag{1.302}$$

it follows that $\xi(\frac{1}{2}+it)$ is real.

(c) We have

$$\xi(s) = (s-1)\zeta(s)\pi^{-\frac{s}{2}}\frac{s}{2}\Gamma\left(\frac{s}{2}\right) = (s-1)\zeta(s)\pi^{-\frac{s}{2}}\frac{s}{2}\Gamma\left(\frac{s}{2}+1\right). \tag{1.303}$$

Since $\lim_{s\longrightarrow 1}(s-1)\zeta(s) = 1$ (by (1.41)) and $\Gamma(\frac{3}{2}) = \frac{\sqrt{\pi}}{2}$, we have $\xi(1) = \frac{1}{2}$, and since $\xi(s) = \xi(1-s)$, we have $\xi(0) = \frac{1}{2}$. □

The following theorem provides information on the location of zeros of the function $\xi(s)$.

Theorem 1.11. (a) *The zeros of $\xi(s)$ (if any) are all in the region $0 \leqslant \sigma \leqslant 1$ and are arranged symmetrically with respect to the lines $\sigma = \frac{1}{2}$ and $t = 0$.*

(b) *The zeros of $\zeta(s)$ are identical (by position and multiplicities) with the zeros of $\xi(s)$, apart from the simple zeros $s = -2, -4, -6, \ldots$.*

(c) *$\xi(s)$ has no zeros on the real axis.*

Proof. (a) We have

$$\xi(s) = \pi^{-\frac{s}{2}}(s-1)\Gamma\left(\frac{s}{2}+1\right)\zeta(s) = h(s)\zeta(s). \tag{1.304}$$

We know that (by (1.9))

$$\zeta(s) \neq 0 \quad (\sigma > 1). \tag{1.305}$$

It is also obvious that

$$h(s) \neq 0 \quad (\sigma > 1), \tag{1.306}$$

so it follows that

$$\xi(s) \neq 0 \quad (\sigma > 1). \tag{1.307}$$

Since $\xi(s) = \xi(1-s)$, we have $\xi(s) \neq 0(\sigma < 0)$. It follows that all zeros are in the region $0 \leq \sigma \leq 1$. Since

$$\xi(\sigma + it) = \overline{\xi(\sigma - it)}, \tag{1.308}$$

if $\sigma + it$ is a zero of the function $\xi(s)$, $\sigma - it$ is also a zero, and these points are symmetric with respect to the real σ-axis.

Let

$$\xi\left(\frac{1}{2} + \sigma + it\right) = 0. \tag{1.309}$$

Then $\xi(1 - (\frac{1}{2} + \sigma + it)) = 0$, i. e., $\xi(\frac{1}{2} - \sigma - it) = 0$. It follows that $\overline{\xi(\frac{1}{2} - \sigma + it)} = 0$, which also means that $\xi(\frac{1}{2} - \sigma + i)t = 0$. Therefore the zeros

$$\frac{1}{2} + \sigma + it, \quad \frac{1}{2} - \sigma + it \tag{1.310}$$

are symmetric with respect to the line $\sigma = \frac{1}{2}$.

(b) The zeros of $\zeta(s)$ differ from the zeros of $\xi(s)$ only where $h(s)$ has zeros or poles. The only zero of the function $h(s)$ is $s = 1$. However, this is not a zero of the function $\xi(s)$, because $\xi(1) = \frac{1}{2}$, nor is it a zero of the function $\zeta(s)$, but it is its pole. The poles of $h(s)$ are simple (since $\Gamma(\frac{s}{2} + 1)$ has simple poles), and these are the points $s = -2, -4, -6, \ldots$. Since at these points, $\xi(s)$ is regular, these points are simple zeros of the function $\zeta(s)$.

(c) Since $\xi(s) \neq 0$ for $\sigma < 0$ and for $\sigma > 1$, it suffices to prove that $\zeta(\sigma) \neq 0$ $(0 < \sigma < 1)$. Since

$$(1 - 2^{1-s})\zeta(s) = (1 - 2^{-s}) + (3^{-s} - 4^{-s}) + \cdots \quad (\sigma > 0) \tag{1.311}$$

and

$$\frac{1}{(2n-1)^{\sigma}} - \frac{1}{(2n)^{\sigma}} > 0, \quad 1 - 2^{1-\sigma} < 0 \quad (0 < \sigma < 1), \tag{1.312}$$

it follows that $\zeta(\sigma) < 0$ $(0 < \sigma < 1)$. □

Definition 1.12. The function $\Xi(t)$ is defined as

$$\Xi(t) = \xi\left(\frac{1}{2} + it\right). \tag{1.313}$$

It is an even, integer, and real function for real t, and its zeros on the real axis correspond to the zeros of the function $\zeta(s)$ on the line $\sigma = \frac{1}{2}$.

Definition 1.13. The function $Z(t)$ is defined as

$$Z(t) = e^{i\theta(t)}\zeta\left(\frac{1}{2} + it\right), \tag{1.314}$$

where

$$e^{i\theta(t)} = \pi^{-i\frac{t}{2}} \frac{\Gamma(\frac{1}{4} + i\frac{t}{2})}{|\Gamma(\frac{1}{4} + i\frac{t}{2})|}. \tag{1.315}$$

It is a real function for real t, and its zeros are the zeros of the function $\zeta(s)$ on the line $\sigma = \frac{1}{2}$.

Proof. We have

$$\overline{Z(t)} = Z(t) \iff \pi^{-it}\Gamma\left(\frac{1}{4} + i\frac{t}{2}\right)\zeta\left(\frac{1}{2} + it\right) = \overline{\Gamma\left(\frac{1}{4} + i\frac{t}{2}\right)\zeta\left(\frac{1}{2} + it\right)}. \tag{1.316}$$

It is known that

$$\begin{aligned}\Gamma\left(\frac{1}{4} + i\frac{t}{2}\right) &= \int_0^\infty x^{(\frac{1}{4}+i\frac{t}{2})-1}e^{-x}\,\partial x = \int_0^\infty x^{\frac{1}{4}-1}e^{-x}e^{i\frac{t}{2}\log x}\,\partial x \\ &= \overline{\int_0^\infty x^{\frac{1}{4}-1}e^{-x}e^{-i\frac{t}{2}\log x}\,\partial x} = \overline{\Gamma\left(\frac{1}{4} - i\frac{t}{2}\right)},\end{aligned} \tag{1.317}$$

$$\zeta\left(\frac{1}{2} + it\right) = \sum_{n=1}^\infty \frac{1}{n^{\frac{1}{2}+it}} = \sum_{n=1}^\infty \frac{1}{n^{\frac{1}{2}}}e^{-\Im m\, t\log x} = \overline{\sum_{n=1}^\infty \frac{1}{n^{\frac{1}{2}}}e^{it\log n}} = \overline{\zeta\left(\frac{1}{2} - it\right)}. \tag{1.318}$$

Based on these relations and the functional equation (1.179) with $s = \frac{1}{2} + it$, we obtain (1.316). □

1.3 Development of the function $\zeta(s)$ into a *Laurent* series around the point $s = 1$

1.3.1 Formulas and estimates for the coefficients of decomposition

Since $\zeta(s)$ is an analytic function in the whole s-plane, except at the point $s = 1$, which is a single pole of order 1, its *Laurent* series around the point $s = 1$ has the following form:

$$\zeta(s) = \frac{1}{s-1} + \sum_{n=0}^\infty \gamma_n(s-1)^n. \tag{1.319}$$

Theorem 1.12. *For the coefficients γ_n ($n = 0, 1, 2, \ldots$), we have*

$$\gamma_n = \frac{(-1)^n}{n!} \lim_{N \longrightarrow \infty} \left(\sum_{m=1}^{N} \frac{\log^n m}{m} - \frac{\log^{n+1} N}{n+1} \right). \tag{1.320}$$

Proof. From (1.41) it follows that

$$\zeta(s) - \frac{s}{s-1} = \zeta(s) - \frac{1}{s-1} - 1 = h(s) = \sum_{n=0}^{\infty} C_n (s-1) = s \int_1^{\infty} \frac{[x] - x}{x^{s+1}} \, \partial x \quad (\sigma > 0), \tag{1.321}$$

where $C_0 = \gamma_0 - 1$, $C_n = \gamma_n$ ($n \geq 1$). Since, according to (1.247),

$$\lim_{s \longrightarrow 1} \left(\zeta(s) - \frac{1}{s-1} \right) = \gamma, \tag{1.322}$$

we have

$$\gamma_0 = \gamma, \quad C_0 = \gamma - 1. \tag{1.323}$$

From (1.42) we know that the integral

$$\int_1^{\infty} \frac{[x] - x}{x^{s+1}} \, \partial x \tag{1.324}$$

can be presented in the form of a uniformly convergent series that determines an analytic function, so that its derivatives can be found by interchanging the signs of the integral and derivatives. Thus

$$\begin{aligned} h^{(k)}(s) &= \left(s \int_1^{\infty} \frac{[x] - x}{x^{s+1}} \, \partial x \right)^{(k)} = \sum_{m=0}^{k} \binom{k}{m} (s)^{(k-m)} \left(\int_1^{\infty} \frac{[x] - x}{x^{s+1}} \, \partial x \right)^{(m)} \\ &= k \left(\int_1^{\infty} \frac{[x] - x}{x^{s+1}} \, \partial x \right)^{(k-1)} + s \left(\int_1^{\infty} \frac{[x] - x}{x^{s+1}} \, \partial x \right)^{(k)} \\ &= s \int_1^{\infty} (-1)^k \frac{[x] - x}{x^{s+1}} \log^k x \, \partial x + k \int_1^{\infty} (-1)^{k-1} \frac{[x] - x}{x^{s+1}} \log^{k-1} x \, \partial x. \end{aligned} \tag{1.325}$$

For $s = 1$, we have

$$h^{(k)}(1) = k! C_k = (-1)^k \left(\int_1^{\infty} \frac{[x] - x}{x^2} \log^k x \, \partial x - k \int_1^{\infty} \frac{[x] - x}{x^2} \log^{k-1} x \, \partial x \right)$$

$$= (-1)^k \lim_{N\longrightarrow\infty}\left(\int_1^N \frac{[x]-x}{x^2}\log^k x\,\partial x - k\int_1^N \frac{[x]-x}{x^2}\log^{k-1} x\,\partial x\right)$$
$$= (-1)^k \lim_{N\longrightarrow\infty}\left(\sum_{m=1}^{N-1}\int_m^{m+1} \frac{[x](\log^k x - k\log^{k-1} x}{x^2}\,\partial x - \int_1^N \frac{\log^k x - k\log^{k-1} x}{x}\,\partial x\right). \tag{1.326}$$

Since

$$\int_m^{m+1} \frac{\log^k x - k\log^{k-1} x}{x^2}\,\partial x = \left(-\frac{\log^k x}{x}\right)_m^{m+1} = \frac{\log^k m}{m} - \frac{\log^k(m+1)}{m+1}, \tag{1.327}$$

it follows that

$$\begin{aligned}\sum_{m=1}^{N-1} m\int_m^{m+1} \frac{\log^k x - k\log^{k-1} x}{x^2}\,\partial x &= \sum_{m=1}^{N-1} m\left(\frac{\log^k m}{m} - \frac{\log^k(m+1)}{m+1}\right)\\ &= \sum_{m=1}^{N-1}\frac{\log^k m}{m} - \frac{N-1}{N}\log^k N.\end{aligned} \tag{1.328}$$

In addition, we have

$$\int_1^N \frac{\log^k x - k\log^{k-1} x}{x}\,\partial x = \frac{\log^{k+1} x}{k+1} - \log^k N, \tag{1.329}$$

so it follows that

$$k!C_k = (-1)^k \lim_{N\longrightarrow\infty}\left(\sum_{m=1}^{N-1}\frac{\log^k m}{m} - \left(1-\frac{1}{N}\right)\log^k N - \frac{\log^{k+1} N}{k+1} + \log^k N\right) = k!\gamma_k, \tag{1.330}$$

which proves formula (1.320). □

By analogy, as in the case of (1.320), for γ_n, we can prove the equality

$$\gamma_n = \frac{(-1)^n}{n!}\cdot\int_1^\infty \frac{\lambda(x)}{x^2}(\log^n x - n\log^{n-1} x)\,\partial x, \tag{1.331}$$

where $\lambda(x) = [x] - x + \frac{1}{2}$. Formula (1.320) is proved by *Briggs and Chowla* [50], whereas formula (1.331) is proved by *Lavrik* [125]. The formula for the coefficients γ_n in the following theorem also belongs to *Lavrik*.

Theorem 1.13. *For $n = 1, 2, \ldots$ we have*

$$\gamma_n = \frac{\xi_n}{12} + \frac{(-1)^{n+1}}{2\pi^2 n!} \sum_{m=1}^{\infty} \frac{1}{m^2} \int_1^{\infty} \frac{\cos 2m\pi x}{x^3} (2\log^n x - 3n\log^{n-1} x + n(n-2)\log^{n-2} x)\, \partial x, \quad (1.332)$$

where

$$\xi_n = \begin{cases} 1, & n = 1, \\ 0, & n > 1. \end{cases} \quad (1.333)$$

Proof. Since for noninteger x, the function $\lambda(x)$ can be developed into the *Fourier* series $\sum_{m=1}^{\infty} \frac{\sin 2m\pi x}{m\pi}$, from (1.331) we have

$$\gamma_n = \frac{(-1)^n}{\pi n!} \sum_{m=1}^{\infty} \frac{1}{m} \int_1^{\infty} \frac{\sin 2m\pi x}{x^2} (\log^n x - n\log^{n-1} x)\, \partial x \quad (1.334)$$

if it is possible to interchange the order of summation and integration. Since the series $\sum_{m=1}^{\infty} \frac{\sin 2m\pi x}{m\pi}$ converges almost everywhere, this is possible on every finite interval. It only remains to prove that

$$\lim_{X \longrightarrow \infty} \sum_{m=1}^{\infty} \frac{1}{m} \int_X^{\infty} \frac{\sin 2m\pi x}{x^2} \log^n x\, \partial x = 0. \quad (1.335)$$

We can write

$$\begin{aligned} \int_X^{\infty} \frac{\sin 2m\pi x}{x^2} \log^n x\, \partial x &= \left(-\frac{\cos 2m\pi x}{2m\pi} \cdot \frac{\log^n x}{x^2} \right)_X^{\infty} + \frac{1}{2m\pi} \int_X^{\infty} \cos 2m\pi x \left(\frac{\log^n x}{x^2} \right)' \partial x \\ &= O\left(\frac{\log^n x}{mx^2} \right) + O\left(\frac{\log^n x}{mx^2} \right) = \left(\frac{\log^n x}{mx^2} \right). \end{aligned} \quad (1.336)$$

Relation (1.335) is obtained from (1.336). In particular, integrating by parts (1.334) for $n = 1$ with $\zeta(2) = \frac{\pi^2}{6}$, we obtain relation (1.333). □

The following result also belongs to *Lavrik* [125].

Theorem 1.14. *For γ_n, we have the estimate*

$$|\gamma_n| \leq \frac{1}{2^{n+1}} \quad (n = 1, 2, \ldots). \quad (1.337)$$

Proof. From (1.333), using $\zeta(2) = \frac{\pi^2}{6}$, it follows that

$$|\gamma_n| \leq \frac{\xi_n}{12} + \frac{1}{2\pi^2 n!} \sum_{m=1}^{\infty} \frac{1}{m^2} \int_1^{\infty} x^{-3} (2\log^n x + 3n\log^{n-1} x + n(n-1)\log^{n-2} x)\, \partial x$$

$$= \frac{\xi_n}{12} + \frac{1}{12n!}(2J_n + 3nJ_{n-1} + n(n-1)J_{n-2}), \tag{1.338}$$

where

$$J_m = \int_1^\infty \frac{\log^m x}{x^3}\,\partial x \quad (m \geq 0). \tag{1.339}$$

It is obvious that

$$\int \frac{\log^m x}{x^3}\,\partial x = \frac{P_m(\log x)}{x^2} + C, \tag{1.340}$$

where $P_m(\log x)$ is a polynomial of degree m with respect to $\log x$. After differentiating and equating the coefficients at the degrees $\log^k x$, we get

$$\int \frac{\log^m x}{x^3}\,\partial x = -\sum_{\nu=0}^{m} \binom{m}{\nu} \frac{\nu!}{2^{\nu+1}} \frac{\log^{m-\nu} x}{x^2} + C, \tag{1.341}$$

and it follows that

$$J_m = \frac{m!}{2^{m+1}}. \tag{1.342}$$

Therefore we also have

$$\begin{aligned} |\gamma_1| &\leqslant \frac{1}{12}(1 + 2J_1 + 3J_0) = \frac{1}{4}, \\ |\gamma_n| &\leqslant \frac{1}{12}\left(\frac{n!}{2^n} + 3\frac{n!}{2^n} + \frac{n!}{2^{n-1}}\right) = \frac{1}{2^{n+1}} \quad (n \geq 2), \end{aligned} \tag{1.343}$$

which proves the theorem. □

The following results concerning γ_n coefficients belong to *Istrail* [155].

Definition 1.14. The *Bernoulli* polynomials $B_n(x)$ [68] are defined by

$$\frac{te^{tx}}{e^t - 1} = \sum_{n=0}^{\infty} \frac{B_n(x)}{n!} t^n \quad (|t| < 1), \tag{1.344}$$

and the periodic *Bernoulli* polynomials $B_n^*(x)$ are given by

$$B_n^*(x) = B_n(x) \quad (0 \leq x < 1), \quad B_n^*(x+1) = B_n^*(x). \tag{1.345}$$

Theorem 1.15. (a) *For $n, k \geq 1$, we have*

$$\gamma_n = \sum_{r=[\frac{n}{2}]+1}^{k} \frac{B_{2r}}{2r} b_{n,2r-1} + \frac{(-1)^{k-1}}{n!(2k)!} \int_1^\infty B_{2k}^*(x) f_n^{(2k)}(x)\,\partial x, \tag{1.346}$$

where

$$b_{j,\nu} = \sum_{r_1=j}^{\nu} \frac{1}{r_1} \sum_{r_2=j-1}^{r_1-1} \frac{1}{r_2} \cdots \sum_{r_j=1}^{r_{j-1}-1} \frac{1}{r_j} \tag{1.347}$$

and

$$f_n(x) = \frac{\log^n(x)}{x}. \tag{1.348}$$

(b) *If the derivatives $f_n^{(2k_0)}(x)$ and $f_n^{(2k_0+2)}(x)$ have the same sign on the interval $N-1 \le x \le N$, then*

$$\gamma_0 = \sum_{m=1}^{N} \frac{1}{m} - \log N - \frac{1}{2N} + \sum_{r=1}^{k-1} \frac{B_{2r}}{2rN^{2r}} + \theta \frac{B_{2k}}{2kN^{2k}}, \tag{1.349}$$

$$\gamma_n = \frac{(-1)^n}{n!} \left\{ \begin{array}{c} \sum_{m=1}^{N} \frac{\log^n m}{m} - \frac{\log^{n+1} N}{n+1} - \frac{\log^n N}{2N} \\ - \sum_{r=1}^{k_0-1} \frac{B_{2r}}{(2r)!} f_n^{(2r-1)}(N) - \theta \frac{B_{2k_0}}{(2k_0)!} f_n^{(2k_0-1)}(N) \end{array} \right\} \quad (n \ge 1), \tag{1.350}$$

where $0 < \theta < 1$.

One of the best analytic estimates for γ_n is given in the afore-mentioned paper of *Istrail*. It is provided in the following theorem.

Theorem 1.16. *For $n \ge 2k$ ($k = 1, 2, \ldots$), we have the following estimate:*

$$|\gamma_n| \le C(k)(2k)^{-n}, \tag{1.351}$$

where

$$C(k) = \frac{|B_{2k}|}{2k} \left\{ 1 + \sum_{j=1}^{2k} b_{j,2k} (2k)^j \right\}, \tag{1.352}$$

and in particular, $C(1) = \frac{1}{2}$, $C(2) = \frac{7}{12}$, $C(3) = \frac{11}{3}$.

Proposition 1.2. *Theorem 1.14 follows from Theorem 1.16 with $k = 1$.*

Before Theorems 1.15 and 1.16 are proved, we give several lemmas that more precisely lead to the proofs.

Lemma 1.14. *For nonnegative integers n, ν, and $b_{j,\nu}$ given in (1.346)–(1.348), provided that $b_{0,\nu} = 1$, we have*

$$f_n^{(\nu)}(x) = \left(\frac{\log^n x}{x} \right)^{(\nu)} = \frac{(-1)^\nu \nu!}{x^{\nu+1}} \sum_{j=0}^{\min(n,\nu)} \frac{(-1)^j n!}{(n-j)!} b_{j,\nu} \log^{n-j} x. \tag{1.353}$$

Proof. First of all, the following relation is easy to prove:

$$b_{j,\nu+1} - b_{j,\nu} = \frac{1}{\nu+1} b_{j-1,\nu} \quad (j \geq 1, \nu \geq 0). \tag{1.354}$$

It will be used later to prove relation (1.353) by applying the induction on ν. □

Lemma 1.15. *Let all the derivatives of $f(x)$ for $x \geq a > 0$ be continuous, and let*

$$\int_a^\infty |f^{(2k)}(x)| \, \partial x < \infty \quad (k \geq 1), \tag{1.355}$$

$$f^{(2k-1)}(x) \longrightarrow 0 \quad (x \longrightarrow \infty). \tag{1.356}$$

Denote

$$F(x) = \int_1^x f(t) \, \partial t, \quad S_k(m) = \sum_{r=1}^k \frac{B_{2r}}{(2r)!} f^{(2r-1)}(m). \tag{1.357}$$

Then the following formula, known as the Euler–Maclaurin summation formula, is valid (for the proof, see the paper by Hardy [47] *(Chapter 4 of the monograph):*

$$\sum_{a \leq m \leq N} f(m) = F(N) + \frac{1}{2} f(N) + S_K(N) + C_K + R_{K,N}, \tag{1.358}$$

where

$$C_K = \frac{1}{2} f(a) - F(a) - S_K(a) - \frac{1}{(2k)!} \int_a^\infty B_{2K}^*(x) f^{(2K)}(x) \, \partial x, \tag{1.359}$$

$$R_{K,N} = \frac{1}{(2K)!} \int_N^\infty B_{2K}^*(x) f^{(2K)}(x) \, \partial x, \tag{1.360}$$

and the Fourier series for $B_{2k}^*(x)$ is

$$B_{2k}^*(x) = \frac{2 \cdot (-1)^{k-1} \cdot (2k)!}{(2\pi)^{2k}} \sum_{m=1}^\infty \frac{\cos 2m\pi x}{m^{2k}}. \tag{1.361}$$

Since $B_{2K}^*(x) = O(1)$, according to (1.356)*, we have $R_{K,N} \longrightarrow 0$ $(N \longrightarrow \infty)$. If conditions* (1.355) *and* (1.356) *are fulfilled for all $k \geqslant K-1$, then*

$$C_{K-1} = \lim_{N \longrightarrow \infty} \left(\sum_{a \leqslant m \leqslant N} f(m) - F(N) - \frac{1}{2} f(N) - S_{K-1}(N) \right). \tag{1.362}$$

According to (1.356), *we have*

$$S_K(N) - S_{K-1}(N) = \frac{B_{2K}}{(2K)!} f^{(2K-1)}(N) \longrightarrow 0 \quad (N \longrightarrow \infty), \tag{1.363}$$

$$C_{K-1} = \lim_{N \longrightarrow \infty} \Bigl(\sum_{a \leqslant m \leqslant N} f(m) - F(N) - \frac{1}{2} f(N) - S_K(N) \Bigr) = C_K. \tag{1.364}$$

It follows that C_K is independent of k for $k \geqslant K - 1$. So, for $k \geqslant K - 1$, C_K does not depend on k and N, i. e., $C = C_K$ only depends on $f(x)$ and $F(X)$, i. e., on $f(x)$ and a. As stated in [96], *there is also so-called another form of the Euler–Maclaurin summation formula, stated in Lemma* 1.16.

Lemma 1.16. *If the derivatives $f^{(2K_0)}(x)$ and $f^{(2K_0+2)}(x)$ have the same sign on the interval $[a, N]$, then*

$$\sum_{a \leqslant m \leqslant N} f(m) = F(N) + \frac{1}{2} f(N) + S_{K_0-1}(N) + C_{K_0} + \theta \cdot \frac{B_{2K_0}}{(2K_0)!} f^{(2K_0-1)}(N), \tag{1.365}$$

$$C_{K_0} = \frac{1}{2} f(a) - F(a) - S_{K_0-1}(a) - \theta \cdot \frac{B_{2K_0}}{(2K_0)!} f^{(2K_0-1)}(a), \tag{1.366}$$

where $0 < \theta < 1$.

We can now pass to proving the already stated Theorems 1.15 and 1.16.

Proof. Suppose $f(x) = f_n(x) = \frac{\log^n x}{x}$. From (1.353) it follows that $f_n^{(K)}(N) = O(\frac{\log^n N}{N^{K+1}})$. From this relation and (1.357) it follows that $S_K(N) = O(\frac{\log^n N}{N^2})$, so $f_n^{(K)}(N)$, $S_K(N)$, $R_{K,N}$, and $f(N)$ tend to zero as $N \longrightarrow \infty$. From (1.358) it follows that

$$C(n) = C_K(n) = \lim_{N \longrightarrow \infty} \Bigl(\sum_{a \leqslant m \leqslant N} \frac{\log^n m}{m} - \frac{\log^{n+1} N}{n+1} \Bigr). \tag{1.367}$$

However, by (1.320), for $a \geqslant 1$, we have

$$C(n) = (-1)^n n! \gamma_n - \sum_{1 \leqslant m \leqslant a-1} \frac{\log^n m}{m}. \tag{1.368}$$

From this relation and (1.359) it follows that for $K \geqslant 1$ and $a \geqslant 1$,

$$\begin{aligned} \gamma_n = \frac{(-1)^n}{n!} \Bigl(& \sum_{1 \leqslant m \leqslant a} \frac{\log^n m}{m} - \frac{\log^{n+1} a}{n+1} - \frac{\log^n a}{2a} \\ & - \sum_{r=1}^{K} \frac{B_{2r}}{(2r)!} f_n^{(2r-1)}(a) - \frac{1}{(2K)!} \int_a^\infty B_{2K}^*(x) f_n^{(2K)}(x)\, \partial x \Bigr). \end{aligned} \tag{1.369}$$

From (1.353) we have

$$f_n^{(2r-1)}(1) = \begin{cases} (-1)^{(2r-1)}(2r-1)!n!b_{n,2r-1}, & n \leqslant 2r-1, \\ 0, & n > 2r-1. \end{cases} \tag{1.370}$$

Equality (1.346) follows from (1.369) with $a = 1$. Next, suppose $n \geq 2K$. From (1.346) it follows that

$$y_n = \frac{(-1)^{K-1}}{n!(2K)!} \int_a^\infty B_{2K}^*(x) f_n^{(2K)}(x)\, \partial x. \tag{1.371}$$

In addition, for $n \geq 2K$, from (1.353) we have

$$f_n^{(2K)}(x) = \frac{(2K)!}{x^{2K+1}} \left(\log^n x + \sum_{j=1}^{2K} \frac{(-1)^j n!}{(n-j)!} b_{j,2K} \log^{n-j} x \right). \tag{1.372}$$

Since, as in the proof of Theorem 1.14,

$$\int_1^\infty \frac{\log^{n-j} x}{x^{2K+1}}\, \partial x = \frac{(n-j)!}{(2K)^{n-j+1}}, \tag{1.373}$$

and by (1.361)

$$|B_{2K}^*(x)| \leqslant 2\frac{(2K)!}{(2\pi)^{2K}} \zeta(2K) \overset{(II.38)}{=} |B_{2K}|, \tag{1.374}$$

from (1.371), (1.372), (1.373), and (1.374) we get

$$\begin{aligned} |y_n| &\leqslant \frac{|B_{2K}|}{n!} \int_1^\infty x^{-2K-1} \left(\log^n x + \sum_{j=1}^{2K} \frac{n!}{(n-j)!} b_{j,2K} \log^{n-j} x \right) \partial x \\ &= \frac{|B_{2K}|}{(2K)^{n+1}} \left\{ 1 + \sum_{j=1}^{2K} b_{j,2K} (2K)^j \right\}. \end{aligned} \tag{1.375}$$

The lemma is proved.

To prove (1.349) and (1.350), given in Theorem 1.15, we will take $f(x) = \frac{\log^n(x)}{x}$ and $a = N-1$ in (1.365). Then from (1.365) and (1.368) it follows that

$$\begin{aligned} \sum_{m=N-1}^{N} \frac{\log^n m}{m} &= \frac{\log^{n+1} N}{n+1} + \frac{\log^n N}{2N} + \sum_{r=1}^{K_0-1} \frac{B_{2r}}{(2r)!} f_n^{(2r-1)}(N) + (-1)^n n! y_n \\ &\quad - \sum_{1 \leqslant m \leqslant N-2} \frac{\log^n m}{m} + \theta \frac{B_{2K_0}}{(2K_0)!} f_n^{(2K_0-1)}(N), \end{aligned} \tag{1.376}$$

from (1.350) directly follows. Now take $f(x) = \frac{1}{x}$ in (1.365). Since $C_k(0) = \gamma_0$ (which follows from (1.368)) and $f_n^{(2r-1)}(N) = -\frac{(2r-1)!}{N^{2r}}$, we get (1.349). □

New results in this area can be found in the paper by Ivić [234].

1.3.2 The relation between the decomposition of the function $\zeta(s)$ around its pole and a major member of the *Dirichlet* divisor problem

The basic idea and main results are given by *Lavrik* [125]. The *Dirichlet* divisor problem refers to the problem of determining asymptotic behavior of the following sum as $x \longrightarrow \infty$:

$$D_k(x) = \sum_{n \leqslant x} d_K(n), \tag{1.377}$$

where $d_K(n)$ is given by formula (1.27). First of all, we will prove the following lemma (*Prachar* [84], p. 427).

Lemma 1.17. *Let the series $f(s) = \sum_{n=1}^{\infty} \frac{a_n}{n^s}$ absolutely converge for $\sigma > 1$, let $|a_n| < b\Psi(n)$ $(b > 0)$, where $\Psi(x)$ is a positive increasing function for large x, and suppose that*

$$\sum_{n=1}^{\infty} \frac{a_n}{n^\sigma} = O\left(\frac{1}{(\sigma - 1)^\alpha}\right) \quad (\alpha > 0,\ \sigma \downarrow 1). \tag{1.378}$$

Let $s = \sigma + it$ be arbitrary, and let $c > 0$, $c + \sigma > 1$, and $T > 0$. Then for noninteger $x > 1$,

$$\begin{aligned}\sum_{n<x} \frac{a_n}{n^s} = \frac{1}{2\pi i} \int_{c-it}^{c+it} f(s+\omega) \frac{x^\omega}{\omega}\, \partial\omega + O\left(\frac{x^c}{T(\sigma + c - 1)^\alpha}\right)\\ + O\left(\frac{\Psi(2x)x^{1-\sigma} \log 2x}{T}\right) + O\left(\frac{\Psi(2x)x^{1-\sigma}}{T|x - N|}\right),\end{aligned} \tag{1.379}$$

where N is the closest natural number to x.

Proof. Applying the *Cauchy* theorem to $(\frac{x}{n})^\omega \frac{1}{\omega}$, for $n < x$ and $U > 0$, we get

$$\frac{1}{2\pi i}\left(\int_{-U-iT}^{c-iT} + \int_{c-iT}^{c+iT} + \int_{c+iT}^{-U+iT} + \int_{-U+iT}^{-U-iT}\right)\left(\frac{x}{n}\right)^\omega \frac{\partial\omega}{\omega} = 1, \tag{1.380}$$

because $\operatorname{res}_{\omega=0}(\frac{x}{n})^\omega \frac{1}{\omega} = 1$. From the inequality

$$\left|\int_{-U+iT}^{-U-iT} \left(\frac{x}{n}\right)^\omega \frac{\partial\omega}{\omega}\right| \leqslant 2T\left(\frac{x}{n}\right)^{-U} \cdot \frac{1}{U} \longrightarrow 0 \quad (U \longrightarrow \infty) \tag{1.381}$$

we have the integrals in the formula

$$\frac{1}{2\pi i}\left(\int_{-\infty-iT}^{c-iT}+\int_{c-iT}^{c+iT}+\int_{c+\Im m\,T}^{-\infty+iT}\right)\left(\frac{x}{n}\right)^{\omega}\frac{\partial\omega}{\omega}=1, \tag{1.382}$$

In addition, we have the relations

$$\int_{-\infty+iT}^{c+iT}\left(\frac{x}{n}\right)^{\omega}\frac{\partial\omega}{\omega}=\left(\frac{(\frac{x}{n})^{\omega}}{\omega\log\frac{x}{n}}\right)_{-\infty+iT}^{c+iT}+\frac{1}{\log\frac{x}{n}}\int_{-\infty+iT}^{c+iT}\left(\frac{x}{n}\right)^{\omega}\frac{\partial\omega}{\omega^2}, \tag{1.383}$$

$$\left|\frac{(\frac{x}{n})^{\sigma+iT}}{(\sigma+iT)\log\frac{x}{n}}\right|=\frac{(\frac{x}{n})^{\sigma}}{\sqrt{\sigma^2+T^2}\log\frac{x}{n}}<\frac{(\frac{x}{n})^{\sigma}}{\sigma\log\frac{x}{n}}\longrightarrow 0\quad(\sigma\longrightarrow-\infty), \tag{1.384}$$

$$\frac{(\frac{x}{n})^{c}}{\sqrt{c^2+T^2}\log\frac{x}{n}}<\frac{(\frac{x}{n})^{c}}{T\log\frac{x}{n}}, \tag{1.385}$$

$$\begin{aligned}\left|\frac{1}{\log\frac{x}{n}}\int_{-\infty+iT}^{c+iT}\frac{(\frac{x}{n})^{\omega}}{\omega^2}\partial\omega\right|&=\left|\frac{1}{\log\frac{x}{n}}\int_{-\infty}^{c}\frac{(\frac{x}{n})^{\sigma+iT}}{(\sigma+iT)^2}\partial\sigma\right|\leqslant\frac{(\frac{x}{n})^{c}}{\log\frac{x}{n}}\int_{-\infty}^{c}\frac{\partial\sigma}{\sigma^2+T^2}\\&<\frac{(\frac{x}{n})^{c}}{\log\frac{x}{n}}\int_{-\infty}^{+\infty}\frac{\partial\sigma}{\sigma^2+T^2}=\frac{\pi}{\log\frac{x}{n}}\cdot\frac{(\frac{x}{n})^{c}}{T}.\end{aligned} \tag{1.386}$$

It follows that

$$\int_{-\infty+iT}^{c+iT}\left(\frac{x}{n}\right)^{\omega}\frac{\partial\omega}{\omega}=O\left(\frac{(\frac{x}{n})^{c}}{T\log\frac{x}{n}}\right). \tag{1.387}$$

Analogously, we have

$$\int_{-\infty-iT}^{c-iT}\left(\frac{x}{n}\right)^{\omega}\frac{\partial\omega}{\omega}=O\left(\frac{(\frac{x}{n})^{c}}{T\log\frac{x}{n}}\right). \tag{1.388}$$

From (1.382) it follows that

$$\frac{1}{2\pi i}\int_{c-iT}^{c+iT}\frac{(\frac{x}{n})^{\omega}}{\omega}\frac{\partial\omega}{\omega}=1+O\left(\frac{(\frac{x}{n})^{c}}{T\log\frac{x}{n}}\right)\quad(x>n). \tag{1.389}$$

When $x<n$, taking as the integration contour the right rectangle with vertices at the points $c-iT$, $U-iT$, $U+iT$, $c+iT$, since the function $(\frac{x}{n})^{\omega}\cdot\frac{1}{\omega}$ is regular at that rectangle, we get

$$\frac{1}{2\pi i}\int_{c-iT}^{c+iT}\frac{(\frac{x}{n})^{\omega}}{\omega}\partial\omega=O\left(\frac{(\frac{x}{n})^{c}}{T|\log\frac{x}{n}|}\right)\quad(x<n). \tag{1.390}$$

Multiplying (1.389) and (1.390) by $\frac{a_n}{n^s}$, after the summation, we obtain

$$\sum_{n=1}^{\infty} \frac{1}{2\pi \mathrm{i}} \int_{c-\mathrm{i}T}^{c+\mathrm{i}T} \frac{a_n}{n^s} \left(\frac{x}{n}\right)^{\omega} \frac{\partial \omega}{\omega} = \sum_{n<1} \frac{a_n}{n^s} + O\left(\frac{x^c}{T} \sum_{n=1}^{\infty} \frac{|a_n|}{n^{\sigma+c} |\log \frac{x}{n}|} \right). \tag{1.391}$$

Since by Lemma 1.12 the series $\sum_{n=1}^{\infty} \frac{a_n}{n^{\omega+s}}$ converges uniformly for $c + \sigma > 1$, in (1.391) the integral and summation symbols can be interchanged. It follows that

$$\frac{1}{2\pi \mathrm{i}} \int_{c-\mathrm{i}T}^{c+\mathrm{i}T} f(\omega + s) \frac{x^{\omega}}{\omega} \partial \omega = \sum_{n<x} \frac{a_n}{n^s} + O\left(\frac{x^c}{T} \sum_{n=1}^{\infty} \frac{|a_n|}{n^{\sigma+c} |\log \frac{x}{n}|} \right). \tag{1.392}$$

If $n < \frac{x}{2}$ or $n > 2x$, then $|\log \frac{x}{n}| > \log 2$. It follows that

$$\sum_{n=1} \frac{|a_n|}{n^{\sigma+c} |\log \frac{x}{n}|} \ll \sum_{n=1}^{\infty} \frac{|a_n|}{n^{\sigma+c}} \ll \frac{1}{(\sigma + c - 1)^{\alpha}}. \tag{1.393}$$

For $\frac{x}{2} \leqslant n \leqslant 2x$, we will consider the following cases: $N < n \leqslant 2x$, $\frac{x}{2} \leqslant n < N$, and $n = N$.

1) $N < n \leqslant 2x$. If $N = [x]$, $n = N + r$, and $[x] < x \leqslant [x] + \frac{1}{2}$, then

$$\log \frac{n}{x} = \log \frac{N+r}{x} \geqslant \log \frac{N+r}{N+\frac{1}{2}} = \log\left(1 + \frac{r - \frac{1}{2}}{N + \frac{1}{2}}\right). \tag{1.394}$$

Since

$$0 < \frac{r - \frac{1}{2}}{N + \frac{1}{2}} < 1, \quad \log(1+t) > \frac{t}{2} \quad (0 < t < 1), \tag{1.395}$$

we have

$$\log \frac{n}{x} > \frac{r - \frac{1}{2}}{2(N + \frac{1}{2})} \geqslant \frac{r}{4(N + \frac{1}{2})} > \frac{r}{4(x + \frac{1}{2})} > \frac{r}{8x} = A_1 \cdot \frac{r}{x}. \tag{1.396}$$

If $N = [x] + 1$, $n = N + r$, and $[x] + \frac{1}{2} \leqslant x < [x] + 1$, then

$$\begin{aligned} \log \frac{n}{x} &= \log \frac{N+r}{x} > \log \frac{N+r}{N} = \log\left(1 + \frac{r}{N}\right) > \frac{r}{2N} \\ &= \frac{r}{2([x+1)} > \frac{r}{2(x + \frac{1}{2})} > \frac{r}{4x} = A_2 \cdot \frac{r}{x}. \end{aligned} \tag{1.397}$$

2) $\frac{x}{2} \leqslant n \leqslant N$. If $N = [x]$, $n = N - r$, and $[x] < x \leqslant [x] + \frac{1}{2}$, then

$$\begin{aligned}\log\frac{x}{n} &= \log\frac{x}{N-r} > \log\frac{N}{N-r} = \log\left(1+\frac{r}{N-r}\right)\\ &> \frac{r}{2(N-r)} > \frac{r}{2N} = \frac{r}{2[x]} > \frac{r}{2x} = A_3\frac{r}{x}.\end{aligned} \tag{1.398}$$

If $N = [x] + 1$, $n = N - r$, and $[x] + \frac{1}{2} \leqslant x < [x] + 1$, then

$$\begin{aligned}\log\frac{x}{n} &= \log\frac{x}{N-r} \geqslant \log\frac{[x]+\frac{1}{2}}{N-r} = \log\frac{N-\frac{1}{2}}{N-r} = \log\left(1+\frac{r-\frac{1}{2}}{N-r}\right)\\ &> \frac{r-\frac{1}{2}}{2(N-r)} > \frac{r}{4(N-r)} > \frac{r}{4N} \geqslant \frac{r}{4(x+\frac{1}{2})} > \frac{r}{8x} = A_1\frac{r}{x}.\end{aligned} \tag{1.399}$$

Therefore for $N < n \leqslant 2x$, we have

$$\begin{aligned}\frac{x^c}{T}\sum_{\substack{N<n\leqslant 2x\\ 1\leqslant r\leqslant x}}\frac{|a_n|}{n^{\sigma+c}|\log\frac{x}{n}|} &\ll \frac{x^c}{T}\sum_{\substack{N<n\leqslant 2x\\ 1\leqslant r\leqslant x}}\frac{x}{r}\cdot\frac{|a_n|}{n^{\sigma+c}} \ll \frac{x^{c+1}}{T}\sum_{\substack{N<n\leqslant 2x\\ 1\leqslant r\leqslant x}}\frac{1}{r}\cdot\frac{\Psi(n)}{n^{\sigma+c}}\\ &\ll \frac{x^{c+1}}{T}\Psi(2x)\sum_{\substack{N<n\leqslant 2x\\ 1\leqslant r\leqslant x}}\frac{1}{x^{\sigma+c}}\cdot\frac{1}{r} \ll \frac{\Psi(2x)}{T}x^{1-\sigma}\log x,\end{aligned} \tag{1.400}$$

since

$$\sum_{1\leqslant r\leqslant x}\frac{1}{r} = O(\log x). \tag{1.401}$$

By the above relations we obtain

$$\frac{x^c}{T}\sum_{\substack{N<n\leqslant 2x\\ 1\leqslant r\leqslant x}}\frac{|a_n|}{n^{\sigma+c}|\log\frac{x}{n}|} = O\left(\frac{\Psi(2x)}{T}x^{1-\sigma}\log x\right). \tag{1.402}$$

For $\frac{x}{2} \leqslant n < N < 2x$, we have

$$\begin{aligned}\frac{x^c}{T}\sum_{\frac{x}{2}<n\leqslant N}\frac{|a_n|}{n^{\sigma+c}|\log\frac{x}{n}|} &\ll \frac{\Psi(N)}{T}x^{1-\sigma}\sum_{\frac{x}{2}<n\leqslant N}\frac{1}{r}\\ &\ll \frac{\Psi(2x)}{T}x^{1-\sigma}\log N \ll \frac{\Psi(2x)}{T}x^{1-\sigma}\log 2x.\end{aligned} \tag{1.403}$$

Relations (1.402) and (1.403) can therefore be presented in the form

$$\frac{x^c}{T}\sum_{\substack{\frac{x}{2}<n\leqslant 2x\\ n\neq N}}\frac{|a_n|}{n^{\sigma+c}|\log\frac{x}{n}|} = O\left(\frac{\Psi(2x)}{T}x^{1-\sigma}\log 2x\right). \tag{1.404}$$

Finally, it follows that

$$\frac{|a_n|}{N^{\sigma+c}|\log\frac{x}{N}|} \ll \frac{|a_n|}{N^{\sigma+c}|\log(1+\frac{x-N}{N})|} \ll \frac{\Psi(2x)x^{1-\sigma-c}}{|x-N|}. \tag{1.405}$$

The lemma follows from all the relations above. □

We will apply Lemma 1.17 to prove the formula given in the following lemma.

Lemma 1.18. *For $c > 1$ and noninteger numbers $x > 1$, we have*

$$D_k(x) = \frac{1}{2\pi i}\int_{c-i\infty}^{c+i\infty} \zeta^k(\omega)\frac{x^\omega}{\omega}\,\partial\omega. \tag{1.406}$$

Proof. Since by (1.28)

$$\zeta^k(s) = \sum_{n=1}^{\infty}\frac{d_k(n)}{n^s} \quad (\sigma > 1), \tag{1.407}$$

in accordance with Lemma 1.17, we will suppose that $a_n = d_k(n)$ and $f(s) = \zeta^k(s)$. For $n = p_1^{m_1}p_2^{m_2}\cdots p_r^{m_r}$, we have

$$d_k(n) = \binom{m_1+k-1}{k-1}\binom{m_2+k-1}{k-1}\cdots\binom{m_r+k-1}{k-1} \tag{1.408}$$

(see: *Vinogradov* [103], p. 32). Then

$$d_k(n) < \binom{n+k-1}{k-1}^n \text{ since } \binom{m_2+k-1}{k-1} < \binom{n+k-1}{k-1} \quad (l < n). \tag{1.409}$$

We have

$$\frac{n!}{m_l!} < \frac{(n+k-1)!}{(m_l+k-1)!}, \quad \text{since} \prod_{u=m_l+1}^{n} u < \prod_{v=m_l+k}^{n+k-1} v. \tag{1.410}$$

It is also obvious that

$$\binom{n+k-1}{k-1}^n < \binom{n+k}{k-1} < \binom{n+k}{k-1}^{n+1}, \tag{1.411}$$

so we can assume that

$$\Psi(n) = \binom{n+k-1}{k-1}^n. \tag{1.412}$$

We have

$$|\zeta^k(s)| \leqslant |\zeta(\sigma)|^k = \sum_{n=1}^{\infty} \frac{d_k(n)}{n^\sigma} = O\left(\frac{1}{(\sigma-1)^k}\right), \tag{1.413}$$

since around the point $\sigma = 1$, we have the following estimate (based on (1.247)):

$$\zeta(\delta) = O\left(\frac{1}{|\sigma-1|}\right). \tag{1.414}$$

Therefore, as $\sigma \downarrow 1$, for $\alpha = k$, the we have following relation:

$$\sum_{n=1}^{\infty} \frac{|a_n|}{n^\sigma} = O\left(\frac{1}{(\sigma-1)^\alpha}\right). \tag{1.415}$$

By the above, (1.406) follows from relation (1.379) for $s = 0$ as $T \longrightarrow \infty$. □

Let us introduce an estimate for $\zeta(s)$, which will lead to another formula for $D_k(x)$.

Lemma 1.19. *We have*

$$\begin{aligned} &\zeta(s) = O(\log|t|) \quad (1 \leqslant \sigma \leqslant 2,\ |t| \geqslant 3), \\ &\zeta(s) = O(|t|^{1-\sigma}\log|t|) \quad \left(\frac{1}{2} \leqslant \sigma \leqslant 1,\ |t| \geqslant 1\right). \end{aligned} \tag{1.416}$$

Proof. The first formula is proved in Lemma 1.5. Let $\frac{1}{2} \leqslant \sigma \leqslant 1$ and $t \geqslant 1$. By (1.40) we have

$$\zeta(s) = \sum_{n=1}^{N+1} \frac{1}{n^s} + \frac{1}{(s-1)(N+1)^{s-1}} - s \sum_{N+1}^{\infty} \frac{\{x\}}{x^{s+1}} dx \quad (\sigma > 0). \tag{1.417}$$

If $[t] = N$, then

$$\begin{aligned} \zeta(s) &= \sum_{n=1}^{t+1} \frac{1}{n^s} + \frac{1}{(s-1)([t]+1)^{s-1}} - s \int_{[t]+1}^{\infty} \frac{\{x\}}{x^{s+1}} \partial x \\ &= \sum_{n=1}^{t+1} \frac{1}{n^s} + \frac{1}{(s-1)([t]+1)^{s-1}} - s \sum_{n=[t]+1}^{[t]+1} \int_{n}^{n+1} \frac{\nu-n}{\nu^{s+1}} \partial \nu. \end{aligned} \tag{1.418}$$

It follows that

$$\begin{aligned} |\zeta(s)| &\leqslant \sum_{n=1}^{t+1} \frac{1}{n^\sigma} + \frac{1}{t}\frac{1}{(t+1)^{\sigma-1}} + (1+t) \sum_{n=[t]+1}^{\infty} \int_{n}^{n+1} \frac{\partial \nu}{\nu^{\sigma+1}} \\ &< \sum_{n=1}^{t+1} \frac{n^{1-\sigma}}{n} + \frac{(t+1)^{1-\sigma}}{t} + 2t \int_{[t]+1}^{\infty} \frac{\partial \nu}{\nu^{\sigma+1}} \end{aligned}$$

$$< (t+1)^{1-\sigma} \sum_{n=1}^{t+1} \frac{1}{n} + + \frac{t+1}{t}(t+1)^{-\sigma} + 2t \int_t^{\infty} \frac{\partial v}{v^{\sigma+1}}$$

$$= (t+1)^{1-\sigma} \sum_{n=1}^{t+1} \frac{1}{n} + \left(1 + \frac{1}{t}\right)\frac{1}{(t+1)^{\sigma}} + \frac{2}{\sigma} t^{1-\sigma}, \tag{1.419}$$

so that

$$\zeta(s) = O(t^{1-\sigma}\log t) + O(t^{-\sigma}) + O(t^{1-\sigma}). \tag{1.420}$$

Hence we have

$$\zeta(s) = O(t^{1-\sigma}\log t). \tag{1.421}$$

For $t \leqslant -1$, we take $[|t|] = N$, and by the same procedure we can conclude that the second formula is true for $t \leqslant -1$, which, together the corresponding expression for $t \geqslant 1$, proves the second formula in the observed region. □

Lemma 1.20. *We have*

$$D_k(x) = xP_k(\log x) + \triangle_k(x), \tag{1.422}$$

where $P_k(\log x)$ *is a polynomial of degree* $k-1$ *with respect to* $\log x$, *and*

$$\triangle_k(x) = \frac{1}{2\pi \mathrm{i}} \int_{c'-\mathrm{i}\infty}^{c'+\mathrm{i}\infty} \zeta^k(\omega)\frac{x^{\omega}}{\omega}\,\partial\omega \quad \left(\frac{k-1}{k} < c' < 1\right). \tag{1.423}$$

Proof. Let us integrate the function $h(\omega) = \zeta^k(\omega)\frac{x^{\omega}}{\omega}$ along the rectangle with vertices at the points $c - \mathrm{i}R$, $c + \mathrm{i}R$, $c' + \mathrm{i}R$, $c' - \mathrm{i}R$, where $c > 1$ and $\frac{k-1}{k} < c' < 1$. From the estimate

$$\left|\int_{c'-\mathrm{i}R}^{c-\mathrm{i}R} \zeta^k(\omega)\frac{x^{\omega}}{\omega}\,\partial\omega\right| = \left|\int_{c'}^{c} \frac{\zeta^k(\sigma-\mathrm{i}R)}{\sigma-\mathrm{i}R} x^{\sigma-\mathrm{i}R}\,\partial\sigma\right| \leqslant \int_{c'}^{c} x^c \frac{|\zeta(\sigma-\mathrm{i}R)|^k}{\sqrt{\sigma^2+R^2}}\,\partial\sigma$$

$$= x^c \int_{c'}^{1} \frac{|\zeta(\sigma-\mathrm{i}R)|^k}{\sqrt{\sigma^2+R^2}}\,\partial\sigma + x^c \int_{1}^{c} \frac{|\zeta(\sigma-\mathrm{i}R)|^k}{\sqrt{\sigma^2+R^2}}\,\partial\sigma$$

$$\leqslant A \cdot \frac{(1-c')x^c}{R} R^{(1-\sigma)k} \log^k R + B(c-1)x^c \frac{\log^k R}{R} \tag{1.424}$$

(for sufficiently large R), which is obtained on the basis of Lemma 1.19, and from $\lim_{R\longrightarrow\infty} \frac{\log^k R}{R} = 0$ and $\lim_{R\longrightarrow\infty} \frac{\log^k R}{R^{1-k+k\sigma}} = 0$ it follows that

$$\lim_{R\longrightarrow\infty} \int_{c'-\mathrm{i}R}^{c-\mathrm{i}R} \zeta^k(\omega)\frac{x^{\omega}}{\omega}\,\partial\omega = 0, \tag{1.425}$$

$$\lim_{R\longrightarrow\infty}\int\limits_{c'+iR}^{c+iR}\zeta^k(\omega)\frac{x^\omega}{\omega}\,\partial\omega = 0. \tag{1.426}$$

From the above relations it follows that

$$\frac{1}{2\pi i}\int\limits_{c-i\infty}^{c+i\infty}\zeta^k(\omega)\frac{x^\omega}{\omega}\,\partial\omega = \operatorname*{res}_{\omega=1}\zeta^k(\omega)\frac{x^\omega}{\omega} + \frac{1}{2\pi i}\int\limits_{c'-i\infty}^{c'+i\infty}\zeta^k(\omega)\frac{x^\omega}{\omega}\,\partial\omega. \tag{1.427}$$

From (1.423) and (1.427) by Lemma 1.18 we have

$$D_k(x) = \operatorname*{res}_{\omega=1}\zeta^k(\omega)\frac{x^\omega}{\omega} + \triangle_k(x). \tag{1.428}$$

Further, we have

$$(\omega-1)\zeta(\omega) = 1+\lambda(\omega) \tag{1.429}$$

and, accordingly,

$$\begin{aligned}
\operatorname*{res}_{\omega=1}\zeta^k(\omega)\frac{x^\omega}{\omega} &= \frac{1}{(k-1)!}\left\{(\omega-1)^k\zeta^k(\omega)\frac{x^\omega}{\omega}\right\}^{(k-1)}_{\omega=1}\\
&= \frac{1}{(k-1)!}\{(1+\lambda(\omega))^k\cdot(\omega^{-1}x^\omega)\}^{(k-1)}_{\omega=1}\\
&= \frac{1}{(k-1)!}\sum_{l=0}^{k-1}\binom{k-1}{l}(1+\lambda(\omega))^{(k-1-l)}_{\omega=1}(\omega^{-1}x^\omega)^{(l)}_{\omega=1}\\
&= \frac{1}{(k-1)!}\sum_{l=0}^{k-1}\binom{k-1}{l}(1+\lambda(\omega))^{(k-1-l)}_{\omega=1}\left(\sum_{p=0}^{l}\binom{l}{p}(-1)^p p!\omega^{-(p+1)}x^\omega(\log x)^{l-p}\right)_{\omega=1}\\
&= \frac{1}{(k-1)!}\sum_{l=0}^{k-1}\binom{k-1}{l}(1+\lambda(\omega))^{(k-1-l)}_{\omega=1}\sum_{p=0}^{l}\binom{l}{p}(-1)^p(l-p)!(\log x)^p\\
&= x\sum_{p=0}^{k-1}\frac{(-1)^{k-1-p}}{p!}\left(\sum_{l=0}^{k-1-p}\frac{(-1)^l}{l!}(1+\lambda(\omega)^{(l)}_{\omega=1})\right)(\log x)^p = xP_k(\log x),
\end{aligned} \tag{1.430}$$

where $P_k(\log x)$ is a polynomial of degree $k-1$ with respect to $\log x$.

The lemma is proved. □

$P_k(\log x)$ is called the main member of the *Dirichlet* divisor problem.

Lemma 1.21. *Let $n = k_1 + 2k_2 + \cdots + nk_n$ with nonnegative integers k_i, and let*

$$A_j(x) = \sum_{k=0}^{\infty} a_k(j)x^k \quad (j = 1,2,\ldots). \tag{1.431}$$

Then

$$\prod_{j=1}^{\infty} A_j(x^j) = \prod_{j=1}^{\infty} a_0(j) + \sum_{n=1}^{\infty} x^n \sum_{\pi(n)} a_{k_1}(1) a_{k_2}(2) \cdots a_{k_n}(n) \prod_{m=n+1}^{\infty} a_0(m), \tag{1.432}$$

where $\sum_{\pi(n)}$ indicates the summation along all decompositions of a number n.

Proof. The coefficient at x^0 in the product on the left-hand side of relation (1.432) is obviously $\prod_{j=1}^{\infty} a_0(j)$. To determine the coefficient at x^n $(n > 0)$, in the expression $A_i(x^i)$, we need to take $k_i + 1$ as a member $(i = 1, 2, \ldots, n)$, where $k_1, k_2, \ldots, k_n$ fulfil the condition $k_1 + k_2 + \cdots + nk_n = n$, which is the only possible decomposition of the number n. □

Lemma 1.22. *Let*

$$Y_0 = 1, \quad Y_n(y_1, y_2, \ldots, y_n) = \sum_{\pi(n)} \frac{n!}{k_1!, k_2!, \ldots, k_n!} \prod_{j=1}^{n} \left(\frac{y_j}{j!}\right)^{k_j}. \tag{1.433}$$

Then

$$\sum_{n=1}^{\infty} Y_n(y_1, y_2, \ldots, y_n) \frac{z^n}{n!} = \exp\left\{\sum_{n=1}^{\infty} y_n \frac{z^n}{n!}\right\}. \tag{1.434}$$

Proof.

$$\exp\left\{\sum_{n=1}^{\infty} y_n \frac{z^n}{n!}\right\} = \prod_{n=1}^{\infty} \exp\left\{y_n \frac{z^n}{n!}\right\} = \prod_{j=1}^{\infty} \left(\sum_{k=0}^{\infty} \frac{y_j^k}{k!(j!)^k} z^{jk}\right) = \prod_{j=1}^{\infty} A_j(z^j), \tag{1.435}$$

where

$$A_j(z) = \sum_{k=0}^{\infty} a_k(j) z^k, \quad a_k(j) = \frac{y_j^k}{k!(j!)^k}, \quad a_0(j) = 1. \tag{1.436}$$

According to Lemma 1.21,

$$\begin{aligned}\prod_{j=1}^{\infty} A_j(z^j) &= 1 + \sum_{n=1}^{\infty} z^n \sum_{\pi(n)} \prod_{j=1}^{n} a_k(j) = \sum_{n=0}^{\infty} z^n \sum_{\pi(n)} \prod_{j=1}^{n} \frac{y^{k_j}}{k_j!(j!)^{k_j}} \\ &= \sum_{n=1}^{\infty} \frac{z^n}{n!} \sum_{\pi(n)} \frac{n!}{k_1!, k_2!, \ldots, k_n!} \prod_{j=1}^{n} \left(\frac{y_j}{j!}\right)^{k_j} = \sum_{n=0}^{\infty} Y_n(y_1, y_2, \ldots, y_n) \frac{z^n}{n!}.\end{aligned} \tag{1.437}$$

□

Lemma 1.23. *For the functions $Y_j(y_1, y_2, \ldots, y_j)$, we have the following relations:*

$$Y_{n+1}(y_1, y_2, \ldots, y_{n+1}) = \sum_{k=0}^{n} Y_{k+1} Y_{n-k}(y_1, y_2, \ldots, y_{n-k}), \tag{1.438}$$

$$\frac{\partial}{\partial y_i}Y_n(y_1,y_2,\ldots,y_n)=\binom{n}{i}Y_{n-i}(y_1,y_2,\ldots,y_{n-i})\quad(i=1,2,\ldots,n),\tag{1.439}$$

$$Y_{n+1}(y_1,y_2,\ldots,y_{n+1})=y_1Y_n(y_1,y_2,\ldots,y_n)+\sum_{k=1}^{n}Y_{k+1}\frac{\partial}{\partial y_k}Y_n(y_1,y_2,\ldots,y_n).\tag{1.440}$$

Proof. Differentiating (1.434) with respect to z (or with respect to y_i), after equating the corresponding coefficients to z, we obtain (1.438) and (1.439). Relation (1.440) is a direct consequence of relations (1.438) and (1.439). □

Lemma 1.24. *Let $g(s)=F(u)$ and $u=\varphi_0(s)$, where F and φ_0 are differentiable functions. Then*

$$g^{(n)}(s)=\sum_{\pi(n)}\frac{n!}{k_1!k_2!\cdots k_n!}F^{(k)}(u)\cdot\prod_{j=1}^{n}\left(\frac{\varphi_0^{(j)}}{j!}\right)^{k_j}\quad(n\geqslant 1),\tag{1.441}$$

where $\pi(n)$ is the set of all solutions to the equation $k_1+2k_2+\cdots+nk_n=n$ in nonnegative integers $k_1,k_2,\ldots,k_n$, and $k=k_1+k_2+\cdots+k_n$. Relation (1.441) *is called Faa di Bruno's relation.*

Proof. We prove formula (1.441) by the induction on n. For $n=1$, the claim is obviously true. Let

$$y_j=F\cdot\varphi_0^{(j)},\quad F^k\longrightarrow F^{(k)}(u).\tag{1.442}$$

Based on the induction assumption for $g^{(n)}(s)$ in (1.441) and (1.432), we have

$$g^{(n)}(s)=Y_n(y_1,y_2,\ldots,y_n)\tag{1.443}$$

and

$$\left[\prod_{j=1}^{n}\left(\frac{\varphi_0^{(j)}}{j!}\right)^{k_j}\right]_s'=\prod_{j=1}^{n}\left(\frac{\varphi_0^{(j)}}{j!}\right)^{k_j}\cdot\sum_{i=1}^{n}k_i\frac{\varphi_0^{(i+1)}}{\varphi_0^{(i)}}.\tag{1.444}$$

Differentiating (1.441) with respect to s, we get

$$\begin{aligned}g^{(n+1)}(s)&=\sum_{\pi(n)}\frac{n!}{k_1!k_2!\cdots k_n!}F^{(k+1)}(u)\varphi_0'\cdot\prod_{j=1}^{n}\left(\frac{\varphi_0^{(j)}}{j!}\right)^{k_j}\\&\quad+\sum_{\pi(n)}\frac{n!}{k_1!k_2!\cdots k_n!}F^{(k)}(u)\cdot\prod_{j=1}^{n}\left(\frac{\varphi_0^{(j)}}{j!}\right)^{k_j}\cdot\sum_{i=1}^{n}k_i\frac{\varphi_0^{(i+1)}}{\varphi_0^{(i)}}.\end{aligned}\tag{1.445}$$

Based on

$$\frac{\partial Y_n}{\partial y_i}=\sum_{\pi(n)}\frac{n!}{k_1!k_2!\cdots k_n!}\prod_{j=1}^{n}\left(\frac{y_j}{j!}\right)^{k_j}\frac{k_i}{y_i},\tag{1.446}$$

$$\sum_{i=1}^{n} y_{i+1}\frac{\partial Y_n}{\partial y_i} = \sum_{\pi(n)} \frac{n!}{k_1!k_2!\cdots k_n!}\prod_{j=1}^{n}\left(\frac{y_j}{j!}\right)^{k_j}\sum_{i=1}^{n} k_i\frac{y_{i+1}}{y_i}$$
$$= \sum_{\pi(n)} \frac{n!}{k_1!k_2!\cdots k_n!}F^{(k)}(u)\prod_{j=1}^{n}\left(\frac{\varphi_0^{(j)}}{j!}\right)^{k_j}\sum_{i=1}^{n} k_i\frac{\varphi_0^{(i+1)}}{\varphi_0^{(i)}}, \tag{1.447}$$

we obtain

$$g^{(n+1)}(s) = y_1Y_n + \sum_{i=1}^{n} y_{i+1}\frac{\partial Y_n}{\partial y_i}. \tag{1.448}$$

From formula (1.440) it follows that $g^{(n+1)}(s) = Y_{n+1}$, which proves formula (1.441) for $n+1$. □

The idea of applying relation (1.441) to estimate the coefficients of the polynomial $P_k(\log x)$ (see (1.422)) was given by *Lavrik* [125]. The following results also belong to him.

Lemma 1.25. *Let*

$$W_{l,n}(s) = \frac{d^n}{ds^n}[(s-1)\zeta(s)]^l, \quad S_{l,n}(\zeta) = \lim_{s\longrightarrow 1} W_{l,n}(s). \tag{1.449}$$

Then

$$S_{l,n}(\zeta) = n!\sum_{\pi(n)}\binom{l}{k}\frac{k!}{k_0!\cdots k_\nu!}\gamma_0^{k_0}\cdots\gamma_\nu^{k_\nu} \quad (1 \leqslant n \leqslant l-1), \tag{1.450}$$

where $\pi(n)$ is the set of all decompositions of the number n in the form

$$n = k_0 + 2k_1 + \cdots + (\nu+1)k_\nu, \quad k_j \geq 0, \tag{1.451}$$

where ν, k_j are nonnegative integers, $n = k_0+k_1+\cdots+k_\nu$, $k_j = 0$ ($n \leqslant j \leqslant \nu$). The coefficients $\gamma_0, \gamma_1, \ldots, \gamma_\nu$ are given by relation (1.320).

Proof. It is known that

$$(s-1)\zeta(s) = 1 + (s-1)\zeta(s), \quad \zeta(s) = \sum_{n=0}^{\infty}\gamma_n(s-1)^n. \tag{1.452}$$

Taking in (1.441)

$$g(s) = [(s-1)\zeta(s)]^l, \quad F(u) = u^l, \quad u = \varphi_0(s) = 1 + (s-1)\zeta(s), \tag{1.453}$$

after differentiating the function $F(u)$, we get

$$W_{l,n}(s) = \sum_{\pi(n)}\frac{n!l(l-1)\cdots(l-k+1)}{k_0!\cdots k_\nu!}u^{l-k}\prod_{j=0}^{\nu}\left(\frac{\varphi_0^{(j+1)}}{(j+1)!}\right)^{k_j} \quad (k \leqslant l), \tag{1.454}$$

whereas

$$\varphi_0^{(j+1)} = (j+1)\varphi^{(j)} + (s-1)\varphi^{(j+1)}, \varphi^{(r)} = r! \sum_{m=r}^{\infty} \binom{m}{r} \gamma_m (s-1)^{m-r}. \tag{1.455}$$

It follows that

$$W_{l,n}(s) = n! \sum_{\pi(n)} \binom{l}{k} \frac{k!}{k_0! \cdots k_\nu!} u^{l-k} \prod_{j=0}^{\nu} \left[\gamma_j + \sum_{m=j+1}^{\infty} \binom{m+1}{j+1} \gamma_m (s-1)^{m-j} \right]^{k_j}. \tag{1.456}$$

Passing to the limits as $s \longrightarrow 1$, from (1.456) we obtain (1.450). □

Theorem 1.17. *Suppose that for $x > 1$ and $l \geqslant 1$,*

$$P_l(\log x) = \sum_{r=1}^{l} a_{l-r} \log^{l-r} x. \tag{1.457}$$

Then

$$a_{l-r} = \frac{(-1)^{r+1}}{(l-r)!} \left[1 + \sum_{n=1}^{r-1} (-1)^n \sum_{\Pi(n)} \frac{l(l-1)\cdots(l-k+1)}{k_0! \cdots k_\nu!} \gamma_0^{k_0} \cdots \gamma_\nu^{k_\nu} \right] \quad (k \leqslant l). \tag{1.458}$$

Proof. Calculating $\operatorname{res}_{s=1} \zeta^l(s) \frac{x^s}{s}$ according to (1.428), we obtain

$$(l-1)! P_l(\log x) = \frac{1}{x} \sum_{n=0}^{l-1} \binom{l-1}{n} \lim_{s \longrightarrow 1} \left(\frac{x^s}{s} \right)^{(l-n-1)} \cdot \lim_{s \longrightarrow 1} \left([(s-1)\zeta(s)]^l \right)^{(n)}. \tag{1.459}$$

Since

$$\lim_{s \longrightarrow 1} (s-1)\zeta(s) = 1, \quad \lim_{s \longrightarrow 1} \left([(s-1)\zeta(s)]^l \right)^{(n)} = S_{l,n}(\zeta) \quad (n \geqslant 1), \tag{1.460}$$

$$\lim_{s \longrightarrow 1} \left(\frac{x}{s} \right)^{(l-n-1)} = x \sum_{r=0}^{l-n-1} (-1)^r r! \binom{l-n-1}{r} \log^{l-n-r-1} x, \tag{1.461}$$

we have

$$\begin{aligned} &(l-1)! P_l(\log x) \\ &= \sum_{r=0}^{l-1} (-1)^r r! \binom{l-1}{r} \log^{l-r-1} x + \sum_{n=1}^{l-1} \sum_{r=0}^{l-n-1} (-1)^r r! \binom{l-1}{n} \binom{l-n-1}{r} \log^{l-n-r-1} x S_{l,n}(\zeta) \\ &= \log^{l-1} x + \sum_{r=1}^{l-1} \log^{l-r-1} x \left\{ (-1)^r r! \binom{l-1}{r} + \sum_{n=1}^{r} (-1)^{r-n} (r-n)! \binom{l-1}{n} \binom{l-n-1}{r-n} S_{l,n}(\zeta) \right\} \\ &= \log^{l-1} x + \sum_{r=1}^{l-1} (-1)^r r! \binom{l-1}{r} \left\{ 1 + \sum_{n=1}^{r} (-1)^n \frac{S_{l,n}(\zeta)}{n!} \right\} \log^{l-n-1} x. \end{aligned} \tag{1.462}$$

It follows that

$$P_l(\log x) = \sum_{r=1}^{l} a_{l-r} \log^{l-r} x, \tag{1.463}$$

where

$$a_{l-r} = \frac{(-1)^{r+1}}{(l-r)!}\left[1 + \sum_{n=1}^{r-1} \frac{(-1)^n}{n!} S_{l,n}(\zeta)\right], \tag{1.464}$$

which leads to (1.458) by (1.450). ☐

Lemma 1.26. *We have the following equation:*

$$\sum_{\pi(n)} \binom{l}{k} \frac{k!}{k_0! k_{1!} \cdots k_\nu!} = \binom{l}{n} \quad (1 \leqslant n \leqslant l-1). \tag{1.465}$$

Proof. Let

$$G(s) = \frac{1}{s-1} + \sum_{m=0}^{\infty} (s-1)^m. \tag{1.466}$$

As in Lemma 1.25, we obtain

$$\lim_{s \longrightarrow 1} \{[(s-1)G(s)]^l\}^{(n)} = n! \sum_{\pi(n)} \binom{l}{k} \frac{k!}{k_0! k_{1!} \cdots k_\nu!}. \tag{1.467}$$

On the other hand, $(s-1)G(s) = \frac{1}{2-s}$ for $|s-1| < 1$. It follows that

$$\{[(s-1)G(s)]^l\}^{(n)} = n!\binom{l}{n}(2-s)^{-l-n}. \tag{1.468}$$

Passing to the limits as $s \longrightarrow 1$ in the last relation, by (1.467) we obtain (1.465). ☐

The following theorem, also proved by *Lavrik* [125], provides an estimate for the coefficients a_{l-r}.

Theorem 1.18. *We have*

$$|a_{l-r}| = \frac{1}{(l-r)!} \sum_{n=0}^{r-1} \gamma^n \binom{l}{n}, \tag{1.469}$$

where γ is Euler's constant.

Proof. From (1.458) and (1.337) (Theorem 1.14) we get

$$|a_{l-r}| \leq \frac{1}{(l-r)!}\left[1 + \sum_{n=1}^{r-1} \sum_{\pi(n)} \binom{l}{k} \frac{k!}{k_0! k_{1!} \cdots k_\nu!} \gamma_0^{k_0} |\gamma_1|^{k_1} \cdots |\gamma_\nu|^{k_\nu}\right]$$

$$\leqslant \frac{1}{(l-r)!}\left[1+\sum_{n=1}^{r-1}\sum_{\pi(n)}\binom{l}{k}\frac{k!}{k_0!k_{1!}\cdots k_\nu!}y_0^{k_0}2^{-2k_1}\dots 2^{-(\nu+1)k_\nu}\right]$$

$$\leqslant |y_0^{k_0}2^{k_0}2^{-n} \leq y_0^n| \leqslant \frac{1}{(l-r)!}\left[1+\sum_{n=1}^{r-1}\sum_{\pi(n)}\binom{l}{k}\frac{k!}{k_0!..k_\nu!}y_0^n\right]$$

$$\overset{(1.465)}{=} \frac{1}{(l-r)!}\sum_{n=0}^{r-1}y^n\binom{l}{n}. \tag{1.470}$$

□

Related to $\triangle_k(x)$ (see Lemma 1.20), we provide the following formula [98]:

$$\int_1^\infty \frac{\triangle_k(u)}{u^2}\,\partial u = a_0^{(k+1)} - \sum_{m=0}^{k-1} m!y_m a_m^{(k)}, \tag{1.471}$$

where

$$P_x(\log x) = a_{k-1}^{(k)}\log^{k-1}x + \cdots + a_1^{(k)}\log x + a_0^{(k)}. \tag{1.472}$$

For the coefficients y_n, *Berndt* [95] gives the following estimates:

$$|y_n| \leqslant \frac{4}{n\pi^n} \quad (n \text{ even}),$$
$$|y_n| \leqslant \frac{2}{n\pi^n} \quad (n \text{ odd}). \tag{1.473}$$

The following relations apply to the coefficients of the polynomials $P_k(\log x)$ and $P_{k+1}(\log x)$:

$$a_k^{(k+1)} = \frac{a_{k-1}^{(k)}}{k},$$
$$a_{k-n}^{(k+1)} = \frac{a_{k-n-1}^{(k)}}{k-n} + \sum_{m=0}^{n} a_{k-n}^{(k)}(n-m)!\binom{k-m}{n-m}y_{n-m} \quad (n = 1, 2, \dots, k-1), \tag{1.474}$$

where the coefficient $a_0^{(k+1)}$ can be determined from (1.458). Formula (1.474) is given in [146]. In relation to the rest of $\triangle_k(x)$, denoting by α_k the infimum a_k for which $\triangle_k(x) = O(x^{a_k+\varepsilon})$ with arbitrarily small $\varepsilon > 0$, it is known that

$$\alpha_k \leqslant 1 - \frac{1}{k} \quad (k \geqslant 2) \;\; [8], \tag{1.475}$$

$$\alpha_k \leqslant 1 - \frac{2}{k+1} \quad (k \geqslant 2) \;\; [21, 26], \tag{1.476}$$

$$\alpha_k \leqslant 1 - \frac{3}{k+2} \quad (k \geqslant 4) \;\; [33]. \tag{1.477}$$

For $\triangle_2(x)$ and $\triangle_3(x)$, *Kolesnik* [160] obtained the following estimates:

$$\triangle_2(x) = O(x^{\frac{35}{108}+\varepsilon}), \tag{1.478}$$

$$\triangle_3(x) = O(x^{\frac{43}{96}+\varepsilon}). \tag{1.479}$$

In 1979, *Heath-Brown* [154] obtained the following estimate:

$$\triangle_k(x) = O(x^{\frac{3k-4}{4k}+\varepsilon}) \quad (4 \leqslant k \leqslant 8), \tag{1.480}$$

whereas formulas for other k are given in [144]:

$$\triangle_9(x) = O(x^{\frac{35}{54}+\varepsilon}), \quad \triangle_{10}(x) = O(x^{\frac{41}{60}+\varepsilon}), \quad \triangle_{11}(x) = O(x^{\frac{7}{10}+\varepsilon}), \tag{1.481}$$

$$\triangle_k(x) = O(x^{\frac{k-2}{k+2}+\varepsilon}), \qquad 12 \leqslant k \leqslant 25, \tag{1.482}$$

$$\triangle_k(x) = O(x^{\frac{k-1}{k+4}+\varepsilon}), \qquad 26 \leqslant k \leqslant 50, \tag{1.483}$$

$$\triangle_k(x) = O(x^{\frac{31k-98}{32k}+\varepsilon}), \quad 51 \leqslant k \leqslant 57, \tag{1.484}$$

$$\triangle_k(x) = O(x^{\frac{7k-34}{7k}+\varepsilon}), \qquad k \geqslant 58. \tag{1.485}$$

For very large k, in 1972, *Karacuba* [100] provided an estimate of the form

$$\triangle_k(x) = O(x^{1-c_1 k^{-\frac{2}{3}}} \cdot (c_2 \log x)^k), \tag{1.486}$$

where c_1 and c_2 are positive absolute constants, and the relation is true uniformly in k.

1.4 *Stirling* formula. *Mellin* transform and $\zeta(s)$

Theorem 1.19. *We have the following relations:*

$$\log \Gamma(n) = \left(n - \frac{1}{2}\right)\log n - n + \log\sqrt{2\pi} + O\left(\frac{1}{n}\right) \quad (n \in \mathbb{N}), \tag{1.487}$$

$$\log \Gamma(s) = \left(s - \frac{1}{2}\right)\log s - s + \log\sqrt{2\pi} + O\left(\frac{1}{|s|}\right) \tag{1.488}$$

in the region $-\pi + \delta \leqslant \arg s \leqslant \pi - \delta$, $(\delta > 0)$, *where* $\log s = 0$ *for* $s = 1$, *and the constant in* O *depends only on* δ.

Proposition 1.3. *Formulas* (1.487) *and* (1.488) *are known as Stirling's formulas* [61]. *Formula* (1.487) *can be presented in the form*

$$(n-1)! = \sqrt{2\pi} \cdot n^{n-\frac{1}{2}} \cdot e^{-n+O(\frac{1}{n})} \tag{1.489}$$

or in the form

$$n! = \sqrt{2\pi} \cdot n^{n+\frac{1}{2}} \cdot e^{-n} \cdot e^{\frac{\theta(n)}{12n}} \quad (0 < \theta(n) < 1). \tag{1.490}$$

Formula (1.487) is a particular case of formula (1.488). It is commonly known, and we will use it in the proof of formula (1.488). Of course, its proof is independent of the following proof of formula (1.488).

Proof. From *Weierstrass'* definition of the function $\Gamma(s)$, given by the formula

$$\frac{1}{\Gamma(s)} = se^{\gamma s} \prod_{n=1}^{\infty}\left(\left(1+\frac{s}{n}\right)e^{-\frac{s}{n}}\right) \quad (\sigma > 0), \tag{1.491}$$

taking the logarithm, we obtain the relation for the main branch of the function $\log\Gamma(s)$:

$$\log\Gamma(s) = -\gamma s - \log s + \sum_{n=1}^{\infty}\left(\frac{s}{n} \log\left(1+\frac{s}{n}\right)\right). \tag{1.492}$$

Next,

$$\begin{aligned}
\int_{0}^{N} \frac{[u]-u+\frac{1}{2}}{u+s}\,\partial u &= \sum_{n=0}^{N-1} \int_{n=0}^{n+1} \frac{n+\frac{1}{2}-u}{u+s}\,\partial u \\
&= \sum_{n=0}^{N-1}\left(\left(n+\frac{1}{2}+s\right)\log(u+s)\right)\Big|_{n}^{n+1} - N \\
&= \sum_{n=1}^{N-1}\left(n-\frac{1}{2}+s-\left(n+\frac{1}{2}+s\right)\right)\log(n+s) \\
&\quad + \left(N-\frac{1}{2}+s\right)\log(N+s) - \left(s+\frac{1}{2}\right)\log s - N \\
&= \left(N-\frac{1}{2}+s\right)\log(N+s) - \left(s+\frac{1}{2}\right)\log s - N - \sum_{n=1}^{N-1}\log(n+s) \\
&= \sum_{n=1}^{N-1}\left(\frac{s}{n} - \log\left(1+\frac{s}{n}\right) - \log n\right) - s\left(1+\frac{1}{2}+\cdots+\frac{1}{N-1}\right) \\
&\quad + \left(N-\frac{1}{2}+s\right)\log(N+s) - \left(s+\frac{1}{2}\right)\log s - N.
\end{aligned} \tag{1.493}$$

The following formulas apply to sufficiently large N:

$$1+\frac{1}{2}+\cdots+\frac{1}{N-1} = \log N + \gamma + O\left(\frac{1}{N}\right), \tag{1.494}$$

$$\log(N+s) = \log N + \frac{s}{N} + O\left(\frac{1}{N^2}\right). \tag{1.495}$$

Indeed, by the definition of the constant γ [74] we have

$$\gamma = \sum_{n=1}^{\infty} \int_0^1 \frac{v}{n(n+v)}\, \partial v = \sum_{n=1}^{\infty} \int_0^1 \left(\frac{1}{n} - \frac{1}{n+v}\right) \partial v$$
$$= 1 + \frac{1}{2} + \cdots + \frac{1}{N-1} - \sum_{n=1}^{N-1} \log \frac{n+1}{n} + \sum_{n=N}^{\infty} \int_0^1 \frac{v}{n(n+v)}\, \partial v$$
$$= 1 + \frac{1}{2} + \cdots + \frac{1}{N-1} - \log N + K_N. \tag{1.496}$$

For the numbers K_N, from the above relation we have the following estimate:

$$K_N \leqslant \frac{1}{N^2} + \frac{1}{(N+1)^2} + \cdots = \left(\frac{1}{N^2} + \frac{1}{(N+1)^2} + \cdots + \frac{1}{(2N-1)^2}\right)$$
$$+ \left(\frac{1}{(2N)^2} + \frac{1}{(2N+1)^2} + \cdots + \frac{1}{(3N-1)^2}\right) + \cdots$$
$$< \frac{1}{N} \cdot \sum_{m=1}^{\infty} \frac{1}{m^2} = \frac{\pi^2}{6} \cdot \frac{1}{N}, \tag{1.497}$$

and formula (1.494) is proved. If $N > |s|$, then

$$\log \frac{N+s}{N} = \log\left(1 + \frac{s}{N}\right)$$
$$= \frac{s}{N} - \frac{1}{2}\left(\frac{s}{N}\right)^2 + \frac{1}{3}\left(\frac{s}{N}\right)^3 - \cdots$$
$$+ (-1)^n \frac{1}{n+1}\left(\frac{s}{N}\right)^{n+1} - \cdots$$
$$\cdots \left|\frac{1}{2}\left(\frac{s}{N}\right)^2 - \frac{1}{3}\left(\frac{s}{N}\right)^3 + \frac{1}{4}\left(\frac{s}{N}\right)^4 - \cdots\right|$$
$$< \left(\frac{|s|}{N}\right)^2 + \left(\frac{|s|}{N}\right)^3 + \cdots = \frac{(\frac{|s|}{N})^2}{1 - \frac{|s|}{N}} < \frac{|s|^2}{\lambda} \cdot \frac{1}{N^2}, \tag{1.498}$$

and formula (1.495) is proved. From (1.493), (1.494), (1.495), and the formula given in item (a) of Theorem 1.19, in which $n = N$, we obtain

$$\int_0^N \frac{[u] - u + \frac{1}{2}}{u+s}\, \partial u = \sum_{n=1}^{N-1} \left\{\frac{s}{n} - \log\left(1 + \frac{s}{n}\right)\right\} - \log\sqrt{2\pi} - \left(s + \frac{1}{2}\right)\log s - \gamma s + s + O\left(\frac{1}{N}\right). \tag{1.499}$$

Taking the limit as $N \longrightarrow \infty$ in (1.499), by (1.492) it follows that

$$\log\Gamma(s) = \left(s - \frac{1}{2}\right)\log s - s + \log\sqrt{2\pi} + \int_0^\infty \frac{[u] - u + \frac{1}{2}}{u + s}\,\partial u. \tag{1.500}$$

Let

$$\varphi(a) = \int_0^a \left([u] - u + \frac{1}{2}\right)\partial u \quad (a \geqslant 0). \tag{1.501}$$

Since

$$\int_n^{n+1} \left([u] - u + \frac{1}{2}\right)\partial u = 0 \tag{1.502}$$

and

$$\int_0^a \left(n - u + \frac{1}{2}\right)\partial u = \frac{1}{2}(a - n)\cdot(1 - (a - n)) \quad (n \leqslant a \leqslant n + 1), \tag{1.503}$$

we get

$$|\varphi(a)| \leqslant \frac{1}{8} \quad (a \geqslant 0). \tag{1.504}$$

If $\alpha = \arg s$, then

$$\begin{aligned}\left|\int_0^\infty \frac{\varphi'(u)}{u + s}\,\partial u\right| &= \left|\int_0^\infty \frac{\varphi(u)}{(u + s)^2}\,\partial u\right| \leqslant \frac{1}{8}\int_0^\infty \frac{\partial u}{|u + s|^2} = \frac{1}{8}\int_0^\infty \frac{\partial u}{u^2 + 2u|s|\cos\alpha + |s|^2} \\ &\leqslant \frac{1}{8}\int_0^\infty \frac{\partial u}{u^2 - 2u|s|\cos\delta + |s|^2} = \left|\begin{matrix} u - |s|\cos\delta \\ = |s|v\sin\delta \end{matrix}\right| = \frac{\pi - \delta}{8\sin\delta}\cdot\frac{1}{|s|}.\end{aligned} \tag{1.505}$$

So (1.488) follows from (1.500) and (1.505). □

Theorem 1.20. *Let the function $f(x)$ be defined and piecewise continuous on the interval $0 < x < \infty$, and suppose that $f(x) = O(x^{\alpha-1})$ as $x \longrightarrow \infty$ and $f(x) = O(x^{\beta-1})$ as $x \longrightarrow 0$ for some $\alpha < \beta$. Then the function*

$$F(s) = \int_0^\infty f(x)x^{-s}\,\partial x \quad (\alpha < \sigma < \beta) \tag{1.506}$$

is analytic in the region $\alpha < \sigma < \beta$, and

$$\frac{1}{2}((x+0)+f(x-0)) = \frac{1}{2\pi i}\int_{\delta-i\infty}^{\delta+i\infty} F(s)x^{s-1}\,\partial s \quad (\alpha < \delta < \beta). \tag{1.507}$$

In particular, if the function $f(x)$ is continuous at x, then

$$f(x) = \frac{1}{2\pi i}\int_{\delta-i\infty}^{\delta+i\infty} F(s)x^{s-1}\,\partial s \quad (\alpha < \delta < \beta). \tag{1.508}$$

Proposition 1.4. *The function $F(s)$ is known as the Mellin transform of the function $f(s)$.*

The proof of the Theorem 1.20 can be found in [81, 83, 85]. Based on Theorem 1.20, we can obtain formulas (1.509), (1.510), and (1.511), where on the basis of *Stirling*'s approximation for $\Gamma(s)$ in (1.488) and *Phragmen–Lindeöf*'s principle, integration can be performed by parallel line. Using Theorem 1.20 and the formulas defining the function $\Gamma(s)$ in (1.161) and (1.148), we obtain

$$e^{-x} = \frac{1}{2\pi i}\int_{\delta-i\infty}^{\delta+i\infty} \Gamma(s)x^{-s}\,\partial s \quad (x > 0,\ \delta > 0), \tag{1.509}$$

$$\frac{1}{e^{x}-1} = \frac{1}{2\pi i}\int_{\delta-i\infty}^{\delta+i\infty} \Gamma(s)\zeta(s)x^{-s}\,\partial s \quad (x > 0,\ \delta > 1), \tag{1.510}$$

$$\frac{1}{e^{x}-1} = \frac{1}{2\pi i}\int_{\delta-i\infty}^{\delta+i\infty} (1-2^{1-s})\Gamma(s)\zeta(s)x^{-s}\,\partial s \quad (x > 0,\ \delta > 0). \tag{1.511}$$

Further, to prove a formula similar to formula (1.509), we will need the following lemma.

Lemma 1.27. *Let z be in the region outside the circles defined by*

$$|z-k| = r, \quad 0 < r < \frac{1}{2}, \quad k = 0, \pm 1, \pm 2, \dots. \tag{1.512}$$

Then there is a constant $M = M(r)$, independent of z, such that

$$\frac{1}{|\sin \pi z|} \leqslant M. \tag{1.513}$$

Proof. First of all, we have

$$\frac{1}{|\sin \pi z|} = \frac{1}{\sqrt{\sin^2 \pi x + \sinh^2 \pi y}}, \tag{1.514}$$

from which we obtain

$$\frac{1}{|\sin \pi z|} \leqslant \frac{1}{\sinh \pi r} \quad (|y| \geqslant b > r), \tag{1.515}$$

$$\frac{1}{|\sin \pi z|} \leqslant \frac{1}{\sinh \frac{\pi r}{\sqrt{2}}} \quad \left(\frac{r}{\sqrt{2}} \leqslant |y| \leqslant b\right), \tag{1.516}$$

$$\frac{1}{|\sin \pi z|} \leqslant \frac{1}{\sin \frac{\pi r}{\sqrt{2}}} \quad \left(\frac{r}{\sqrt{2}} \leqslant |x-k| \leqslant \frac{1}{2}\right), \quad k \in \mathbb{Z}, \tag{1.517}$$

and the lemma follows. □

We will prove the following formula (based on the already stated equations):

$$e^{-w} = \frac{1}{2\pi i} \int_{\delta - i\infty}^{\delta + i\infty} \Gamma(s) x^{-s}\, \partial s \quad \left(|\arg \omega| < \frac{\pi}{2},\ \omega \neq 0,\ \delta > 0\right). \tag{1.518}$$

Proof. Let

$$I = \frac{1}{2\pi i} \int_{\delta - i\infty}^{\delta + i\infty} \Gamma(s) \omega^{-s}\, \partial s = \frac{1}{2\pi i} \int_{-\delta - i\infty}^{-\delta + i\infty} \Gamma(-s) \omega^{s}\, \partial s. \tag{1.519}$$

Since

$$\Gamma(-s)\Gamma(1+s) = \frac{-\pi}{\sin \pi s}, \tag{1.520}$$

it follows that

$$I = -\frac{1}{2\pi i} \int_{-\delta - i\infty}^{-\delta + i\infty} \frac{\pi \omega^{s}}{\Gamma(s+1) \sin \pi s}\, \partial s. \tag{1.521}$$

We will integrate the function $f(s) = \frac{\pi \omega^s}{\Gamma(s+1)\sin \pi s}$ along the contour defined in Figure 1.6.

Let $s = \mathrm{Re}^{\theta i}$, $-\pi < \theta < \pi$. Then

$$\left|\frac{1}{\Gamma(s+1)}\right| = \left|\frac{1}{s\Gamma(s)}\right| \backsim \left|\begin{matrix} -\pi + \lambda \leqslant \theta \leqslant \pi - \lambda,\ \lambda > 0 \\ \Gamma(s) \backsim \sqrt{2\pi} s^{s-\frac{1}{2}} e^{-s} \end{matrix}\right|$$
$$\backsim \frac{1}{\sqrt{2\pi}} \left|e^{s-(s+\frac{1}{2})\log s}\right| = \frac{1}{\sqrt{2\pi}} e^{(1-\log R)R\cos\theta + \theta R \sin\theta - \frac{1}{2}\log R}. \tag{1.522}$$

If $\omega = \rho e^{\phi i}$, $-\frac{\pi}{2} < \phi < \frac{\pi}{2}$, $\phi \neq 0$, and $\theta \neq 0$, then

$$\left|s \frac{\pi \omega^s}{\Gamma(s+1)\sin \pi s}\right| \backsim \sqrt{2\pi} e^{-R(\log R - \log \rho - 1)\cos\theta + \frac{1}{2}\log R + (\theta - \phi \mp \pi) R \sin\theta} = \sqrt{2\pi} \cdot L, \tag{1.523}$$

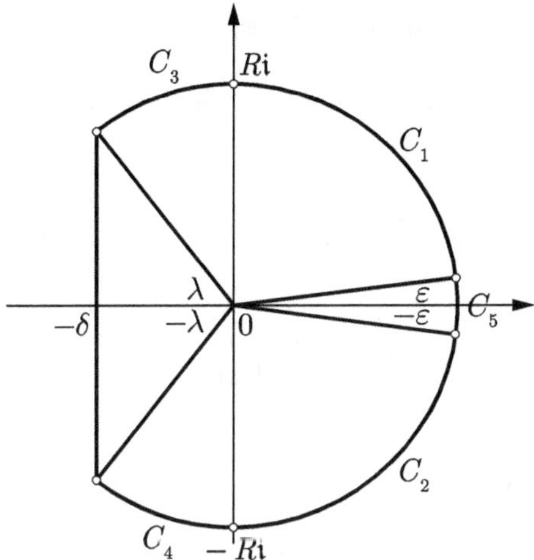

Figure 1.6: The contour of integration of the function $f(s) = \frac{\pi\omega^s}{\Gamma(s+1)\sin \pi s}$.

where in the formula, the sign "–" is used if $\sin\theta > 0$ and "+" otherwise. In relation to the given ρ, let us choose R such that $\log R - \log \rho - 1 > 0$. If the point s is on the curve C_1, then $\varepsilon \leqslant \theta \leqslant \frac{\pi}{2}$, that is,

$$L < R^{\frac{1}{2}} \cdot e^{(\theta-\phi-\pi)R\sin\theta} \leqslant R^{\frac{1}{2}} \cdot e^{(\theta-\phi-\pi)R\sin\theta} \leqslant \frac{R^{\frac{1}{2}}}{e^{(\frac{\pi}{2}+\phi)R\sin\varepsilon}}, \tag{1.524}$$

It follows that $L \longrightarrow 0$ $(R \longrightarrow \infty)$ uniformly in s on C_1. Then for the points s on the curve C_2, we have the estimate

$$L < R^{\frac{1}{2}} \cdot e^{(\theta-\phi-\pi)R\sin\theta} \leqslant R^{\frac{1}{2}} \cdot e^{(-\theta+\phi-\pi)R\sin\theta} \leqslant \frac{R^{\frac{1}{2}}}{e^{(\frac{\pi}{2}-\phi)R\sin\varepsilon}}, \tag{1.525}$$

so that $L \longrightarrow 0$ $(R \longrightarrow \infty)$ uniformly in s on C_2. For the points s on the curve C_3, we have the following estimates:

$$-R(\log R - \log\rho - 1)\cos\theta \leqslant \delta(\log R - \log\rho - 1), \tag{1.526}$$

$$-\pi \cdot R\sin\theta < (\theta - \phi - \pi)R\sin\theta < (\theta - \phi - \pi)R\sin\lambda \leqslant -(\lambda + \phi)R\sin\lambda, \tag{1.527}$$

where the second relations hold when $\lambda + \phi > 0$, which is true for a sufficiently large R. Therefore, for a constant P_1,

$$L \leqslant P_1 \frac{R^{\delta+\frac{1}{2}}}{e^{(\lambda+\phi)R\sin\lambda}} \longrightarrow 0 \quad (R \longrightarrow \infty) \tag{1.528}$$

uniformly in s on C_3. For the points s on the curve C_4, by analogy we have the estimates

$$-R(\log R - \log\rho - 1)\cos\theta \leqslant \delta(\log R - \log\rho - 1), \tag{1.529}$$

$$-\pi R\sin(-\theta) < (-\theta+\phi-\pi)R\sin(-\theta) < (-\theta+\phi-\pi)R\sin\lambda \leqslant -(\lambda-\phi)R\sin\lambda, \tag{1.530}$$

where the second of these relations, (1.530), is valid when $\lambda - \phi > 0$, which is true for a sufficiently large R. Therefore, for a constant P_2,

$$L \leqslant P_2 \frac{R^{\delta+\frac{1}{2}}}{e^{(\lambda-\phi)R\sin\lambda}} \longrightarrow 0 \quad (R \longrightarrow \infty) \tag{1.531}$$

uniformly in s on C_4. Now let the point s be on the arc of C_5, and let $|s| = m + \frac{1}{2}$ $(m \leq N)$. Then the curve C_5 is inside the region in which Lemma 1.27 is valid, and hence

$$\frac{1}{|\sin\pi s|} \leqslant 2M. \tag{1.532}$$

It follows that for sufficiently large m, we can write

$$\begin{aligned} \left| s\frac{\pi\omega^s}{\Gamma(s+1)\sin\pi s} \right| &\leqslant M\cdot\sqrt{2\pi}\cdot e^{R_m(-(\log R_m-\log\rho-1)\cos\varepsilon+|\theta-\phi|\sin\varepsilon)+\frac{1}{2}\log R_m} \\ &\leqslant |\log x < 2x| \\ &\leqslant M\sqrt{2\pi}\cdot e^{R_m(-(\log R_m-\log\rho-1)\cos\varepsilon+1+|\theta-\phi|\sin\varepsilon)} \longrightarrow 0 \quad (m\longrightarrow\infty) \end{aligned} \tag{1.533}$$

uniformly in s on the curve C_5, where $R_m = m + \frac{1}{2}$. So we can conclude that the integral of the function $f(s)$ along the curve C tends to zero as $R \longrightarrow \infty$. Since

$$\lim_{s\longrightarrow -n} f(s) = \lim_{s\longrightarrow -n} \frac{\pi\omega^s}{s\Gamma(s)\sin\pi s} = \lim_{s\longrightarrow -n} \frac{\omega^s\Gamma(1-s)}{s} = -\frac{n!}{n\omega^n} \quad (n \in \mathbb{N}), \tag{1.534}$$

the point $s = -n$ is a regular point of the function $f(s)$, so by the *Cauchy* theorem it follows that

$$I = \sum_{n=0}^{\infty} \operatorname*{res}_{s=n} \frac{\pi\omega^s}{\Gamma(s+1)\sin\pi s} = \sum_{n=0}^{\infty} \frac{(-\omega)^n}{n!} = e^{-\omega}. \tag{1.535}$$

Formula (1.518) is proved. □

2 *Riemann* hypothesis on the function $\zeta(s)$

2.1 The first *Riemann* hypothesis on the function $\zeta(s)$

The first *Riemann* hypothesis is as follows:

The number $N(T)$ of zeros of the function $\zeta(s)$ in the rectangle $0 \leqslant \sigma \leqslant 1, 0 \leqslant t \leqslant T$ is given by the formula

$$N(T) = \frac{T}{2\pi} \log \frac{T}{2\pi} - \frac{T}{2\pi} + O(\log T) \quad (T \longrightarrow \infty). \tag{2.1}$$

Formula (2.1) had been stated in *Riemann*'s work [9], dated in 1859, and was first proved by *von Mangoldt* [22] in 1905. That is why the following statement is known as the *Riemann–Mangoldt* theorem.

Theorem 2.1. *The number $N(T)$ of zeros of the function $\zeta(s)$ in the rectangle $0 \leqslant \sigma \leqslant 1$, $0 \leqslant t \leqslant T$, is given by formula* (2.1).

It is important to point out the following immediate consequence of this formula.

Proposition 2.1. *Since*

$$\lim_{T \longrightarrow \infty} \frac{1 + \log 2\pi}{2\pi} T \left(\frac{\frac{T}{2\pi} + \frac{O(\log T)}{\log T}}{\frac{1+\log 2\pi}{2\pi} T} \log T - 1 \right) = \infty, \tag{2.2}$$

the number of zeros of the function $\zeta(s)$ in the "critical" region, i. e. in the region $0 \leqslant \sigma \leqslant 1$, is infinite.

Before providing the proof of Theorem 2.1, we will prove the following two lemmas.

Lemma 2.1. *In the region $\sigma \geqslant 1$, $t \geqslant 2$, we have the following estimate:*

$$|\zeta(s)| < c \log t, \tag{2.3}$$

where $c > 0$ is an absolute constant. In the region $\sigma \geqslant \delta$ $(0 < \delta < 1)$, $t \geqslant 1$, we have the following estimate:

$$|\zeta(s)| < c(\delta) t^{1-\delta}, \tag{2.4}$$

where $c(\delta) > 0$ is a constant depending only on δ.

Proof. The proof of the lemma is similar to that of Lemma 1.5. In this case, we have the following estimates:

$$\left| \sum_{n=1}^{N} \frac{1}{n^s} \right| \leqslant 3 \log t \quad (N = [t]), \tag{2.5}$$

https://doi.org/10.1515/9783112233276-002

$$\left|\frac{N^{1-s}}{s-1}\right| \leqslant \frac{1}{2}, \tag{2.6}$$

$$\left|s\int_N^\infty \frac{\{x\}}{x^{s+1}}\,\partial x\right| \leqslant (\sigma+\lambda)\int_N^\infty \frac{\partial x}{x^{\sigma+1}} = \frac{\sigma+\lambda}{\sigma}\cdot\frac{1}{N^\sigma} \leqslant \frac{\sigma+t}{\sigma}\cdot\frac{1}{N} \leqslant \frac{1+t}{N} \leqslant \frac{t+1}{t-1} \leqslant 3, \tag{2.7}$$

from which it follows that

$$|\zeta(s)| \leqslant 3\log t + \frac{1}{2} + 3 < 10\log t. \tag{2.8}$$

Thus the first estimate given in Lemma 2.1 is proved. When $0 < \delta < 1, \sigma \geqslant \delta, t \geqslant 1$, for $N = [t]$, we have

$$\begin{aligned}\left|\sum_{n=1}^{[t]}\frac{1}{n^s}\right| &\leqslant \sum_{n=1}^{[t]}\frac{1}{n^\sigma} \leqslant \sum_{n=1}^{[t]}\frac{1}{n^\delta} < 1 + \int_1^{[t]}\frac{\partial x}{x^\delta} = 1 - \frac{1}{1-\delta} + \frac{[t]^{1-\delta}}{1-\delta} \leqslant 1 + \frac{t^{1-\delta}-1}{1-\delta} \\ &< t^{1-\delta} + \frac{t^{1-\delta}+t^{1-\delta}}{1-\delta} = \left(1+\frac{2}{1-\delta}\right)t^{1-\delta},\end{aligned} \tag{2.9}$$

$$\left|\frac{N^{1-s}}{s-1}\right| = \frac{N^{1-\sigma}}{|s-1|} \leqslant \frac{N^{1-\sigma}}{t} \leqslant N^{1-\sigma} \leqslant t^{1-\delta}, \tag{2.10}$$

$$\begin{aligned}\left|s\int_N^\infty \frac{\{x\}}{x^{s+1}}\,\partial x\right| &\leqslant \sqrt{\sigma^2+t^2}\int_N^\infty \frac{\partial x}{x^{\sigma+1}} \leqslant \frac{\sigma+t}{\sigma}\cdot\frac{1}{N^\sigma} \leqslant \left(1+\frac{t}{\delta}\right)N^{-\delta} \\ &\leqslant \left(1+\frac{t}{\delta}\right)\frac{2^\delta}{t^\delta} < \left(1+\frac{t}{\delta}\right)\frac{2}{t^\delta} \leqslant \left(2+\frac{2}{\delta}\right)t^{1-\delta}.\end{aligned} \tag{2.11}$$

The second estimate in the lemma directly follows from these inequalities. □

Lemma 2.2. *Let $f(s)$ be a regular function on a circle $|s-s_0| \leqslant R$ $(R > 0)$ and have the highest number n of zeros in the region $|s-s_0| \leqslant r < R$ (multiple zeros are counted with their multiplicity). If $f(s_0) \neq 0$, then*

$$\left(\frac{R}{r}\right)^n \leqslant \frac{M}{|f(s_0)|}, \tag{2.12}$$

where $M = \max\{|f(s)| : |s-s_0| = R\}$.

Proof. We can assume that $s_0 = 0$; otherwise, we can make the replacement $s = s_0 + s'$ and observe the new variable s'. Let $a_1, a_2, \ldots, a_n$ be zeros of the function $f(s)$ in the region $|s| \leqslant r$. Then

$$f(s) = \varphi(s)\prod_{k=1}^{n}\frac{R(s-a_k)}{R^2-\overline{a_k}s}, \tag{2.13}$$

where the function $\varphi(s)$ is regular in the region $|s| \leqslant R$ (because $R^2 - \overline{a_k}s \neq 0$ for every s in the region $|s| \leqslant R$, since $\frac{R^2}{\overline{a_k}}$ is the point inverse to the point a_k with respect to the circle $|s| = R$). Since

$$\left|\frac{R(s-a_k)}{R^2-\overline{a_k}s}\right| = \frac{R}{|s|}\cdot\left|\frac{s-a_k}{\overline{s-a_k}}\right| = 1 \quad (|s| = R), \tag{2.14}$$

we have

$$|\varphi(s)| = |f(s)| \leqslant M \quad (|s| = R). \tag{2.15}$$

By the maximum modulus principle it follows that $|\varphi(o)| \leqslant M$ and thus

$$|f(0)| = |\varphi(0)|\prod_{k=1}^{n}\frac{|a_k|}{R} \leqslant M\left(\frac{r}{R}\right)^n, \tag{2.16}$$

which for $|f(0)| \neq 0$ gives the result. □

Now we give the proof of Theorem 2.1, following the proof that was first presented by *Backlund* [30] in 1918.

Proof. According to Theorems 1.4, 1.10, and 1.11, $N(T)$ is equal to the number of zeros $\rho = \beta + \gamma\mathrm{i}$ of the function $\zeta(s)$ in the region $0 < \beta < 1$, $0 < \gamma \leqslant T$. Suppose $T > 3$ and suppose it is different from any γ. Since $\xi(s)$ is regular in the rectangle $0 < \beta < 1$, $0 < \gamma \leqslant T$, in this region, it has infinitely many zeros. Let R be a rectangle with the vertices $2 + \mathrm{i}T$, $-1 + \mathrm{i}T$, $-1 - \mathrm{i}T$, $2 - \mathrm{i}T$ in the stated order. According to Theorems 1.10 and 1.11, the function $\xi(s)$ within R has $2N(T)$ zeros, and there is no zero on the boundary. By the formula for the meromorphic function $f(s)$ in D we have

$$N - P = \frac{1}{2\pi\mathrm{i}}\int_D \frac{f'(s)}{f(s)}\,\partial s, \tag{2.17}$$

where N is the number of zeros, and P is the number of poles of the function $f(s)$ within D. It follows that

$$N(T) = \frac{1}{4\pi}I_m\left(\int_R \frac{\xi'(s)}{\xi(s)}\,\partial s\right). \tag{2.18}$$

Since

$$\xi(s) = \frac{1}{2}s(s-1)\eta(s), \quad \eta(s) = \pi^{-\frac{s}{2}}\Gamma\left(\frac{s}{2}\right)\zeta(s), \tag{2.19}$$

we have

$$\frac{\xi'(s)}{\xi(s)} = \frac{1}{s} + \frac{1}{s-1} + \frac{\eta'(s)}{\eta(s)}, \tag{2.20}$$

$$\Im\mathrm{m}\left(\int_R\left(\frac{1}{s} + \frac{1}{s-1}\right)\partial s\right) = 4\pi. \tag{2.21}$$

In addition, we have

$$\eta(s) = \eta(1-s), \quad \eta(\sigma - \mathrm{i}t) = \overline{\eta(\sigma + \mathrm{i}t)}, \tag{2.22}$$

and so

$$\mathfrak{Im}\left(\int_R \frac{\eta'(s)}{\eta(s)}\,\partial s\right) = 4\,\mathfrak{Im}\left(\int_L \frac{\eta'(s)}{\eta(s)}\,\partial s\right), \tag{2.23}$$

where L is a part of the boundary R from the point $s = 2$ to the point $s = 2 + \mathrm{i}T$ (part L_1) and from the point $s = 2 + \mathrm{i}T$ to the point $s = \frac{1}{2} + \mathrm{i}T$ (part L_2). In addition, we have

$$\begin{aligned}\mathfrak{Im}\left(\int_L \frac{\eta'(s)}{\eta(s)}\,\partial s\right) &= \mathfrak{Im}\left(\int_L\left(-\frac{\log\pi}{2}\right)\partial s\right) + \mathfrak{Im}\left(\int_L \frac{\frac{1}{2}\Gamma'(\frac{s}{2})}{\Gamma(\frac{s}{2})}\,\partial s + \int_L \frac{\zeta'(s)}{\zeta(s)}\,\partial s\right)\\ &= -\frac{T}{2}\log\pi + \mathfrak{Im}\log\Gamma\left(\frac{1}{4} + \frac{T}{2}\mathrm{i}\right) + \mathfrak{Im}\left(\int_L \frac{\zeta'(s)}{\zeta(s)}\,\partial s\right).\end{aligned} \tag{2.24}$$

To evaluate $\mathfrak{Im}\log\Gamma(\frac{1}{4} + \frac{T}{2}\mathrm{i})$, we will apply Theorem 1.19. Specifically, we will prove the following relation, which immediately follows from Theorem 1.19:

$$\log\Gamma(s+\alpha) = \left(s + \alpha - \frac{1}{2}\right)\log s - s + \log\sqrt{2\pi} + O\left(\frac{1}{|s|}\right) \tag{2.25}$$

at $|s| \longrightarrow \infty$ uniformly in the angle $|\arg s| \leqslant \pi - \varepsilon < \pi$ for a limited α. Indeed, from Theorem 1.19 it follows that

$$\begin{aligned}\log\Gamma(s+\alpha) &= \left(s+\alpha-\frac{1}{2}\right)\log(s+\alpha) - (s+\alpha) + \log\sqrt{2\pi} + O\left(\frac{1}{|s+\alpha|}\right)\\ &= \left(s+\alpha-\frac{1}{2}\right)\log s - s + \log\sqrt{2\pi} + \left(s+\alpha-\frac{1}{2}\right)\log\left(1+\frac{\alpha}{s}\right) - \alpha + O\left(\frac{1}{|s|}\right).\end{aligned} \tag{2.26}$$

Since

$$\left|\log\left(1+\frac{\alpha}{s}\right) - \frac{\alpha}{s}\right| \leqslant \sum_{n=2}^{\infty}\frac{(\frac{|\alpha|}{|s|})^n}{n} < \sum_{n=2}^{\infty}\frac{(\frac{|\alpha|}{|s|})^n}{2} = \frac{1}{2}\frac{(\frac{|\alpha|}{|s|})^2}{1-\frac{|\alpha|}{|s|}} < \frac{|\alpha|^2}{|s|^2}, \tag{2.27}$$

i. e.,

$$\log\left(1+\frac{\alpha}{s}\right) = \frac{\alpha}{s} + O\left(\frac{1}{|s|^2}\right), \tag{2.28}$$

(2.25) directly follows. From (2.25) we have

$$\mathfrak{Im}\log\Gamma\Big(\frac{1}{4}+\frac{T}{2}\mathrm{i}\Big)=\mathfrak{Im}\Big(\Big(-\frac{1}{4}+\frac{T}{2}\mathrm{i}\Big)\log\Big(\frac{T}{2}\mathrm{i}\Big)-\frac{T}{2}\mathrm{i}+\log\sqrt{2\pi}\Big)+O\Big(\frac{1}{T}\Big)$$
$$=\frac{T}{2}\log\frac{T}{2}-\frac{\pi}{8}-\frac{T}{2}+O\Big(\frac{1}{T}\Big). \tag{2.29}$$

From (2.29), (2.24), (2.23), (2.21), (2.20), and (2.18) we get

$$N(T)=\frac{T}{2\pi}\log\frac{T}{2\pi}-\frac{T}{2\pi}+\frac{7}{8}+\frac{1}{\pi}\mathfrak{Im}\Big(\int_L\frac{\zeta'(s)}{\zeta(s)}\,\partial s\Big)+O\Big(\frac{1}{T}\Big). \tag{2.30}$$

Let m be the number of the different points s' on L, excluding its ends, such that $\mathfrak{Re}\,\zeta(s')=0$. On any of $m+1$ segments of L, to which it is divided by zeros s', $\mathfrak{Re}\,\zeta(s)$ does not change the sign. Let S be one of those segments. Then

$$\mathfrak{Im}\int_S\frac{\zeta'(s)}{\zeta(s)}\,\partial s=\mathfrak{Im}\int_{\zeta(S)}\frac{\partial z}{z}, \tag{2.31}$$

where $\zeta(S)$ is the image of S under the function ζ. It is obvious that $\zeta(S)$ is inside the semiplane $\mathfrak{Re}\,z\geqslant 0$ or $\mathfrak{Re}\,z\leqslant 0$. If the curve has the starting point a and the end point b, and if $[a,b]$ denotes the straight line connecting a and b, then by the *Cauchy* theorem on contour deformation we have

$$\int_{\zeta(S)}\frac{\partial z}{z}=\int_{K(a,b)}\frac{\partial z}{z}, \tag{2.32}$$

where $K(a,b)$ is a semicircle with diameter $[a,b]$ in the semiplane in which $\zeta(S)$ lies. It follows that

$$\Big|\mathfrak{Im}\int_S\frac{\zeta'(s)}{\zeta(s)}\,\partial s\Big|=\Big|\mathfrak{Im}\int_{K(a,b)}\frac{\partial z}{z}\Big|\leqslant\pi, \tag{2.33}$$

which further gives

$$\Big|\mathfrak{Im}\int_L\frac{\zeta'(s)}{\zeta(s)}\,\partial s\Big|\leqslant(m+1)\pi. \tag{2.34}$$

Since

$$\mathfrak{Re}\,\zeta(2+\mathrm{i}T)=\sum_{n=1}^{\infty}\frac{1}{n^2}\cos(t\log n)\geqslant 1-\sum_{n=1}^{\infty}\frac{1}{n^2}>1-\frac{1}{2^2}-\int_2^{\infty}\frac{\partial v}{v^2}=\frac{1}{4}, \tag{2.35}$$

the points s' are not on L_1. So m is the number of different points s' lying on L_2 such that $\mathfrak{Re}\,\zeta(s')=0$, i. e., for them, $\frac{1}{2}<\sigma<2$, $\mathfrak{Re}\,\zeta(\sigma+\mathrm{i}T)=0$. Let us look at the function $g(s)=\frac{1}{2}\{\zeta(s+\mathrm{i}T)+\zeta(s-\mathrm{i}T)\}$. It is obvious that

$$g(\sigma)=\mathfrak{Re}\,\zeta(\sigma+\mathrm{i}T), \tag{2.36}$$

because $\zeta(\sigma - \mathrm{i}T) = \overline{\zeta(\sigma + \mathrm{i}T)}$, so m is the number of different zeros on the σ-axis ($\frac{1}{2} < \sigma < 2$) of the function $g(s)$. Since $g(s)$ is regular in R, except for $s = 1 \pm \mathrm{i}T$, m is finite. We will apply Lemma 2.2 to the function $g(s)$ to estimate m from above in relation to T. For that purpose, we will consider the circles $|s-2| \leqslant \frac{7}{4}$ and $|s-2| \leqslant \frac{3}{2}$. Since we assume that $T > 3$, $g(s)$ is regular on the larger circle. Since $g(2) \neq 0$ (because $g(2) = \mathfrak{Re}\,\zeta(2+\mathrm{i}T) > \frac{1}{4}$), taking $\delta = \frac{1}{4}$, for s such that $|s-2| = \frac{7}{4}$, by (2.4) we have

$$|g(s)| < \frac{1}{2}c(|t+T|^{\frac{3}{4}} + |t-T|^{\frac{3}{4}}\} < Cc(2+T)^{\frac{3}{4}}, \tag{2.37}$$

since in that case, $\sigma \geqslant \frac{1}{4}$, $1 \leqslant |t \pm T| < 2 + T$, $(|t \pm T| < |t| + T \leqslant \frac{7}{4} + T < 2 + T$, $T \pm t > 3 \pm t \geqslant 3 - \frac{7}{4} > 1)$. From (2.12) it follows that

$$\left(\frac{7}{6}\right)^m < \frac{M}{g(2)} < \frac{c_1(2+T)^{\frac{3}{4}}}{\frac{1}{4}} < T \quad (T > T_0 \geqslant 3), \tag{2.38}$$

$$m < c_2 \log T \quad (T > T_0,\ c_2 > 0\text{-const.}). \tag{2.39}$$

From (2.39), (2.34), and (2.30) we obtain (2.1) as $T \longrightarrow \infty$, assuming that T is different from any γ. If T is equal to some γ, then we can take $T' > T$, where T' is different from any γ, and let $T' \downarrow T$. The theorem is proved. □

After it was proven that the function $\zeta(s)$ has infinitely many zeros in the "critical region", in 1914, *Hardy* obtained a stronger result. This result is given in the following theorem.

Theorem 2.2. *The function $\zeta(s)$ has infinitely many zeros on the line $\sigma = \frac{1}{2}$.*

Proof. The proof follows the proof given by *Hardy* [27]. We know that (see (1.518))

$$\mathrm{e}^{-y} = \frac{1}{2\pi\mathrm{i}} \int_{k-\mathrm{i}\infty}^{k+\mathrm{i}\infty} \Gamma(v) y^{-v}\,\partial v \quad \left(k > 0,\ |\arg y| < \frac{\pi}{2},\ y \neq 0\right). \tag{2.40}$$

For $k > \frac{1}{2}$, it follows that

$$\frac{1}{2\pi\mathrm{i}} \int_{k-\mathrm{i}\infty}^{k+\mathrm{i}\infty} \Gamma(v)(\pi x)^{-v}\zeta(2v)\,\partial v = \sum_{n=1}^{\infty} \mathrm{e}^{-n^2\pi x} \quad \left(|\arg y| < \frac{\pi}{2},\ x \neq 0\right), \tag{2.41}$$

because $\zeta(2v) = \sum_{n=1}^{\infty} \frac{1}{n^{2v}}$ for $v = k + \mathrm{i}t$ $(-\infty < t < +\infty)$ is a uniformly convergent series for $k > \frac{1}{2}$. Next, with

$$\varphi(x) = \sum_{n=1}^{\infty} \mathrm{e}^{-n^2\pi x} \quad \left(|\arg x| < \frac{\pi}{2},\ x \neq 0\right) \tag{2.42}$$

it follows that

$$\varphi(x) = \frac{1}{2\pi i} \int_{k-i\infty}^{k+i\infty} \frac{1}{2}\Gamma\left(\frac{v}{2}\right)\pi^{-\frac{v}{2}}\zeta(v)x^{-\frac{v}{2}}\,\partial v \quad (k > 1). \tag{2.43}$$

We will integrate the function $f(s) = \frac{1}{2}\pi^{-\frac{s}{2}}\Gamma(\frac{s}{2})\zeta(s)x^{-\frac{s}{2}}$ over the rectangle with vertices $k-iT, k+iT, \frac{1}{2}+iT, \frac{1}{2}-iT$. From (1.188) we conclude that $f(s)$ has a unique singularity at $s = 1$, a pole of order 1. By (1.247) it follows that

$$\operatorname*{res}_{s=1} f(s) = \frac{1}{2}\lim_{s\longrightarrow 1} \pi^{-\frac{s}{2}}x^{-\frac{s}{2}}\Gamma\left(\frac{s}{2}\right)(s-1)\zeta(s) = \frac{1}{2}x^{-\frac{1}{2}}. \tag{2.44}$$

Let us now evaluate the integrals of $f(s)$ on the horizontal sides of the rectangle. According to the *Stirling* formula (1.488),

$$\left|\Gamma\left(\frac{s}{2}\right)\right| = \sqrt{2\pi}\left|e^{\frac{s-1}{2}\log\frac{s}{2}-\frac{s}{2}+O(\frac{1}{|s|})}\right| \leqslant K_1 \cdot e^{\frac{\sigma-1}{2}\log\frac{|s|}{2}-\frac{T}{2}\lambda} \leqslant K_2 \cdot T^{\frac{\sigma-1}{2}} \cdot e^{-\frac{T}{2}\lambda} \tag{2.45}$$

for sufficiently large T, where $\lambda = \arg s$. Next,

$$\left|x^{-\frac{s}{2}}\right| \leqslant K_3 \cdot e^{\frac{T}{2}\alpha} \quad (\alpha = \arg x). \tag{2.46}$$

Let T be large enough so that $c = \frac{\lambda-\alpha}{2} > 0$. By Lemma 1.19 it follows that

$$\zeta(s) = O(T^{1-\sigma}\cdot\log T) \quad \left(\frac{1}{2} \leqslant \sigma \leqslant 1\right), \tag{2.47}$$

and by Lemma 2.1

$$\zeta(s) = O(\log T) \quad (\sigma \geqslant 1). \tag{2.48}$$

In addition, for sufficiently large T, we have

$$\frac{T^{\frac{1-\sigma}{2}}\log T}{e^{cT}} < \frac{T^{\frac{3-\sigma}{2}}}{e^{cT}} \longrightarrow 0\ (T \longrightarrow \infty) \quad \left(\frac{1}{2} \leqslant \sigma \leqslant 1\right), \tag{2.49}$$

$$\frac{T^{\frac{\sigma-1}{2}}\log T}{e^{cT}} < \frac{T^{\frac{\sigma+1}{2}}}{e^{cT}} \longrightarrow 0\ (T \longrightarrow \infty) \quad (\sigma \geqslant 1). \tag{2.50}$$

Therefore the integral on the top of the rectangle tends to 0 as $T \longrightarrow \infty$. By analogy the integral on the bottom side of the rectangle tends to 0 as $T \longrightarrow \infty$. By the *Cauchy* theorem and (2.43) it follows that

$$\varphi(x) = \frac{1}{2\pi i}\left(2\pi i\cdot\frac{1}{2}x^{-\frac{1}{2}} + \int_{\frac{1}{2}-i\infty}^{\frac{1}{2}+i\infty} \frac{1}{2}\Gamma\left(\frac{v}{2}\right)\pi^{-\frac{v}{2}}\zeta(v)x^{-\frac{v}{2}}\,\partial v\right), \tag{2.51}$$

which, together with the definition of the functions $\xi(\lambda)$ and $\Xi(t)$, leads to

$$\varphi(x) = \frac{1}{2}x^{-\frac{1}{2}} - \frac{1}{2\pi}\int\limits_0^\infty \frac{x^{-\frac{1}{4}}(x^{-i\frac{t}{2}} + x^{i\frac{t}{2}})}{t^2 + \frac{1}{4}}\Xi(t)\,\partial t. \tag{2.52}$$

Taking $x = e^{\alpha i}$ ($|\alpha| < \frac{\pi}{2}$, $\alpha \neq 0$) in (2.52), we obtain the following relation for $\varphi(x)$:

$$\int\limits_0^\infty \frac{\cosh\frac{\alpha t}{2}}{t^2 + \frac{1}{4}}\Xi(t)\,\partial t = \pi\left\{\frac{1}{2}e^{-\frac{\alpha}{4}i} - e^{\frac{\alpha}{4}i}\varphi(e^{\alpha i})\right\}. \tag{2.53}$$

Based on the parity of the subintegral function in (2.53) with respect to α and the functional equation $2\varphi(x) + 1 = \frac{1}{\sqrt{x}}(2\varphi(\frac{1}{x}) + 1)$ for $\varphi(x)$, we have

$$\int\limits_0^\infty \frac{\cosh\frac{\alpha t}{2}}{t^2 + \frac{1}{4}}\Xi(t)\,\partial t = \pi\cos\frac{\alpha}{4} - \frac{\pi}{2}e^{\frac{\alpha}{4}i}\{1 + 2\varphi(e^{\alpha i})\} \quad \left(|\alpha| < \frac{\pi}{2},\ \alpha \neq 0\right). \tag{2.54}$$

By applying the same functional equation for $\varphi(x)$ it turns out that

$$\begin{aligned}\varphi(i + \delta) &= \sum_{n=1}^\infty e^{-n^2\pi(i+\delta)} = \sum_{n=1}^\infty (-1)^n e^{-n^2\pi\delta} = 2\varphi(4\delta) - \varphi(\delta)\\ &= \frac{1}{\sqrt{\delta}}\varphi\left(\frac{1}{4\delta}\right) - \frac{1}{\sqrt{\delta}}\varphi\left(\frac{1}{\delta}\right) - \frac{1}{2},\end{aligned} \tag{2.55}$$

i. e.,

$$\frac{1}{2} + \varphi(i + \delta) = \frac{1}{\sqrt{\delta}}\varphi\left(\frac{1}{4\delta}\right) - \frac{1}{\sqrt{\delta}}\varphi\left(\frac{1}{\delta}\right). \tag{2.56}$$

In addition, we have the following equations:

$$e^{-\frac{n^2\pi}{\delta}} \leqslant k!\left(\frac{n^2\pi}{\delta}\right)^{-k} \quad (k \geqslant 1), \tag{2.57}$$

$$\lim_{\alpha\to\frac{\pi}{2}} \frac{d^{2n}}{d\alpha^{2n}}\{1 + 2\varphi(e^{\alpha i})\} = 2\lim_{\delta\to 0}\left\{\frac{1}{\sqrt{\delta}}\varphi\left(\frac{1}{4\delta}\right) - \frac{1}{\sqrt{\delta}}\varphi\left(\frac{1}{\delta}\right)\right\}^{(2n)} = 0. \tag{2.58}$$

Further, by the *Stirling* formula

$$\left|\Gamma\left(\frac{1}{4} + i\frac{t}{2}\right)\right| \leqslant K_1 t^{-\frac{1}{4}} \cdot e^{-\frac{\pi}{4}t}, \tag{2.59}$$

and by Lemma 1.19

$$\left|\zeta\left(\frac{1}{2} + it\right)\right| \leqslant K_2 t^{\frac{1}{2}} \cdot \log t. \tag{2.60}$$

In addition,

$$\cosh\frac{\alpha}{2}t \leqslant K_3 \mathfrak{e}^{(\frac{\pi}{4}-\frac{\delta}{2})t} \quad (\delta > 0). \tag{2.61}$$

From (1.313) we conclude that

$$\int_T^\infty \frac{\cosh\frac{\alpha t}{2}}{t^2+\frac{1}{4}}\Xi(t)\,\mathfrak{d}t \longrightarrow 0 \quad (T \longrightarrow \infty) \tag{2.62}$$

uniformly in α, so the upper integral in (2.54) can be differentiated over α under the integral sign, which with (2.58) leads to

$$\lim_{\alpha\longrightarrow\frac{\pi}{2}} \int_0^\infty \frac{t^{2n}\cosh\frac{\alpha t}{2}}{t^2+\frac{1}{4}}\Xi(t)\,\mathfrak{d}t = \frac{(-1)^n\pi}{2^{2n}}\cos\frac{\pi}{8}. \tag{2.63}$$

It is well known that zeros of the function $\Xi(t)$ on the real axis correspond to zeros of the function $\zeta(s)$ on the line $\sigma = \frac{1}{2}$. To prove that the function $\zeta(s)$ has infinitely many zeros on the line $\sigma = \frac{1}{2}$, we must first prove that the function $\Xi(t)$ has infinitely many real zeros, that is, there is no T after which $\Xi(t)$ does not change the sign. If we assume the opposite, i. e., that $\Xi(t) > 0$ $(t \geqslant T)$, then

$$\lim_{\alpha\longrightarrow\frac{\pi}{2}} \int_T^\infty \frac{t^{2n}\cosh\frac{\alpha t}{2}}{t^2+\frac{1}{4}}\Xi(t)\,\mathfrak{d}t = L \quad (L > 0). \tag{2.64}$$

Therefore

$$\int_T^{T'} \frac{t^{2n}\cosh\frac{\alpha t}{2}}{t^2+\frac{1}{4}}\Xi(t)\,\mathfrak{d}t \leqslant L \quad \left(0 < \alpha < \frac{\pi}{2},\ T' > T\right). \tag{2.65}$$

For $\alpha \longrightarrow \frac{\pi}{2}$, we have

$$\int_T^{T'} \frac{t^{2n}\cosh\frac{\pi t}{4}}{t^2+\frac{1}{4}}\Xi(t)\,\mathfrak{d}t \leqslant L. \tag{2.66}$$

The integral $\int_0^\infty \frac{t^{2n}\cosh\frac{\pi t}{4}}{t^2+\frac{1}{4}}\Xi(t)\mathfrak{d}t$ converges. Accordingly, the integral $\int_T^\infty \frac{t^{2n}\cosh\frac{\alpha t}{2}}{t^2+\frac{1}{4}}\Xi(t)\mathfrak{d}t$ uniformly converges for $0<\alpha<\frac{\pi}{2}$. This, together with (2.63), leads to

$$\int_0^\infty \frac{t^{2n}\cosh\frac{\pi t}{4}}{\lambda^2+\frac{1}{4}}\Xi(t)\,\mathfrak{d}t = \frac{(-1)^n\pi\cos\frac{\pi}{8}}{2^{2n}} \quad (\forall n). \tag{2.67}$$

Let n be odd. Since in that case the right-hand side of relation (2.67) is negative, we have

$$\int_T^\infty \frac{t^{2n}\cosh\frac{\pi t}{4}}{t^2+\frac{1}{4}}\Xi(t)\,\partial t < -\int_0^T \frac{t^{2n}\cosh\frac{\pi}{4}}{t^2+\frac{1}{4}}\Xi(t)\,\partial t < K\cdot T^{2n}, \tag{2.68}$$

where K is independent of n. Since $\Xi(t) > 0$ $(t \geqslant T)$, we have $m = m(T)$, so that

$$\frac{\Xi(t)}{t^2+\frac{1}{4}} \geqslant m \quad (2T \leqslant t \leqslant 2T+1). \tag{2.69}$$

It follows that

$$\int_T^\infty \frac{t^{2n}\cosh\frac{\pi t}{4}}{t^2+\frac{1}{4}}\Xi(t)\,\partial t \geqslant \int_{2T}^{2T+1} mt^{2n}\,\partial t \geqslant m(2T)^{2n}. \tag{2.70}$$

From (2.68) and (2.70) it follows that $m\cdot 2^{2n} < K$ $(\forall n)$, which does not hold for a sufficiently large n. The same conclusion is obtained if $\Xi(t) > 0$ $(t \geqslant T)$, since it is sufficient to take $\Xi(t) = \Phi(t)$. The theorem is proved. □

Remark 2.1. *Davenport* in [93] (p. 110), states that, to the best of his knowledge, up to 1967, the estimate

$$S(T) = \frac{1}{\pi}\Im\mathrm{m}\left(\int_L \frac{\zeta'(s)}{\zeta(s)}\,\partial s\right) = O(\log T),$$

obtained by *von Mangold* in 1904, has not been improved in the 20th century.

Remark 2.2. Regarding the *Hardy* theorem, *Landau* [28] stated:

> One of the most significant achievements of mathematics of this era is the article by Mr. *E. H. Hardy* "On zeros of the *Riemann* ζ function".

2.2 II, III, and IV *Riemann* hypotheses on the function $\zeta(s)$ and the region where the function has no zeros

These hypotheses are mentioned in *Riemann*'s memoir on the function $\zeta(s)$ from 1859 [9].

II hypothesis. Let ρ pass all nontrivial zeros of the $\zeta(s)$ function. Then the series $\sum_\rho \frac{1}{|\rho|^2}$ converges, whereas the series $\sum_\rho \frac{1}{|\rho|}$ diverges.

III hypothesis. The function $\xi(s)$ can be presented in the following form:

$$\xi(s) = ae^{bs}\prod_\rho\left(1-\frac{s}{\rho}\right)e^{\frac{s}{\rho}}, \tag{2.71}$$

where a, b are constants, and ρ passes all nontrivial zeros of the function $\zeta(s)$.

IV hypothesis. If

$$P(x) = \sum_{n \leqslant x} \frac{\wedge(n)}{\log n}, \quad P_0(x) = \frac{1}{2}[P(x+0) + P(x-0)], \tag{2.72}$$

then for $x \geqslant 1$, we have

$$P_0(x) = \operatorname{li} x - \sum_{\rho} \operatorname{li} x^{\rho} + \int_{x}^{\infty} \frac{\partial u}{(u^2-1)\log u} - \log 2, \tag{2.73}$$

where

$$\operatorname{li} x = \int_{2}^{x} \frac{\partial u}{\log u}, \quad \operatorname{li} e^{\omega} = \int_{-\infty+iv}^{u+iv} \frac{e^z}{z} \partial z \quad (\omega = u + iv,\ v \gtrless 0). \tag{2.74}$$

Proposition 2.2. *Hypotheses II and III were proved in 1893 by Hadamard* [15], *where in hypothesis III,* $a = \frac{1}{2}$, $b = \log 2 + \frac{1}{2}\log \pi - 1 - \frac{\gamma}{2}$, *and* γ *is Euler's constant.*

Proposition 2.3. *Hypothesis IV was proved by von Mangoldt* [16] *in 1894, and in* [17] *in 1895, there was given an important analogue for*

$$\Psi(x) = \sum_{n \leqslant x} \wedge(n). \tag{2.75}$$

If

$$\Psi_0(x) = \frac{1}{2}[\Psi(x+0) + \Psi(x-0)], \tag{2.76}$$

then for $x \geqslant 1$,

$$\Psi_0(x) = x - \sum_{\rho} \frac{x^{\rho}}{\rho} - \frac{\zeta'(0)}{\zeta(0)} - \frac{1}{2}\log\left(1 - \frac{1}{x^2}\right), \tag{2.77}$$

where ρ passes all nontrivial zeros of the function $\zeta(s)$; in the sum $\sum_{\rho} \frac{x^{\rho}}{\rho}$, the elements with ρ and $\overline{\rho}$ are considered jointly, and the sum itself is given by the relation

$$\sum_{\rho} \frac{x^{\rho}}{\rho} = \lim_{T \to \infty} \sum_{|\gamma| \leqslant T} \frac{x^{\rho}}{\rho} \quad (\rho = \beta + i\gamma). \tag{2.78}$$

If x is an integer, then the last element $\wedge(x)$ in the relation for $\Psi(x)$ is replaced by the element $\frac{1}{2} \wedge (x)$.

Remark 2.3. It can be directly concluded from the above and from Definition 2.4 that there is a direct link between the distribution of prime numbers and the distribution of nontrivial zeros of the function $\zeta(s)$.

The function $\xi(s)$ is known to be integer (Theorem 1.10), e. g., regular in any finite region of the s-plane.

Definition 2.1. It is said that the integer function $f(s)$ is of finite order if there is a constant λ such that

$$\left|f(s)\right| < e^{r^{\lambda}} \quad (|s| = r > r_0). \tag{2.79}$$

According to the *Liouville* theorem, for a nonconstant function $f(s)$ of finite order, $\lambda > 0$. If inequality (2.79) is satisfied for a given λ, then it also holds for $\lambda' > \lambda$. So there are infinitely many numbers $\lambda' > 0$ that satisfy (2.79).

Definition 2.2. The infimum of the set of all such λ' (that satisfy the above inequality) is called the order of the function $f(s)$ and is denoted by ρ.

Then for $\varepsilon > 0$, there exists r_0 such that

$$\left|f(s)\right| < e^{r^{\rho+\varepsilon}} \quad (|s| = r > r_0). \tag{2.80}$$

Definition 2.3. Let

$$M(r) = \max_{|s|=r}\left|f(s)\right| \quad (r > 0). \tag{2.81}$$

Suppose

$$M(r) < e^{r^{\rho+\varepsilon}} \quad (r > r_0), \tag{2.82}$$

and

$$M(r) < e^{r^{\rho-\varepsilon}} \tag{2.83}$$

for infinitely many values of r that $\to \infty$. Logarithmizing these two inequalities twice, we get

$$\frac{\log\log M(r)}{\log r} < \rho + \varepsilon \tag{2.84}$$

and

$$\frac{\log\log M(r)}{\log r} > \rho - \varepsilon \tag{2.85}$$

for infinitely many values of r that $\to \infty$. It follows that

$$\rho = \overline{\lim_{r\to\infty}}\frac{\log\log M(r)}{\log r}. \tag{2.86}$$

Lemma 2.3. *The function $\xi(s)$ is a whole function of the first order.*

Proof. Let

$$M(r) = \max_{|s|=r} |\xi(s)|. \tag{2.87}$$

Let us first prove that

$$\log M(r) \backsim \frac{1}{2} r \log r \ (r \to \infty). \tag{2.88}$$

Since

$$\left| \frac{1}{2} s(s-1)\pi^{-\frac{s}{2}} \right| < e^{C_1|s|} \quad \left(\sigma \geqslant \frac{1}{2}\right), \tag{2.89}$$

by Lemma 2.1 (for $\delta = \frac{1}{2}$) it follows that

$$|\zeta(s)| < C_2|s|^{\frac{1}{2}} \quad \left(\sigma \geqslant \frac{1}{2},\ |s| > 2\right). \tag{2.90}$$

In the region $\sigma \geqslant \frac{1}{2}$, $|s| > 3$, we can apply the *Stirling* formula to $\Gamma(\frac{s}{2})$, which leads to the estimate

$$\left| \Gamma\left(\frac{s}{2}\right) \right| < e^{\frac{|s|}{2} \log |s| + C_3|s|}, \tag{2.91}$$

since $\log \frac{s}{2} = \log |s| - \log 2 + i \arg s$. By the definition of $\xi(s)$ (Definition 1.11), from (2.89), (2.90), and (2.91) we have

$$|\xi(s)| < e^{\frac{r}{2} \log r + C_4 r} \quad \left(\sigma \geqslant \frac{1}{2},\ |s| = r > 3\right). \tag{2.92}$$

From the relation $\xi(s) = \xi(1-s)$ and (2.92) it follows that

$$|\xi(s)| < e^{\frac{|1-s|}{2} \log |1-s| + C_4|1-s|} < e^{\frac{r}{2} \log r + C_5 r} \quad \left(\sigma \leqslant \frac{1}{2},\ |s| = r > 4\right). \tag{2.93}$$

Based on the relations above, we can conclude that

$$M(r) < e^{\frac{r}{2} \log r + C_6 r} \quad (r > 4). \tag{2.94}$$

On the other hand,

$$M(r) \geqslant \xi(r) > \pi^{-\frac{r}{2}} \Gamma\left(\frac{r}{2}\right) > e^{\frac{r}{2} \log r - C_7 r} \quad (r > 2). \tag{2.95}$$

From (2.94) and (2.95) it follows that

$$\frac{r}{2}\log r - C_7 r < \log M(r) < \frac{r}{2}\log r + C_6 r \quad (r > 4), \tag{2.96}$$

and, accordingly,

$$\log M(r) \backsim \frac{r}{2}\log r \quad (r \to \infty), \tag{2.97}$$

$$\lim_{r\to\infty}\frac{\log\log M(r)}{\log r} = \lim_{r\to\infty}\frac{\log(\frac{r}{2}\log r)}{\log r} = 1, \tag{2.98}$$

and thus $\rho = 1$. The lemma is proved. □

Based on the facts regarding the functions of finite order [53], we can conclude that the function $\xi(s)$ has infinitely many zeros $\rho_1, \rho_2, \ldots$ such that the series $\sum_{\rho_n}\frac{1}{|\rho_n|^{1+\varepsilon}}$ converges for all $\varepsilon > 0$, whereas the series $\sum_{\rho_n}\frac{1}{|\rho_n|}$ diverges. Since $\xi(s)$ is a first-order function, it can be represented in the form

$$\xi(s) = ae^{bs}\prod_{\rho}\left(1-\frac{s}{\rho}\right)e^{\frac{s}{\rho}}. \tag{2.99}$$

Since the zeros of $\xi(s)$ are nontrivial zeros of the function $\zeta(s)$, we have the following theorem.

Theorem 2.3. *The function $\zeta(s)$ in the region $0 \leqslant \sigma \leqslant 1$ has infinitely many zeros $\rho_1, \rho_2, \ldots$, for which $\sum_{\rho_n}\frac{1}{|\rho_n|^{1+\varepsilon}} < \infty$ for all $\varepsilon > 0$, whereas $\sum_{\rho_n}\frac{1}{|\rho_n|} = \infty$.*

A closer characterization of the function $\xi(s)$ is given by the following theorem.

Theorem 2.4. *In formula* (2.99)*:*

$$a = \frac{1}{2}, \quad b = \log 2 + \frac{1}{2}\log\pi - \frac{\gamma}{2} - 1 \quad (\gamma \textit{ is Euler's constant } [11, 29, 200]); \tag{2.100}$$

$$\frac{\xi'(s)}{\xi(s)} = -1 - \frac{\gamma}{2} + \frac{1}{2}\log 4\pi + \sum_{\rho}\left(\frac{1}{s-\rho} + \frac{1}{\rho}\right); \tag{2.101}$$

$$\frac{\zeta'(s)}{\zeta(s)} = -1 - \frac{\gamma}{2} + \log 2\pi - \frac{1}{s-1} - \frac{1}{2}\cdot\frac{\Gamma'(\frac{s}{2}+1)}{\Gamma(\frac{s}{2}+1)} + \sum_{\rho}\left(\frac{1}{s-\rho} + \frac{1}{\rho}\right). \tag{2.102}$$

Proof. Logarithmizing and differentiating (2.99), we obtain

$$\frac{\xi'(s)}{\xi(s)} = b + \sum_{\rho}\left(\frac{1}{s-\rho} + \frac{1}{\rho}\right). \tag{2.103}$$

Since

$$\xi(s) = \pi^{-\frac{s}{2}}(s-1)\Gamma\left(\frac{s}{2}+1\right)\zeta(s), \tag{2.104}$$

logarithmizing and differentiating this relation, by (2.103) we obtain

$$\frac{\xi'(s)}{\xi(s)} = b + \frac{1}{2}\log\pi - \frac{1}{s-1} - \frac{1}{2}\cdot\frac{\Gamma'(\frac{s}{2}+1)}{\Gamma(\frac{s}{2}+1)} + \sum_{\rho}\left(\frac{1}{s-\rho} + \frac{1}{\rho}\right). \tag{2.105}$$

Formulas (2.99) and (2.102) follow from (2.103) and (2.105) when we determine b. From the formula

$$\frac{1}{s\Gamma(s)} = e^{\gamma s}\prod_{n=1}^{\infty}\left(1+\frac{s}{n}\right)e^{-\frac{s}{n}}, \tag{2.106}$$

taking $\frac{s}{2}$ instead of s and using the relation $\frac{s}{2}\Gamma(\frac{s}{2}) = \Gamma(\frac{s}{2}+1)$, after logarithmization and differentiation, we get

$$-\frac{1}{2}\cdot\frac{\Gamma'(\frac{s}{2}+1)}{\Gamma(\frac{s}{2}+1)} = \frac{\gamma}{2} + \sum_{n=1}^{\infty}\left(\frac{1}{s+2n} - \frac{1}{2n}\right). \tag{2.107}$$

From (2.99) (the term for $\xi(s)$) and (1.251) it follows that

$$a = \xi(0) = \xi(1) = \frac{1}{2}\pi^{-\frac{1}{2}}\Gamma\left(\frac{1}{2}\right)\lim_{s\to1}(s-1)\zeta(s) = \frac{1}{2}. \tag{2.108}$$

From (2.103) and from the relation $\xi(s) = \xi(1-s)$ we obtain

$$b = \frac{\xi'(0)}{\xi(0)} = -\frac{\xi'(1)}{\xi(1)}. \tag{2.109}$$

Since

$$\frac{\xi'(s)}{\xi(s)} = \frac{\zeta'(s)}{\zeta(s)} + \frac{1}{s-1} - \frac{1}{2}\log\pi + \frac{1}{2}\cdot\frac{\Gamma'(\frac{s}{2}+1)}{\Gamma(\frac{s}{2}+1)}, \tag{2.110}$$

from (2.109) and (2.110) it follows that

$$b = -\lim_{s\to1}\left(\frac{\zeta'(s)}{\zeta(s)} + \frac{1}{s-1}\right) + \frac{1}{2}\log\pi - \frac{1}{2}\cdot\frac{\Gamma'(\frac{3}{2})}{\Gamma(\frac{3}{2})}. \tag{2.111}$$

Based on (2.107), we can write the following equation

$$-\frac{1}{2}\cdot\frac{\Gamma'(\frac{3}{2})}{\Gamma(\frac{3}{2})} = \frac{\gamma}{2} + \sum_{n=1}^{\infty}\left(\frac{1}{2n+1} - \frac{1}{2n}\right) = \frac{\gamma}{2} - 1 + \log 2, \tag{2.112}$$

and, accordingly,

$$b = -\lim_{s\to1}\left(\frac{\zeta'(s)}{\zeta(s)} + \frac{1}{s-1}\right) + \frac{\gamma}{2} - 1 + \frac{1}{2}\log 4\pi. \tag{2.113}$$

According to (1.41),

$$\zeta(s) = \frac{s}{s-1} - sf(s), \quad f(s) = \int_1^\infty \frac{x-[x]}{x^{s+1}}\,\partial x. \tag{2.114}$$

It follows that

$$\lim_{s\to 1}\left(\frac{\zeta'(s)}{\zeta(s)} + \frac{1}{s-1}\right) = 1 - f(1). \tag{2.115}$$

According to (1.246),

$$f(1) = \int_1^\infty \frac{x-[x]}{x^2}\,\partial x = 1-\gamma. \tag{2.116}$$

It finally follows that

$$b = -\frac{\gamma}{2} - 1 + \frac{1}{2}\log 4\pi. \qquad \square$$

Proposition 2.4. *From formulas* (2.109) *and* (2.110) *it follows that* $\Gamma'(1) = -\gamma$, $\zeta(0) = -\frac{1}{2}$ *(as well as from* (1.266)*), and from numeric values of b we have*

$$\zeta'(0) = -\frac{1}{2}\log 2\pi. \tag{2.117}$$

2.2.1 A region with no zeros of the function

Theorem 2.5 (Karatsuba [170]). *Let* $\rho_n = \beta_n + i\gamma_n$, $n = 1, 2, \ldots$, *be all nontrivial zeros of the function* $\zeta(s)$. *Then for* $T \geqslant 2$, *we have*

(a)

$$\sum_{n=1}^\infty \frac{1}{1+(T-\gamma_n)^2} \leqslant c\log T. \tag{2.118}$$

(b) *The number of zeros of the function* $\zeta(s)$ *for which* $T \leqslant |\mathfrak{Im}\,\rho_n| \leqslant T+1$ *does not exceed*

$$C_1 \log T. \tag{2.119}$$

(c)

$$\sum_{|T-\gamma_n|>1} \frac{1}{|T-\gamma_n|^2} = O(\log T). \tag{2.120}$$

(d) *In the region* $-1 \leqslant \sigma \leqslant 2$, $|t| \geqslant 2$,

$$\frac{\zeta'(s)}{\zeta(s)} = \sum_{|t-\gamma_n|\leqslant 1} \frac{1}{s-\rho_n} + O(\log|t|). \tag{2.121}$$

Proof. (a) Let $s = 2 + iT$. From (2.102) and (2.107) it follows that

$$\frac{\zeta'(s)}{\zeta(s)} = -\frac{1}{s-1} + \sum_{n=1}^{\infty}\left(\frac{1}{s-\rho_n} + \frac{1}{\rho_n}\right) + \sum_{n=1}^{\infty}\left(\frac{1}{s+2n} - \frac{1}{2n}\right) - 1 + \log 2\pi. \tag{2.122}$$

However,

$$\begin{aligned}\left|\sum_{n=1}^{\infty}\left(\frac{1}{s+2n} - \frac{1}{2n}\right)\right| &\leqslant \sum_{n\leqslant T}\left|\frac{1}{s+2n} - \frac{1}{2n}\right| + \sum_{n>T}\left|\frac{s}{2n(s+2n)}\right| \\ &\leqslant \sum_{n\leqslant T}^{\infty}\left(\frac{1}{2n} + \frac{1}{2n}\right) + \sum_{n>T}\frac{|s|}{4n^2} \leqslant C_0 \log T,\end{aligned} \tag{2.123}$$

since according to the proof of (1.494), it follows that

$$\sum_{n\leqslant T}\frac{1}{n} = O(\log T), \tag{2.124}$$

$$\sum_{n>T}\frac{1}{n^2} = O\left(\frac{1}{T}\right). \tag{2.125}$$

From (2.122) and (2.123) we have

$$\begin{aligned}-\Re\mathfrak{e}\,\frac{\zeta'(s)}{\zeta(s)} &= \Re\mathfrak{e}\left(\frac{1}{s-1} + 1 - \log 2\pi - \sum_{n=1}^{\infty}\left(\frac{1}{s+2n} - \frac{1}{2n}\right)\right) - \Re\mathfrak{e}\left(\sum_{n=1}^{\infty}\left(\frac{1}{s-\rho_n} + \frac{1}{\rho_n}\right)\right) \\ &\leqslant C_2 \log T - \Re\mathfrak{e}\left(\sum_{n=1}^{\infty}\left(\frac{1}{s-\rho_n} + \frac{1}{\rho_n}\right)\right).\end{aligned} \tag{2.126}$$

Since

$$\left|\frac{\zeta'(s)}{\zeta(s)}\right| = \left|\sum_{n=1}^{\infty}\frac{\Lambda(n)}{n^{2+iT}}\right| \leqslant \sum_{n=1}^{\infty}\frac{\Lambda(n)}{n^2} \leqslant \sum_{n=1}^{\infty}\frac{\log n}{n^2} = -\zeta'(2), \tag{2.127}$$

from (2.126) we can conclude that

$$\Re\mathfrak{e}\left(\sum_{n=1}^{\infty}\left(\frac{1}{s-\rho_n} + \frac{1}{\rho_n}\right)\right) \leqslant C_3 \log T. \tag{2.128}$$

In addition, we have the following relations:

$$\mathfrak{Re}\,\frac{1}{\rho_n} = \frac{\beta_n}{\beta_n^2+\gamma_n^2} \geqslant 0, \tag{2.129}$$

$$\mathfrak{Re}\,\frac{1}{s-\rho_n} = \mathfrak{Re}\,\frac{1}{2-\beta_n+\mathrm{i}(T-\gamma_n)} = \frac{2-\beta_n}{(2-\beta_n)^2+(T-\gamma_n)^2} \geqslant \frac{\frac{1}{2}}{1+(T-\gamma_n)^2}, \tag{2.130}$$

because $1 \leqslant 2-\beta_n \leqslant 2$. Therefore (a) follows from (2.128).

(b) Let $T \leqslant \gamma_n \leqslant T+1$, which can be assumed because the zeros of $\zeta(s)$ are symmetric with respect to the σ-axis. Assuming that for every $C_4 > 0$, $N(T+1)-N(T) > C_4 \log T$, we have

$$\sum_{T\leqslant\gamma_n\leqslant T+1} \frac{1}{1+(T-\gamma_n)^2} \geqslant \frac{1}{2}(N(T+1)-N(T)) > \frac{1}{2}C_4\log T, \tag{2.131}$$

which for $C_4 = 2c$ leads to the result that contradicts the result obtained in (a).

(c) The result follows from the following inequalities:

$$\sum_{|T-\gamma_n|>1} \frac{1}{(T-\gamma_n)^2} \leqslant 2\sum_{|T-\gamma_n|>1}\frac{1}{1+(T-\gamma_n)^2} \leqslant 2\sum_{n=1}^{\infty}\frac{1}{1+(T-\gamma_n)^2} \leqslant 2c\log T. \tag{2.132}$$

(d) Estimate in (2.123) is also valid for $s = \sigma+it$, $|t| \geqslant 2$, $-1 \leqslant \sigma \leqslant 2$, and therefore

$$\frac{\zeta'(s)}{\zeta(s)} = \sum_{n=1}^{\infty}\left(\frac{1}{s-\rho_n}+\frac{1}{\rho_n}\right) + O(\log|t|). \tag{2.133}$$

Subtracting relation (2.133) with $s = \sigma+it$ from it with $s = 2+it$, we get:

$$\frac{\zeta'(s)}{\zeta(s)} = \sum_{n=1}^{\infty}\left(\frac{1}{s-\rho_n}-\frac{1}{2+it-\rho_n}\right) + O(\log|t|). \tag{2.134}$$

For $|\gamma_n - t| > 1$,

$$\left|\frac{1}{\sigma+\mathrm{i}t-\rho_n}-\frac{1}{2+\mathrm{i}t-\rho_n}\right| \leqslant \frac{2-\sigma}{(t-\gamma_n)^2} \leqslant \frac{3}{(t-\gamma_n)^2}, \tag{2.135}$$

and for $|\gamma_n - t| \leqslant 1$,

$$\sum_{|t-\gamma_n|\leqslant1}\frac{1}{|2+it-\rho_n|} \leqslant \sum_{|t-\gamma_n|\leqslant1}\frac{1}{\sqrt{1+(t-\gamma_n)^2}} \leqslant \sqrt{2}\sum_{|t-\gamma_n|\leqslant1}\frac{1}{1+(t-\gamma_n)^2}. \tag{2.136}$$

This, together with (a), (c), and (2.134), provides the proof of proposition (d). □

Theorem 2.6 (*de la Vallée Poussin* [20] on the region in which the function $\zeta(s)$ has no zeros). *There is an absolute constant $c > 0$ such that in the region of the s-plane*

$$\sigma \geqslant 1 - \frac{c}{\log(|t|+2)}, \tag{2.137}$$

there are no zeros of the function $\zeta(s)$.

Proof. The proofs for Theorems 2.5 and 2.6 follow the proof presented in [170]. Since $s = 1$ is a pole of the function $\zeta(s)$, then in the region $|s-1| \leqslant \gamma_0$, for some $\gamma_0 > 0$, $\zeta(s)$ has no zeros. Let $\rho_n = \beta_n + i\gamma_n$ be a zero of the function $\zeta(s)$ for which $|\gamma_n| > \gamma_0$. For $\sigma > 1$, we have

$$-\frac{\zeta'(s)}{\zeta(s)} = \sum_{n=1}^{\infty} \frac{\wedge(n)}{n^s} = \sum_{n=1}^{\infty} \wedge(n) \cdot n^{-\sigma} \cdot \mathfrak{e}^{-it\log n} \tag{2.138}$$

and

$$-\mathfrak{Re}\,\frac{\zeta'(s)}{\zeta(s)} = \sum_{n=1}^{\infty} \wedge(n) \cdot n^{-\sigma} \cdot \cos(t\log n). \tag{2.139}$$

As in the proof of Lemma 1.6, we use the polynomial

$$3 + 4\cos\varphi + \cos 2\varphi = 2(1+\cos\varphi)^2 \geqslant 0. \tag{2.140}$$

It follows that

$$3\left(-\frac{\zeta'(\sigma)}{\zeta(\sigma)}\right) + 4\left(-\mathfrak{Re}\,\frac{\zeta'(\sigma+it)}{\zeta(\sigma+it)}\right) + \left(-\mathfrak{Re}\,\frac{\zeta'(\sigma+2it)}{\zeta(\sigma+2it)}\right) \geqslant 0. \tag{2.141}$$

For $1 < \sigma \leqslant 2$, we have

$$\left|\sum_{n=1}^{\infty}\left(\frac{1}{\sigma+2n} - \frac{1}{2n}\right)\right| < \frac{1}{2}\sum_{n=1}^{\infty}\frac{1}{n^2} = \frac{\pi^2}{12} = K_1. \tag{2.142}$$

In addition,

$$\begin{aligned}\left|\sum_{n=1}^{\infty}\left(\frac{1}{\sigma-\rho_n} + \frac{1}{\rho_n}\right)\right| &\leqslant \sum_{n=1}^{\infty}\frac{\sigma}{|\rho_n|\cdot|\sigma-\rho_n|}\\ &< \left|\begin{matrix}\text{because } |\sigma-\rho_n| > |\gamma_n|,\\ |\rho_n| > |\gamma_n|\end{matrix}\right| < 2\sum_{n=1}^{\infty}\frac{1}{|\gamma_n|^2}\\ &< \left|\begin{matrix}|\rho_n| < \sqrt{1+\gamma_n^2} < 1 + |\gamma_n| < 2|\gamma_n|\\ \text{with } |\gamma_n| > 1;\ |\gamma_n| > 12 \text{ (see [53])}\end{matrix}\right| < 8\sum_{n=1}^{\infty}\frac{1}{|\rho_n|^2} = K_2,\end{aligned} \tag{2.143}$$

because the series $\sum_{n=1}^{\infty}\frac{1}{|\rho_n|^2}$ converges. From (2.122), for $s = \sigma, 1 < \sigma \leqslant 2$, it follows that

$$-\frac{\zeta'(\sigma)}{\zeta(\sigma)} < \frac{1}{\sigma-1} + K, \tag{2.144}$$

where $K > 0$ is an absolute constant. For $1 < \sigma \leqslant 2$, $|t| > \gamma_0$, from (2.122) we obtain the equation

$$-\mathfrak{Re}\,\frac{\zeta'(s)}{\zeta(s)} < L\log(|t|+2) - \sum_{k=1}^{\infty}\mathfrak{Re}\left(\frac{1}{s-\rho_k} + \frac{1}{\rho_k}\right), \tag{2.145}$$

where $L > 0$ is an absolute constant. However,

$$\mathfrak{Re}\,\frac{1}{s-\rho_k} = \frac{\sigma-\beta_k}{(\sigma-\beta_k)^2+(t-\gamma_k)^2} > 0, \tag{2.146}$$

since $\sigma > 1 \geqslant \beta_k \geqslant 0$, and also

$$\mathfrak{Re}\,\frac{1}{\rho_k} = \frac{\beta_k}{\beta_k^2+\gamma_k^2} \geqslant 0, \tag{2.147}$$

which, together with (2.145), leads to

$$-\mathfrak{Re}\,\frac{\zeta'(\sigma+it)}{\zeta(\sigma+it)} < L\log(|t|+2) - \frac{\sigma-\beta_k}{(\sigma-\beta_k)^2+(t-\gamma_k)^2} \tag{2.148}$$

for all $n \geqslant 1$. From (2.148) it follows that

$$-\mathfrak{Re}\,\frac{\zeta'(\sigma+2it)}{\zeta(\sigma+2it)} < L\log(2|t|+2) < L_1\log(|t|+2). \tag{2.149}$$

Relations (2.144), (2.148), and (2.149), together with relation (2.141), give

$$\frac{3}{\sigma-1} - 4\frac{\sigma-\beta_k}{(\sigma-\beta_n)^2+(t-\gamma_n)^2} + C\log(|t|+2) \geqslant 0, \tag{2.150}$$

where $C > 1$ is an absolute constant. The last inequality is satisfied for all $1 < \sigma \leqslant 2$ and $|t| > \gamma_0$. Let

$$t = \gamma_n, \quad \sigma = 1 + \frac{1}{2C\log(|t|+2)}. \tag{2.151}$$

Then by (2.150) it follows that

$$\beta_n \leqslant 1 - \frac{1}{14C\log(|t|+2)}, \tag{2.152}$$

where C does not depend on n. The theorem is proved. □

In 1899, *de la Vallée Poussin* [20] proved the following theorem.

Theorem 2.7. *There is a constant $c > 0$ such that $\zeta(s) \neq 0$ in the region*

$$\sigma \geqslant 1 - \frac{c}{\log^{\alpha}(|t|+2)} \quad (\alpha \geqslant 1,\ const.). \tag{2.153}$$

Therefore Theorem 2.6 is a particular case of this result. The result given in formula (2.153) *is the first nontrivial result regarding the region in which there are no zeros of the function* $\zeta(s)$.

In relation to (2.153), the best result of this form is obtained by *Vinogradov* [67] in 1958.

Theorem 2.8. *There is* $c > 0$ *such that* $\zeta(s) \neq 0$ *in the region*

$$\sigma \geqslant 1 - \frac{c}{\log^{\alpha}(|t| + 2)} \quad \left(\alpha > \frac{2}{3}\right). \tag{2.154}$$

The result of this form is somehow connected to $\zeta(1+it)$. In addition to (2.154), in the same paper, *Vinogradov* proved the asymptotic equality given in the following theorem.

Theorem 2.9. *We have the following estimate:*

$$\zeta(1 + it) = O(\log^{\frac{2}{3}} |t|) \quad (|t| \geqslant 2). \tag{2.155}$$

Related to the result given above is the following theorem, cited in [170] on p. 98.

Theorem 2.10. *There is an absolute constant* $\gamma_1 > 0$ *such that in the region,*

$$\sigma \geqslant 1 - \frac{\gamma_1}{\log^{\frac{2}{3}} |t|}, \quad |t| \geqslant 2, \tag{2.156}$$

we have

$$\zeta(\sigma + it) = O(\log^{\frac{2}{3}} |t|). \tag{2.157}$$

The result (2.157) is directly related to the formula for $\pi(x)$ (which was proved by *de la. Vallée Poussin* [20]). In 1958, *Vinogradov* [67] obtained the following formula.

Theorem 2.11.

$$\pi(x) = \int_{2}^{x} \frac{\partial u}{\log u} + O(xe^{-\alpha \log^{\frac{3}{5}} x}), \tag{2.158}$$

where $\alpha > 0$ *is a constant.*

De la Vallée Poussin [20] proved the following:

Theorem 2.12. *For sufficiently large R, the function* $\zeta(s)$ *has no zeros in the region*

$$\sigma \geqslant 1 - \frac{1}{R \log t}, \quad t \geqslant 12. \tag{2.159}$$

Relying on the idea of the introduction of the polynomial $3 + 4\cos\varphi + \cos 2\varphi$, which was used in the proof of Theorem 2.6, *Stechkin* [91, 92] introduced the class of polynomials

$$p_n(\varphi) = a_0 + a_1 \cos\varphi + \cdots + a_n \cos n\varphi \quad (n \geqslant 2) \tag{2.160}$$

such that $p_n(\varphi) \geqslant 0$ for all φ, $a_k \geqslant 0$ $(k = 0, 1, \ldots, n)$, and $a_0 < a_1$. Among other things, he improved the result (2.159). One of his results is given in the following theorem.

Theorem 2.13. *The function $\zeta(s)$ has no zeros in the region*

$$\sigma \geqslant 1 - \frac{1}{R\log t}, \quad t \geqslant 12, \ R = 9.65. \tag{2.161}$$

Rosser and *Schoenfeld* [121] proved the following:

Theorem 2.14. *The function $\zeta(s)$ has no zeros in the region*

$$\sigma \geqslant 1 - \frac{1}{R_1 \log|\frac{t}{17}|}, \quad t \geqslant 21, \ R_1 = 9.645908801. \tag{2.162}$$

In 1895, in relation to the “critical region”, *von Mangoldt* [17] proved the following:

Theorem 2.15. *We have the following inequality:*

$$\zeta(\sigma + \mathrm{i}t) \neq 0 \quad (0 < |t| \leqslant 12). \tag{2.163}$$

2.3 The fifth *Riemann*'s hypothesis and nonnumerical results related to the first and fifth *Riemann* hypotheses

Hypothesis. The fifth *Riemann* hypothesis on the function $\zeta(s)$ states that

All nontrivial zeros of $\zeta(s)$ are on the line $\sigma = \frac{1}{2}$, and they are all simple.

As the remaining four *Riemann* hypotheses on the function have been proven, the only one that remains to be proved is the fifth *Riemann* hypothesis. In 2018, the British scientist *Michael Ati* offered a proof of the hypothesis, but it has not yet been verified. Advances regarding the *Riemann* hypothesis, in relation to *Hardy*'s result (Theorem 2.2), were made in 1921 by *Hardy–Littlewood* [32], given in the form of the following theorem.

Theorem 2.16. *For the total number $N_0(T)$ [114] of zeros of the function $\zeta(s)$ in the region $\sigma = \frac{1}{2}$, $0 < t \leq T$, we have*

$$N_0(T) > cT, \tag{2.164}$$

where c is an absolute constant.

This result was improved only in 1942 by *Selberg* [45], who proved the following theorem.

Theorem 2.17. *We have the following inequality:*

$$N_0(T) > cT\log T \tag{2.165}$$

for sufficiently large T, where c is an absolute constant.

From this inequality it follows that the number of zeros of the function $\zeta(s)$ on the line $\sigma = \frac{1}{2}$, $0 < t \leqslant T$ is finite. Indeed,

$$\begin{aligned}\frac{N_0(T)}{N(T)} &> \frac{cT\log T}{\frac{T}{2\pi}\log\frac{T}{2\pi} - \frac{T}{2\pi} + \alpha(T)\log T} \\ &= \frac{c}{\frac{1}{2\pi} - \frac{\log 2\pi}{2\pi}\cdot\frac{1}{\log T} + \frac{\alpha(T)}{T}} > \frac{c}{\frac{1}{2\pi} + \delta_1} = \delta\end{aligned} \tag{2.166}$$

for sufficiently large T, i. e.,

$$N_0(T) > \delta N(T) \quad \text{(for sufficiently large } T\text{)}. \tag{2.167}$$

If the *Riemann* hypothesis were correct, then we would have

$$N_0(T) = N(T) \backsim \frac{T}{2\pi}\log T \quad (T \longrightarrow \infty). \tag{2.168}$$

An improvement of *Selberg*'s result (because in (2.167) the constant δ is very small) was made by *Levinson* [116]:

Theorem 2.18. *For sufficiently large T, we have*

$$N_0(T) > \frac{1}{3}N(T). \tag{2.169}$$

As $T \longrightarrow \infty$, we immediately arrive at the following theorem.

Theorem 2.19. *At least one-third of all zeros of $\zeta(s)$ in the region $0 < \sigma < 1$ is on the line $\sigma = \frac{1}{2}$.*

Levinson, who came to this result and the proof in [120, 184], also proved the fact that directly leads to (2.169), which itself gives more information about the number of zeros of the function $\zeta(s)$ than relation (2.169). He proved the following:

Theorem 2.20. *For $L = \log\frac{T}{2\pi}$ and $U = \frac{T}{L^{10}}$, we have*

$$N_0(T+U) - N_0(T) > 0.34\frac{U}{2\pi}\log\frac{T}{2\pi} \quad (T \text{ sufficiently large}). \tag{2.170}$$

Further, by Theorem 2.1 we have

$$\begin{aligned} &N_0(T+U) - N(T) \\ &= \frac{T+U}{2\pi}\log\frac{T+U}{2\pi} - \frac{T+U}{2\pi} + \alpha(T+U)\log(T+U) - \frac{T}{2\pi}\log\frac{T}{2\pi} + \frac{T}{2\pi} - \beta(T)\log T \\ &= \frac{U}{2\pi}\log\frac{T}{2\pi}\left\{\begin{matrix} \frac{T+U}{U}\frac{\log\frac{T+U}{2\pi}}{\log\frac{T}{2\pi}} - \frac{T+U}{U}\cdot\frac{1}{\log\frac{T}{2\pi}} + \frac{\alpha(T+U)\log(T+U)}{\frac{U}{2\pi}\log\frac{T}{2\pi}} \\ -\frac{T}{U} + \frac{T}{U\log\frac{T}{2\pi}} - \frac{\beta(T)}{\frac{U}{2\pi}}\cdot\frac{\log T}{\log\frac{T}{2\pi}} \end{matrix}\right\} \end{aligned} \tag{2.171}$$

and

$$\frac{\log\frac{T+U}{2\pi}}{\log\frac{T}{2\pi}} = 1 + \frac{\log\frac{T+U}{T}}{\log\frac{T}{2\pi}}, \quad \log\frac{T+U}{T} = \frac{U}{T} + \gamma(T)\frac{U^2}{T^2}. \tag{2.172}$$

After rearranging this relation, it turns out that

$$N(T+U) - N(T) = \frac{U}{2\pi}\log\frac{T}{2\pi}\left\{\begin{matrix} 1 - \frac{1}{\log\frac{T}{2\pi}} + \frac{T+U}{U}\cdot\frac{1}{\log\frac{T}{2\pi}}\cdot(\frac{U}{T} + \gamma(T)\frac{U^2}{T^2}) \\ + \frac{\alpha(T+U)}{\frac{U}{2\pi}}\cdot\frac{\frac{U}{T}+\gamma(T)\frac{U^2}{T^2}+\log T}{\log\frac{T}{2\pi}} - \frac{\beta(T)}{\frac{U}{2\pi}}(1 + \frac{\log 2\pi}{\log\frac{T}{2\pi}}) \end{matrix}\right\}. \tag{2.173}$$

Since $\lim_{T\longrightarrow\infty}\frac{U}{T} = 0$, for sufficiently large T, we have

$$N(T+U) - N(T) < (1+\delta)\frac{U}{2\pi}\log\frac{T}{2\pi}. \tag{2.174}$$

Choosing T large enough to satisfy the condition $1+\delta < \frac{0.340}{0.335}$, from (2.170) and (2.174) we arrive at the following theorem.

Theorem 2.21.

$$N_0(T+U) - N(T) > 0.335\cdot(N(T+U) - N(T)). \tag{2.175}$$

Further development of (2.175) implies that

$$\begin{aligned} N_0(T_0+U_0) - N_0(T_0) &> 0.335\cdot(N(T_0+U_0) - N(T_0)), \\ N_0(T_1+U_1) - N_0(T_1) &> 0.335\cdot(N(T_1+U_1) - N(T_1)), \\ &\vdots \\ N_0(T_k+U_k) - N_0(T_k) &> 0.335\cdot(N(T_k+U_k) - N(T_k)) \end{aligned} \tag{2.176}$$

with $T_{l+1} = T_l + U_l$, where U_l is defined as in (2.170). It follows that

$$N_0(T_k+U_k) - N_0(T_0) > 0.335\cdot(N(T_k+U_k) - N(T_0)), \tag{2.177}$$

that is,

$$N_0(T_0+U)-N_0(T_0)>0.335\cdot(N(T_0+U)-N(T_0)), \tag{2.178}$$

where U is obtained after finally many applications of formula (2.176) using (2.170). From Theorem 2.1, for fixed T_0, we have

$$N_0(T_0)=\alpha\cdot N(T_0), \quad (\alpha \text{ fixed}, \ 0<\alpha\leqslant 1), \tag{2.179}$$

$$N_0(T_0)=\lambda\cdot N(T_0+U), \quad (\lambda \text{ negligible compared to the large } U). \tag{2.180}$$

Based on this relation, from (2.178) we have

$$N_0(T_0+U)>\{0.335+\lambda(\alpha-0.335)\}(T_0+U). \tag{2.181}$$

If $\alpha>0.335$, then it immediately follows that

$$N_0(T_0+U)>0.335\cdot N(T_0+U), \tag{2.182}$$

and if $\alpha<0.335$, then for large U,

$$N_0(T_0+U)>(0.335-\delta)\cdot N(T_0+U). \tag{2.183}$$

Therefore, for sufficiently large U, we get

$$N_0(T_0+U)>\frac{1}{3}N(T_0+U). \tag{2.184}$$

Since for a sufficiently large T, the function $f(T)=T+\frac{T}{(\log\frac{T}{2\pi})^{10}}$ is continuous and increasing, U does not have to be as defined in Theorem 2.20. Based on (2.184), for arbitrary U and for sufficiently large T, we have the following inequality:

$$N_0(T)>\frac{1}{3}N(T). \tag{2.185}$$

In 1975, *Levinson* [120] proved the following:

Theorem 2.22. *With the notations of Theorem* 2.20, *for sufficiently large T, we have*

$$N_0(T+U)-N_0(T)>0.3474\frac{U}{2\pi}\log\frac{T}{2\pi}. \tag{2.186}$$

Since the zeros of the function $\xi(s)$ defined in Section 1.2.4 in the region $0<\sigma<1$ are exactly the zeros of the function $\zeta(s)$ in this region, $\xi(\frac{1}{2}+it)$ is real, and $\xi(s)$ is the whole function, according to *Rolle*'s theorem, between two consecutive zeros on the line $\sigma=\frac{1}{2}$, there is at least one zero of the function $\xi'(s)$. Based on the above and Theorem 2.18, we arrive at the following theorem.

Theorem 2.23. *At least one-third of the zeros of the function $\xi'(s)$ in the region $0<\sigma<1$ lies on the line $\sigma=\frac{1}{2}$.*

In 1974, *Levinson* [115] proved the following theorem.

Theorem 2.24. *The number of zeros of the function $\xi'(s)$ in the region $\sigma = \frac{1}{2}$, $T < t < T+U$ for $U = \frac{T}{\log^{10} T}$ is greater than $0.7172 \cdot \frac{U}{2\pi} \log \frac{T}{2\pi}$.*

Using the same methodology as in (2.170)–(2.175), we get the following result.

Theorem 2.25. *The number of zeros of the function $\xi'(s)$ in the region $\sigma = \frac{1}{2}$, $T < t < T+U$ for $U = \frac{T}{\log^{10} T}$ is greater than $\frac{7}{10}$ of the total number of zeros of the function $\xi'(s)$ in the region $0 < \sigma < 1$, $T < t < T + U$.*

Definition 2.4. Let $N_1(T)$ denote the number of simple zeros of the function $\zeta(s)$ in the region $\sigma = \frac{1}{2}$, $0 < t \leqslant T$.

Selberg and *Heath-Brown* [137], using *Levinson*'s result (2.186), proved the following fact.

Theorem 2.26. *For sufficiently large T and $U = \frac{T}{\log^{10} T}$, we have*

$$N_1(T + U) - N_1(T) > 0.3474(N(T + U) - N(T)). \tag{2.187}$$

The following theorem is the first and immediate consequence of Theorem 2.26 with respect to Theorem 2.1 regarding the fact that the function $\zeta(s)$ has infinitely many zeros in the region $0 < \sigma < 1$:

Theorem 2.27. *The function $\zeta(s)$ has infinitely many simple zeros in the region $0 < \sigma < 1$* [165]. *For $U = \frac{T}{(\log \frac{T}{2\pi})^{10}}$ and sufficiently large T, we have the following inequality:*

$$N_1(T + U) - N_1(T) > 0.3532 \cdot (N(T + U) - N(T)). \tag{2.188}$$

In the same paper, the following evaluation was obtained:

$$N_0(T) > 0.36 \cdot N(T). \tag{2.189}$$

Since the method used by *Selberg* and *Heath-Brown* is such that only simple zeros can be counted, from relations (2.169) and (2.186) we arrive at the following theorem.

Theorem 2.28. *At least one-third of zeros of the function $\zeta(s)$ in the region $0 < \sigma < 1$ are simple zeros and lie on the line $\sigma = \frac{1}{2}$.*

In 1972, *Montgomery* [101], assuming the *Riemann* hypothesis, proved that

$$N_1(T) \geqslant \left(\frac{2}{3} + o(1) \cdot N(T)\right) \quad (T \longrightarrow \infty). \tag{2.190}$$

We will here present some results concerning the zeros of odd order in short intervals of the line $\sigma = \frac{1}{2}$, together with appropriate estimates for the distance between two consecutive zeros of the function $\zeta(s)$ on the line $\sigma = \frac{1}{2}$. In 1918, *Hardy* and *Littlewood* [31] proved the following theorem.

Theorem 2.29. *For arbitrary $\varepsilon > 0$, there is $T_0 = T_0(\varepsilon) > 0$ such that for $T \geqslant T_0$ and $H \geqslant T^{\frac{1}{4}+\varepsilon}$, the interval $(T, T+H)$ contains a zero of odd order of the function $\zeta(\frac{1}{2}+it)$.*

Definition 2.5. Let $N^0(T)$ denote the number of zeros of odd order of the function $\zeta(\frac{1}{2}+it)$ in the interval $(0, T)$.

In 1921, *Hardy* and *Littlewood* [19] proved the following theorem on the number of zeros of odd order of the function $\zeta(\frac{1}{2}+it)$ in the interval $(T, T+H)$:

Theorem 2.30. *For arbitrary $\varepsilon > 0$, there are $T_0 = T_0(\varepsilon) > 0$ and $c = c(\varepsilon) > 0$ such that for $T \geqslant T_0$ and $H = T^{\frac{1}{4}+\varepsilon}$, the number of zeros of odd order of the function $\zeta(\frac{1}{2}+it)$ in the interval $(T, T+H)$ is not smaller than cH, that is,*

$$N^0(T+H) - N^0(T) \geqslant cH. \tag{2.191}$$

Selberg [44] significantly improved Theorem 2.30 by the following result.

Theorem 2.31. *For arbitrary $\varepsilon > 0$, there are $T_0 = T_0(\varepsilon) > 0$ and $c = c(\varepsilon) > 0$ such that for $T \geqslant T_0$ and $H = T^{\frac{1}{4}+\varepsilon}$, we have the following inequality:*

$$N^0(T+H) - N^0(T) \geqslant cH \log T. \tag{2.192}$$

Since, according to (2.174),

$$N(T+H) - N(T) < dH \log T \tag{2.193}$$

(because in (2.174) the conclusion is only drawn from the *Riemann–Mangoldt* formula for $N(T)$ and from $\lim_{T\longrightarrow\infty} \frac{U}{T} = 0$), we have

$$N^0(T+H) - N^0(T) \leqslant N(T+H) - N(T) < dH \log T. \tag{2.194}$$

This, together with (2.192), leads to the conclusion that the result expressed in that inequality cannot be amplified, in the sense that H is replaced by H^λ $(\lambda > 1)$ or $\log T$ by $(\log T)^\mu$ $(\mu > 1)$. Thus (2.192) provides a proper estimate of the number of zeros of the function $\zeta(s)$ in short intervals of the line $\sigma = \frac{1}{2}$:

$$N(T+H) - N(T) \geqslant N_0(T+H) - N_0(T) \geqslant cH \log T. \tag{2.195}$$

In relation to inequality (2.192), we can also present the following *Selberg* hypothesis.

Hypothesis. For arbitrary $\varepsilon > 0$, there are $T_0 = T_0(\varepsilon) > 0$ and $c = c(\varepsilon) > 0$ such that for $T \geqslant T_0$ and $H = T^{a+\varepsilon}$, where a is a fixed positive number smaller than $\frac{1}{2}$, we have the following inequality:

$$N^0(T+H) - N^0(T) \geqslant cH \log T. \tag{2.196}$$

Moser In [139, 171] proved the following theorems.

Theorem 2.32. *For $T \geqslant T_0 > 0$ and $H \geqslant T^{\frac{1}{6}} \log^6 T$, the interval $(T, T+H)$ contains a zero of odd order of the function $\zeta(\frac{1}{2}+it)$.*

Theorem 2.33. *For $T \geqslant T_0 > 0$ and $H \geqslant T^{\frac{5}{12}} \log^3 T$, we have the following inequality:*

$$N^0(T+H) - N^0(T) \geqslant cH \quad (c > 0 \textit{ absolute constant}). \tag{2.197}$$

Karacuba [156, 178] proved the following theorems.

Theorem 2.34. *For $T \geqslant T_0 > 0$ and $H \geqslant T^{a} \log^2 T$, $a = \frac{5}{32}$, the interval $(T, T+H)$ contains a zero of odd order of the function $\zeta(\frac{1}{2}+it)$.*

Theorem 2.35. *For $T \geqslant T_0 > 0$ and $H \geqslant T^{\frac{27}{82}} \log^2 T$, we have the following inequality:*

$$N^0(T+H) - N^0(T) \geqslant cH \quad (c > 0 \textit{ absolute constant}). \tag{2.198}$$

For the function $Z(t)$ (see (1.314), we have the following hypothesis: The interval $(T, T+H)$ for $H = T^{\varepsilon}$, where $\varepsilon > 0$ is arbitrarily small fixed number, and $T \geqslant T_0 = T_0$, contains a zero of odd order of the function $Z(t)$. It is known that it does not follow from the *Riemann* hypothesis. *Karacuba* [156] proved the following theorem.

Theorem 2.36. *Let k be a natural number such that $T \geqslant T_0(k) > 0$, and let $H \geqslant T^{\frac{1}{6k+6}} \log T$. Then the interval $(T, T+H)$ contains a zero of odd order of the function $Z^{(k)}(t)$.*

Moser's Theorem 2.33 was the first step toward obtaining a positive solution of *Selberg*'s hypothesis. In 1984, *Karacuba* [178] gave a result on *Selberg*'s hypothesis, presented in the following theorem.

Theorem 2.37. *Let ε be an arbitrary positive number that does not exceed 0.001, $T \geqslant T_0$, and $H = T^{\frac{27}{82}+\varepsilon}$. Then there is a positive constant $c = c(\varepsilon)$ such that*

$$N^0(T+H) - N^0(T) \geqslant cH. \tag{2.199}$$

Ivic [233] explicitly proved that the sign of *Hardy*'s function $Z(t)$ depends on the parity of the function $N(T)$ (the number of zeros). Two definitions of this function are analyzed, and some related issues are discussed.

3 Integrals and sums in connection with the function $\zeta(s)$ and the density of the zeros distribution

3.1 The density of the distribution of zeros of the function $\zeta(s)$ in the critical zone $0 \leqslant \sigma \leqslant 1$. *Lindelöf* hypothesis. Numerical results regarding the zeros of the function $\zeta(s)$

3.1.1 The density of the distribution of zeros of the function $\zeta(s)$ in the critical zone

Definition 3.1. In the region $0 \leqslant \sigma \leqslant 1$, $T \geqslant 2$, the number $N(\sigma, T)$ denotes the total number of zeros of the function $\zeta(s)$ in the rectangle $|\,\mathfrak{Im}\, s| \leqslant T, \sigma \leqslant \mathfrak{Re}\, s \leqslant 1$.

The problem of the density of the distribution of zeros of the function $\zeta(s)$ in the region $0 \leqslant \sigma \leqslant 1$ refers to determining the best possible estimate for $N(\sigma, T)$. The results obtained regarding this problem are listed in [90, 99, 132, 144, 169, 177]. Here we list only some of them. In 1972, *Huxley* [99] proved the following estimate:

$$N(\sigma, T) \leqslant T^{\frac{12}{5}(1-\sigma)} \log^c T \quad \left(\sigma \geqslant \frac{1}{2}\right), \tag{3.1}$$

where $c > 0$ is a constant. In 1983, *Jutila* [169] proved that

$$N(\sigma, T) \leqslant_\varepsilon T^{1-(1-\varepsilon)(\sigma-\frac{1}{2})} \log T \quad \left(\sigma \geqslant \frac{1}{2}\right) \tag{3.2}$$

for each fixed $\varepsilon > 0$, where $\leqslant_\varepsilon$ indicates that the constant in O depends on ε. In 1984, *Ivić* [177] obtained the following result:

$$N(\sigma, T) \leqslant T^{\frac{3(1-\sigma)}{7\sigma-4}+\varepsilon} \quad \left(\frac{3}{4} \leqslant \sigma \leqslant \frac{10}{13}\right), \tag{3.3}$$

$$N(\sigma, T) \leqslant T^{\frac{9(1-\sigma)}{8\sigma-4}+\varepsilon} \quad \left(\frac{10}{13} \leqslant \sigma \leqslant 1\right). \tag{3.4}$$

In this area, it still remains unproven the following "hypothesis on the density of the distribution of zeros of the function $\zeta(s)$ in $0 \leqslant \sigma \leqslant 1$":

$$N(\sigma, T) \leqslant T^{(2+\varepsilon)(1-\sigma)}, \quad T \geqslant T_0(\varepsilon) \quad (\varepsilon > 0). \tag{3.5}$$

3.1.2 Lindelöf hypothesis

Lindelöf's hypothesis says that

$$\zeta\left(\frac{1}{2} + it\right) = O(t^\varepsilon) \tag{3.6}$$

for each $\varepsilon > 0$.

https://doi.org/10.1515/9783112233276-003

This hypothesis has not yet been proven. *Titchmarsh* [50], p. 283, states that the *Lindelöf* hypothesis follows from the *Riemann* hypothesis. This fact was first proved by *J. E. Litttlewood.* Based on the one of the hypotheses equivalent to the *Lindelöf* hypothesis (see [50], p. 280), we can see that the *Lindelöf* hypothesis does not imply the *Riemann* hypothesis. The results reached so far regarding formula (3.6) are still very far from the expected estimate. *Kolesnik* [160, 186] (1982, 1985, respectively) gives the following estimates:

$$\zeta\left(\frac{1}{2}+it\right)=O(t^{\frac{35}{216}+\varepsilon}), \tag{3.7}$$

$$\zeta\left(\frac{1}{2}+it\right)=O(t^{\frac{139}{858}+\varepsilon}), \tag{3.8}$$

which still remain the best results of this kind.

If the *Lindelöf* hypothesis were correct, we would have the following result [50]:

$$N(\sigma,T)=O(T^{2(1-\sigma)+\varepsilon}), \tag{3.9}$$

that is [170],

$$N(\sigma,T)=O(T^{(2+\varepsilon)(1-\sigma)}\log^c T). \tag{3.10}$$

The problems concerning $N(\sigma,T)$ and the *Lindelöf* hypothesis have been extensively dealt with in [50, 90, 122].

3.1.3 Numerical results regarding zeros of the function $\zeta(s)$

The ordinates of the first six zeros of the $\zeta(s)$ function on the line $\sigma=\frac{1}{2}$ are $t_1=14.13$; $t_2=21.02$; $t_3=25.01$; $t_4=30.42$; $t_5=32.93$; $t_6=37.58$. The earliest attempt to join the distribution of the zeros of the function $\zeta(s)$ on the line $\sigma=\frac{1}{2}$ with the corresponding regularity of this type, such that to each "observed" interval, there corresponds exactly one zero of the function $\zeta(s)$, was formalized by *Gram*'s "law" (1902–1904).

Let τ_n, $n\in N$, be a real positive root of the equation

$$\pi^{-1}\Im\mathrm{m}\left(\log\Gamma\left(\frac{1}{4}+\pi\tau\mathrm{i}\right)\right)-\tau\log\pi=n. \tag{3.11}$$

The number τ_n is called the nth *Gram* point, and the interval $I_n=(\tau_n,\tau_{n+1})$ is called the nth *Gram* interval. *Gram*'s "law" states that the function $\zeta(\frac{1}{2}+2\pi\tau\mathrm{i})$ has a single zero in I_n. This "law" has been proven to be wrong. Part of the research was devoted to the problem of examining deviations from *Gram*'s "law".

Numerical tests can reveal possible inaccuracies of the *Riemann* hypothesis if some of zeros of the function $\zeta(s)$ are outside the line $\sigma=\frac{1}{2}$ and in the domain that can be covered by a computer. The more zeros are detected on the line $\sigma = \frac{1}{2}$ (though it has no significant effect on the accuracy of the *Riemann* hypothesis), the better the estimate of all those values that depend on the number of zeros on the line $\sigma = \frac{1}{2}$ $(t \leqslant T)$.

The results of *van de Lune, te Riele*, and *Winter* [188] in 1985 (the first serious results of this type, resulting from the refinement of algorithms on computer platforms) were obtained on the computers of type CYBER 205 and CDC CYBER 175/750 in *Centrum voor Wiskunde en Informatica*, Amsterdam (Center for Mathematics and Computer Science) and led to the following: The first 1,500,000,001 zeros of the *Riemann* zeta function in the critical zone lie on the line $\sigma = \frac{1}{2}$, and they all are simple. This confirms the *Riemann* hypothesis in the region $0 \leqslant \sigma \leqslant 1$, $0 \leqslant t < 545{,}439{,}823{,}215$ [63, 64, 89, 121, 134, 188].

3.2 Some integrals and some sums in which the function $\zeta(s)$ occurs

In this section, we first determine the values of the integrals of the form

$$\int_0^1 t^n \log \Gamma(t)\, \partial t, \quad \int_0^1 t^n \log \sin \pi t \, \partial t \tag{3.12}$$

for negative integers n. Then we calculate the sums

$$\sum_{k=2}^{\infty} (-1)^k \frac{\zeta(k)}{n+k}, \tag{3.13}$$

for negative integers n. It is known that the integral

$$I_n = \int_0^1 t^n \log \Gamma(t)\, \partial t \tag{3.14}$$

is calculated for $n = 0$ (the so-called *Raabe* integral) [94]:

$$I_0 = \int_0^1 \log \Gamma(t)\, \partial t = \frac{1}{2} \log 2\pi. \tag{3.15}$$

For the integrals

$$\int_0^1 t^n \log \sin \pi t \, \partial t, \tag{3.16}$$

the values for $n = 0$ and $n = 1$ were found in [94]:

$$I_0 = \int_0^1 \log \sin \pi t \, \partial t = -\log 2, \tag{3.17}$$

$$I_1 = \int_0^1 t \log \sin \pi t \, \partial t = -\frac{1}{2} \log 2. \tag{3.18}$$

The series

$$K_n = \sum_{k=2}^{\infty} (-1)^k \frac{\zeta(k)}{n+k} \tag{3.19}$$

was calculated for $n = 0$ [72]:

$$K_0 = \sum_{k=2}^{\infty} (-1)^k \frac{\zeta(k)}{n+k} = \gamma, \tag{3.20}$$

and for $n = 1$ [118, 172]:

$$K_1 = \sum_{k=2}^{\infty} (-1)^k \frac{\zeta(k)}{n+k} = 1 + \frac{\gamma}{2} - \frac{1}{2} \log 2\pi. \tag{3.21}$$

In formulas (3.20) and (3.21), γ is *Euler*'s constant. For the series

$$L_n = \sum_{k=1}^{\infty} \frac{\zeta(2k) - 1}{n+k} \tag{3.22}$$

the sum is obtained for $n = 0$ [173]:

$$L_0 = \sum_{k=1}^{\infty} \frac{\zeta(2k) - 1}{k} = \log 2. \tag{3.23}$$

The series

$$M_n = \sum_{k=1}^{\infty} \frac{\zeta(2k+1) - 1}{n+k} \tag{3.24}$$

was determined for $n = 1$ [173]:

$$M_1 = \sum_{k=1}^{\infty} \frac{\zeta(2k+1) - 1}{n+k} = \log 2 - \gamma. \tag{3.25}$$

The integrals I_n and J_n and the sums K_n, L_n, and M_n will be calculated below for all nonnegative integers n, and the sum M_n for $n \geqslant 1$. The following formula applies to the integers $n \geqslant 0$:

$$I_n = \int_0^1 x^n \log \Gamma(x) \, \partial x = \frac{\log 2\pi}{2(n+1)} - \frac{n(\gamma + \log 2\pi)}{2(n+1)(n+2)}$$

$$+\frac{1}{2}\sum_{k=1}^{[\frac{n}{2}]}(-1)^{k+1}\frac{n!}{(n-2k+1)!}\cdot\frac{\zeta(2k+1)}{(2\pi)^{2k}}$$
$$+\frac{1}{\pi}\sum_{k=1}^{[\frac{n+1}{2}]}(-1)^{k+1}\frac{n!}{(n-2k+2)!}\cdot\frac{\zeta'(2k)}{(2\pi)^{2k-1}}; \tag{3.26}$$

in particular, for $n = 1$,

$$I_1 = \int_0^1 x\log\Gamma(x)\,\partial x = \frac{\zeta'(2)}{2\pi^2} - \frac{\gamma}{12} + \frac{\log 2\pi}{6}. \tag{3.27}$$

Proof. To prove formula (3.26), we start from the *Kummer* series for $\log\Gamma(x)$ [54],

$$\log\Gamma(x) = \frac{1}{2}\log 2\pi + \sum_{m=1}^{\infty}\left(\frac{1}{2m}\cos 2m\pi x + (\gamma + \log 2m\pi)\frac{\sin 2m\pi x}{m\pi}\right) \quad (0 < x < 1) \tag{3.28}$$

and the following formulas:

$$\int_0^1 x^n \cos 2m\pi x\,\partial x = \sum_{k=1}^{[\frac{n}{2}]}(-1)^{k+1}\frac{n!}{(n-2k+1)!}\cdot\frac{1}{(2m\pi)^{2k}}, \tag{3.29}$$

$$\int_0^1 x^n \sin 2m\pi x\,\partial x = \sum_{k=1}^{[\frac{n+1}{2}]}(-1)^{k+1}\frac{n!}{(n-2k+2)!}\cdot\frac{1}{(2m\pi)^{2k-1}}, \tag{3.30}$$

$$\zeta'(s) = -\sum_{m=1}^{\infty}\frac{\log m}{m^s} \quad (\Re e(s) > 1). \tag{3.31}$$

Since I_n exists because of the existence of I_0, and since *Fourier* series (3.28) multiplied by x^n may be integrated term-by-term ([149], p. 431), we get the following formula:

$$I_n = \int_0^1 x^n\log\Gamma(x)\,\partial x = \frac{\log 2\pi}{2(n+1)} + 2(\gamma + \log 2\pi)\cdot\sum_{k=1}^{[\frac{n+1}{2}]}(-1)^k\frac{n!}{(n-2k+2)!}\cdot\frac{\zeta(2k)}{(2\pi)^{2k}}$$
$$+\frac{1}{2}\sum_{k=1}^{[\frac{n}{2}]}(-1)^{k+1}\frac{n!}{(n-2k+1)!}\cdot\frac{\zeta(2k+1)}{(2\pi)^{2k}} + \frac{1}{\pi}\sum_{k=1}^{[\frac{n+1}{2}]}(-1)^{k+1}\frac{n!}{(n-2k+2)!}\cdot\frac{\zeta'(2k)}{(2\pi)^{2k-1}}. \tag{3.32}$$

According to (1.297),

$$\zeta(2k) = \frac{(-1)^{k-1}2^{2k-1}\pi^{2k}}{(2k)!}B_{2k}. \tag{3.33}$$

The second sum on the right-hand side of (3.33) is equal to

$$-\frac{1}{2(n+1)(n+2)}\sum_{k=1}^{[\frac{n+1}{2}]}\binom{n+2}{2k}B_{2k}. \tag{3.34}$$

Formula (3.26) follows from (3.32) if we show that

$$\sum_{k=1}^{[\frac{n+1}{2}]}\binom{n+2}{2k}B_{2k}=\frac{n}{2}. \tag{3.35}$$

The last relation follows from the equality

$$\sum_{k=0}^{l}\binom{l}{k}B_k=B_l, \tag{3.36}$$

because $B_0 = 1$, $B_1 = -\frac{1}{2}$, and $B_{2k+1} = 0$ $(k \geqslant 1)$. For I_n $(n \geqslant 0)$, we have the following formula:

$$\begin{aligned} I_n = {} & \frac{\log\pi}{(n+1)(n+2)} - \frac{n-(\gamma+\log 2)}{2(n+1)(n+2)} - \frac{1}{2}\int_0^1 x^n \log\sin\pi x\,\partial x \\ & + \frac{1}{\pi}\sum_{k=1}^{[\frac{n+1}{2}]}(-1)^{k+1}\frac{n!}{(n-2k+2)!}\cdot\frac{\zeta'(2k)}{(2\pi)^{2k-1}}. \end{aligned} \tag{3.37}$$

Formula (3.37) follows from (3.30), (3.31), and the following *Kummer* series for the function $\log\Gamma(x)$ [54]:

$$\begin{aligned} \log\Gamma(x) = {} & \left(\frac{1}{2}-x\right)(\gamma+\log 2)+(1-x)\log\pi \\ & -\frac{1}{2}\log\sin\pi x+\sum_{m=1}^{\infty}\frac{\log m}{m\pi}\sin 2m\pi x \quad (0<x<1). \end{aligned} \tag{3.38}$$

Further, from (3.26) and (3.37) we obtain the formula for J_n $(n \geqslant 0)$:

$$J_n=\int_0^1 x^n\log\sin\pi x\,\partial x=-\frac{\log 2}{n+1}+\sum_{k=1}^{[\frac{n}{2}]}(-1)^k\frac{n!}{(n-2k+1)!}\cdot\frac{\zeta(2k+1)}{(2\pi)^{2k}}; \tag{3.39}$$

In particular,

$$J_2=\int_0^1 x^2\log\sin\pi x\,\partial x=-\frac{\log 2}{3}-\frac{\zeta(3)}{2\pi^2}, \tag{3.40}$$

$$J_3 = \int_0^1 x^3 \log \sin \pi x \, \partial x = -\frac{\log 2}{4} - \frac{3\zeta(3)}{4\pi^2}. \tag{3.41}$$

For I_n, we also have

$$I_n = \frac{\gamma}{(n+1)(n+2)} + \frac{1}{(n+1)^2} + \frac{1}{n+1} \sum_{m=2}^{\infty} (-1)^{m+1} \frac{\zeta(m)}{n+1+m}. \tag{3.42}$$

□

Proof. To prove (3.39), from the expression for the function $\frac{\Gamma'(x)}{\Gamma(x)}$ it follows that

$$\frac{\Gamma'(x)}{\Gamma(x)} = -\gamma - \sum_{m=1}^{\infty} \left(\frac{1}{x+m} - \frac{1}{m} \right) \quad (x > 0). \tag{3.43}$$

Since the series on the right-hand side converges uniformly, integrating by parts, we obtain

$$\begin{aligned} I_n &= -\frac{1}{n+1} \int_0^1 x^{n+1} \frac{\Gamma'(x)}{\Gamma(x)} \partial x \\ &= \frac{\gamma}{(n+1)(n+2)} + \frac{1}{(n+1)^2} + \frac{1}{n+1} \sum_{m=1}^{\infty} \int_0^1 x^{n+1} \left(\frac{1}{x+m} - \frac{1}{m} \right) \partial x. \end{aligned} \tag{3.44}$$

Since

$$x^{n+1} \left(\frac{1}{x+m} - \frac{1}{m} \right) = -x^{n+2} \frac{1}{1+\frac{x}{m}} \cdot \frac{1}{m^2} = -\frac{1}{m^2} \sum_{p=1}^{\infty} (-1)^{p+1} \frac{x^{n+p+1}}{m^{p-1}}, \tag{3.45}$$

it follows that

$$\begin{aligned} \sum_{m=1}^{\infty} \int_0^1 x^{n+1} \left(\frac{1}{x+m} - \frac{1}{m} \right) \partial x &= \sum_{m=1}^{\infty} \frac{1}{m^2} \sum_{p=1}^{\infty} (-1)^p \frac{1}{m^{p-1}(n+p+2)} \\ &\overset{*}{=} \sum_{p=1}^{\infty} \frac{(-1)^p}{n+p+2} + \sum_{p=1}^{\infty} (-1)^p \frac{\zeta(p+1) - 1}{n+p+2} \overset{!}{=} \sum_{p=1}^{\infty} (-1)^p \frac{\zeta(p+1)}{n+p+2}. \end{aligned} \tag{3.46}$$

The equality $*$ in (3.46) is valid because of the absolute convergence of the double series for $m \geqslant 2$:

$$\left|\sum_{m=1}^{\infty}\frac{1}{m^2}\sum_{p=1}^{\infty}(-1)^p\frac{1}{m^{p-1}(n+p+2)}\right| \leqslant \sum_{m=2}^{\infty}\frac{1}{m^2}\sum_{p=1}^{\infty}(-1)^p\frac{1}{m^{p-1}(n+p+2)}$$
$$< \frac{1}{n+3}\sum_{m=2}^{\infty}\sum_{p=1}^{\infty}\frac{1}{m^{p+1}} = \frac{1}{n+3}\sum_{m=2}^{\infty}\frac{1}{(m-1)m} = \frac{1}{n+3}. \tag{3.47}$$

Since the series after the equality sign ! in (3.46) is conditionally convergent, it follows that the series that determines K_n in (3.19) conditionally converges. Relation (3.42) follows from the relations obtained above. From (3.26) and (3.42) it follows that K_n can be expressed as

$$K_n = \sum_{k=2}^{\infty}(-1)^k\frac{\zeta(k)}{n+k} = \frac{1}{n} + \frac{\gamma}{2} - \frac{\log 2\pi}{n+1}$$
$$+ \frac{1}{2}\sum_{m=1}^{[\frac{n-1}{2}]}(-1)^m\frac{n!}{(n-2m)!}\frac{\zeta(2m+1)}{(2\pi)^{2m}}$$
$$+ \frac{1}{\pi}\sum_{m=1}^{[\frac{n}{2}]}(-1)^m\frac{n!}{(n-2m+1)!}\frac{\zeta'(2m)}{(2\pi)^{2m-1}}. \tag{3.48}$$

In particular,

$$K_2 = \sum_{k=2}^{\infty}(-1)^k\frac{\zeta(k)}{k+2} = \frac{1+\gamma}{2} - \frac{\log 2\pi}{3} - \frac{\zeta'(2)}{\pi^2}, \tag{3.49}$$

$$K_3 = \sum_{k=2}^{\infty}(-1)^k\frac{\zeta(k)}{k+3} = \frac{2+3\gamma}{6} - \frac{\log 2\pi}{4} - \frac{3\zeta(3)}{4\pi^2} - \frac{3\zeta'(2)}{2\pi^2}. \tag{3.50}$$

For J_n $(n \geqslant 0)$, we have

$$J_n = \frac{\log\pi}{n+1} - \frac{1}{(n+1)^2} - \sum_{k=1}^{\infty}\frac{\zeta(2k)}{k(n+2k+1)}. \tag{3.51}$$

Formula (3.51) follows from

$$\log\frac{\sin t}{t} = \sum_{k=1}^{\infty}(-1)^k\frac{2^{2k-1}}{k(2k)!}B_{2k}t^{2k} \quad (|t| < \pi) \tag{3.52}$$

and from the expression

$$J_n = \frac{1}{\pi^{n+1}}\int_0^{\pi} t^n \log\sin t\,\partial t = \frac{1}{\pi^{n+1}}\int_0^{\pi} t^n\left(\log\frac{\sin t}{t} + \log t\right)\partial t \tag{3.53}$$

after the integration and using the expression for $\zeta(2k)$ (see (1.294)). Further, from (3.51) and (3.39) it follows that

$$\sum_{k=1}^{\infty}\frac{\zeta(2k)}{k(n+2k)}=\frac{\log\pi}{n}-\frac{1}{n^2}+\sum_{m=1}^{[\frac{n-1}{2}]}(-1)^{m+1}\frac{(n-1)!}{(n-2m)!}\frac{\zeta(2m+1)}{(2\pi)^{2m}}\quad(n\geqslant 1).\tag{3.54}$$

For $n=2$, using the equality $L_0=\log 2$, we obtain

$$L_1=\sum_{k=1}^{\infty}\frac{\zeta(2k)-1}{k+1}=\frac{3}{2}-\log\pi.\tag{3.55}$$

Taking $2n$ instead of n in (3.54), we get

$$\sum_{k=1}^{\infty}\frac{\zeta(2k)}{k(k+n)}=\frac{\log\pi}{n}-\frac{1}{2n^2}+2\sum_{m=1}^{n-1}(-1)^{m+1}\frac{(2n-1)!}{(2n-2m)!}\frac{\zeta(2m+1)}{(2\pi)^{2m}}\quad(n\geqslant 1).\tag{3.56}$$

Further, from the known formula for the function $\Psi(s)=\frac{\Gamma'(s)}{\Gamma(s)}$,

$$\Psi(1+s)=-\gamma+s\sum_{k=1}^{\infty}\frac{1}{k(k+s)}\quad(\Re\mathfrak{e}(s)>0),\tag{3.57}$$

it follows that

$$\sum_{k=1}^{\infty}\frac{1}{k(n+2k)}=\frac{1}{n}\left\{\Psi\left(1+\frac{n}{2}\right)+\gamma\right\}.\tag{3.58}$$

From this and from (3.56) we obtain the following expression for L_n:

$$\begin{aligned}L_n=\sum_{k=1}^{\infty}\frac{\zeta(2k)-1}{n+k}&=1+\frac{1}{2}+\frac{1}{3}+\cdots+\frac{1}{n}-\log\pi+\frac{1}{2n}\\&+\sum_{m=1}^{n-1}(-1)^m\frac{(2n)!}{(2n-2m)!}\frac{\zeta(2m+1)}{(2\pi)^{2m}}\quad(n\geqslant 1).\end{aligned}\tag{3.59}$$

In particular,

$$L_2=\sum_{k=1}^{\infty}\frac{\zeta(2k)-1}{k+2}=\frac{7}{24}-\log\pi-\frac{3\zeta(3)}{\pi^2}.\tag{3.60}$$

Taking $2n+1$ instead of n in (3.54), we have

$$\begin{aligned}\sum_{k=1}^{\infty}\frac{\zeta(2k)-1}{2n+2k+1}&=1+\frac{1}{3}+\cdots+\frac{1}{2n+1}-\log 2-\frac{1}{2}\log\pi+\frac{1}{2}\frac{1}{2n+1}\\&+\frac{1}{2}\sum_{m=1}^{n}(-1)^m\frac{(2n+1)!}{(2n+1-2m)!}\frac{\zeta(2m+1)}{(2\pi)^{2m}}\quad(n\geqslant 0).\end{aligned}\tag{3.61}$$

In particular, for $n = 0, 1$,

$$\sum_{k=1}^{\infty} \frac{\zeta(2k) - 1}{2k + 1} = \frac{3}{2} - \log 2 - \frac{1}{2} \log \pi, \tag{3.62}$$

$$\sum_{k=1}^{\infty} \frac{\zeta(2k) - 1}{2k + 3} = \frac{3}{2} - \log 2 - \frac{1}{2} \log \pi - \frac{3\zeta(3)}{4\pi^2}. \tag{3.63}$$

Further, from (3.48) it follows that

$$\begin{aligned}\sum_{k=2}^{\infty} (-1)^k \frac{\zeta(k) - 1}{n + k} &= \frac{1}{n} + \frac{\gamma}{2} - \frac{\log 2\pi}{n + 1} \\ &\quad + (-1)^n \left(\log 2 - \left(1 - \frac{1}{2} + \frac{1}{3} - \cdots + (-1)^n \frac{1}{n + 1} \right) \right) \\ &\quad + \frac{1}{2} \sum_{m=1}^{[\frac{n-1}{2}]} (-1)^m \frac{n!}{(n - 2m)!} \frac{\zeta(2m + 1)}{(2\pi)^{2m}} \\ &\quad + \frac{1}{\pi} \sum_{m=1}^{[\frac{n}{2}]} (-1)^m \frac{n!}{(n - 2m)!} \frac{\zeta'(2m)}{(2\pi)^{2m-1}}.\end{aligned} \tag{3.64}$$

Based on equations (3.61) and (3.64), we can derived the following expression for the series M_n:

$$\begin{aligned}M_n = \sum_{k=1}^{\infty} \frac{\zeta(2k + 1) - 1}{n + k} &= 1 + \frac{1}{2} + \cdots + \frac{1}{n} - \gamma - \log \pi + \frac{\log 2\pi}{n} \\ &\quad - \frac{1}{2n - 1} + \frac{2}{\pi} \sum_{m=1}^{n-1} (-1)^{m+1} \frac{(2n - 1)!}{(2n - 2m)!} \frac{\zeta'(2m)}{(2\pi)^{2m-1}} \quad (n \geqslant 1).\end{aligned} \tag{3.65}$$

In particular, for $n = 1, 2$,

$$M_1 = \sum_{k=1}^{\infty} \frac{\zeta(2k + 1) - 1}{k + 1} = \log 2 - \gamma, \tag{3.66}$$

$$M_2 = \sum_{k=1}^{\infty} \frac{\zeta(2k + 1) - 1}{k + 2} = \frac{7}{6} - \gamma - \frac{1}{2} \log \frac{\pi}{2} + \frac{\zeta'(2)}{\pi^2}. \tag{3.67}$$

From (3.54) and (3.64) we get

$$\sum_{k=1}^{\infty} \frac{\zeta(2k + 1) - 1}{k + 2} = \frac{13}{12} - \frac{2 \log 2}{3} - \frac{\log \pi}{6} - \frac{\gamma}{2} + \frac{\zeta'(2)}{\pi^2}. \tag{3.68}$$

Finally, introducing the function

$$F_n(x) = \int_0^x t^n \log \Gamma(t) \, \partial t \quad (0 < x \leqslant 1) \tag{3.69}$$

and applying for $x = \frac{1}{2}$ the same method, for $n = 1, 2$, we obtain

$$\sum_{k=1}^{\infty}(-1)^k \frac{\zeta(k)}{2^k(k+1)} = 1 - \frac{2}{3}\log 2 - \frac{1}{4}\log\pi + \frac{3\zeta'(2)}{2\pi^2}, \tag{3.70}$$

$$\sum_{k=2}^{\infty}(-1)^k \frac{\zeta(k)}{2^k(k+2)} = \frac{1}{4} - \frac{\gamma}{6} - \frac{7}{12}\log 2 - \frac{1}{4}\log\pi + \frac{7\zeta(3)}{8\pi^2} + \frac{\zeta'(2)}{2\pi^2}, \tag{3.71}$$

and for $n \geqslant 3$,

$$\begin{aligned}\sum_{k=2}^{\infty}(-1)^k \frac{\zeta(k)}{2^k(n+k)} &= \frac{\gamma}{2n(n+1)} + \frac{1}{n^2} - \frac{\log 2}{2n} - \frac{(n-1)!\cos\frac{n\pi}{2}}{2\pi^n}\zeta(n+1)\\ &\quad - \frac{(n-1)!\sin\frac{n\pi}{2}}{\pi^{n+1}}(\gamma + \log 2\pi)\zeta(n+1) + \frac{(n-1)!\sin\frac{n\pi}{2}}{\pi^{n+1}}\zeta'(n+1)\\ &\quad + \frac{\gamma + \log 4\pi}{n(n+1)}\left(B_{n+1} - \frac{n-1}{2}\right) - \frac{\gamma + \log 2\pi}{n(n+1)}\sum_{m=1}^{[\frac{n+1}{2}]}\binom{n+1}{2m}2^{2m-1}B_{2m}\\ &\quad + \frac{1}{2}\sum_{m=1}^{[\frac{n}{2}]}(-1)^{m+1}\frac{(n-1)!}{(n-2m)!}\frac{1-2^{-2m}}{\pi^{2m}}\zeta(2m+1)\\ &\quad + \sum_{m=1}^{[\frac{n+1}{2}]}(-1)^{m+1}\frac{(n-1)!}{(n-2m+1)!}\frac{1-2^{-2m+1}}{\pi^{2m}}\zeta'(2m).\end{aligned} \tag{3.72}$$

Putting $2n$ instead of n in this relation and using the formula

$$\sum_{m=1}^{n}\binom{2n+1}{2m}2^{2m-1}B_{2m} = n \tag{3.73}$$

(see [35], p. 154, formula 25), after rearranging, we get

$$\begin{aligned}\sum_{k=2}^{\infty}(-1)^k \frac{\zeta(k)}{2^k(k+2n)} &= \frac{1-2n}{2n(2n+1)}\gamma + \frac{1}{4n^2} + \frac{1-8n}{4n(2n+1)}\log 2\\ &\quad + \frac{1-4n}{4n(2n+1)}\log\pi + (-1)^{n-1}\frac{(2n-1)!}{2\pi^{2n}}\zeta(2n+1)\\ &\quad + \frac{1}{2}\sum_{m=1}^{n}(-1)^{m+1}\frac{(2n-1)!}{(2n-2m)!}\frac{1-2^{-2m}}{\pi^{2m}}\zeta(2m+1)\\ &\quad + \sum_{m=1}^{n}(-1)^{m+1}\frac{(2n-1)!}{(2n-2m+1)!}\frac{1-2^{-2m+1}}{\pi^{2m}}\zeta'(2m).\end{aligned} \tag{3.74}$$

Putting $2n+1$ instead of n in (3.72) and using the formula

$$\sum_{m=1}^{n}\binom{2n+2}{2m}2^{2m-1}B_{2m} = \frac{2n+1}{2} - (2^{2n+1}-1)B_{2n+2} \tag{3.75}$$

(see [35], p. 154, formula 24), after rearranging, we get

$$\begin{aligned}\sum_{k=2}^{\infty}(-1)^k\frac{\zeta(k)}{2^k(k+2n)} &= \frac{-n}{(n+1)(2n+1)}\gamma+\frac{1}{(2n+1)^2}\\ &-\frac{4n+1}{4(n+1)(2n+1)}\log\pi\\ &+\frac{2B_{2n+2}-(8n+3)}{4(n+1)(2n+1)}\log 2+(-1)^n\frac{(2n)!}{\pi^{2n+2}}\zeta'(2n+2)\\ &+\frac{1}{2}\sum_{m=1}^{n}(-1)^{m+1}\frac{(2n)!}{(2n-2m+1)!}\frac{1-2^{-2m}}{\pi^{2m}}\zeta(2m+1)\\ &+\sum_{m=1}^{n+1}(-1)^{m+1}\frac{(2n)!}{(2n-2m+2)!}\frac{1-2^{-2m+1}}{\pi^{2m}}\zeta'(2m).\end{aligned} \tag{3.76}$$

□

Various authors dealt with this issue throughout the entire 20th century [25, 42, 69, 73]. An interesting relation between the numbers $\zeta(2n)$ and $\zeta'(2n)$ was obtained by *Richmond* and *Szekeres* [158] in 1981:

$$\sum_{n=1}^{\infty}\frac{\zeta(2n)}{n^2}-\sum_{n=1}^{\infty}\frac{\zeta'(2n)}{n}=\frac{\pi^2}{3}. \tag{3.77}$$

We also have a similar relation of the form

$$\sum_{n=2}^{\infty}\frac{\zeta(n)}{n^2}+\sum_{n=1}^{\infty}\frac{1}{n}\sum_{k=2}^{\infty}(-1)^k\frac{\zeta(k)}{n+k}=\frac{\pi^2}{6}. \tag{3.78}$$

Proof. To prove (3.78), let us start with

$$\begin{aligned}I=\int_0^1\Psi(1-x)\log\partial dx &= \int_0^1\left(-\gamma+\sum_{n=1}^{\infty}\left(\frac{1}{n}-\frac{1}{n-x}\right)\right)\log x\,\partial x\\ &= -\gamma(x\log x-x)|_0^1+\sum_{n=1}^{\infty}\int_0^1\left(\frac{1}{n}-\frac{1}{n-x}\right)\log x\,\partial x\\ &= -\gamma-\sum_{n=1}^{\infty}\frac{1}{n}\int_0^1\frac{\log x}{n-x}\partial x=\gamma-\sum_{n=1}^{\infty}\frac{1}{n^2}\sum_{m=0}^{\infty}\frac{1}{n^m}\int_0^1 x^{m+1}\log x\,\partial x\\ &= \gamma+\sum_{n=1}^{\infty}\frac{1}{n^2}\sum_{m=0}^{\infty}\frac{1}{n^m(m+2)^2}=\gamma+\sum_{n=2}^{\infty}\frac{\zeta(n)}{n^2}.\end{aligned} \tag{3.79}$$

On the other hand,

$$I = \int_0^1 \Psi(x)\log(1-x)\,\partial x = \int_0^1 \frac{\Gamma'(x)}{\Gamma(x)}\log(1-x)\,\partial x = \int_0^1 \frac{\log\Gamma(x)}{1-x}\,\partial x$$

$$= \sum_{n=0}^{\infty}\int_0^1 x^n \log\Gamma(x)\,\partial x)$$

$$\overset{(3.42)}{=} \sum_{n=0}^{\infty}\left(\frac{\gamma}{(n+1)(n+2)} + \frac{1}{(n+1)^2} + \frac{1}{n+1}\sum_{k=2}^{\infty}(-1)^{k+1}\frac{\gamma(k)}{n+1+k}\right)$$

$$= \gamma + \frac{\pi^2}{6} - \sum_{n=1}^{\infty}\frac{1}{n}\sum_{k=2}^{\infty}(-1)^k\frac{\zeta(k)}{n+k}. \tag{3.80}$$

Equality (3.78) follows directly from (3.79) and (3.80). □

3.3 *Mordell*'s sums

Let

$$S_k = \sum_{m=1}^{\infty}\sum_{n=1}^{\infty}\frac{1}{m^k n^k (m+n)^k} \quad (k \in \mathbb{N}). \tag{3.81}$$

In 1958, *Mordell* [66] calculated the sums

$$S_1 = 2\zeta(3), \quad S_2 = \frac{\pi^6}{2835}. \tag{3.82}$$

He also proved that $S_k = \pi^{3k}p_k$, where p_k is a rational number if k is even. He asked the question "how to find" an appropriate result if k is odd. Although *Apéry* in 1978 proved that $\zeta(3)$ is an irrational number (a simple proof is cited by *Beukers* [136] in 1979); the relationship between $\zeta(3)$ and π^9 is so far unknown. This means that *Mordell*'s question still remains unanswered. We will call S_k the *Mordell* sums. In this section, derive auxiliary formulas and determine S_3. The authors believe that in this way the sum S_{2k+1} is determined analytically for the first time, after that was done for S_1.

Theorem 3.1. *Let*

$$F(a,b) = \sum_{m=1}^{\infty}\frac{1}{m^a}\sum_{n=1}^{m}\frac{1}{n^b} \quad (a, b \in \mathbb{N},\ a \geqslant 2). \tag{3.83}$$

Then

$$F(a,b) = \left[1 + (-1)^{b+1}\binom{a+b-2}{b}\right]\zeta(a+b)$$

$$+(-1)^{b+1}\binom{a+b-2}{b-1}F(a+b-1,1)$$
$$+\sum_{k=2}^{b}(-1)^{b-k}\binom{a+b-1-k}{a-1}\zeta(a+b-k)\zeta(k)$$
$$+(-1)^{b}\sum_{l=2}^{a}\binom{a+b-1-l}{b-1}F(l,a+b-l), \tag{3.84}$$

that is,

$$S_k=2\sum_{j=1}^{k}\binom{2k-1-j}{k-1}F(3k-j,j)-2\binom{2k-1}{k-1}\zeta(3k), \tag{3.85}$$

$$S_3=20\zeta(9)-12\zeta(7)\zeta(2). \tag{3.86}$$

Proof. To prove (3.84), let

$$\zeta_k(x)=\sum_{m=1}^{\infty}\frac{x^m}{m^k}\quad(k\in\mathbb{N}). \tag{3.87}$$

It is obvious that $\zeta_k(1)=\zeta(k)$. Further, it follows that

$$F_{(a,b)}(x)=\sum_{m=1}^{\infty}\frac{x^m}{m^a}\sum_{n=1}^{m}\frac{x^n}{n^b}\quad(0\leqslant x\leqslant 1, a,b\in N, a\geqslant 2). \tag{3.88}$$

Then

$$\begin{aligned}F_{(a,b)}(x)&=\zeta_{a+b}(x^2)+\sum_{m=2}^{\infty}\frac{x^m}{m^a}\sum_{n=1}^{m-1}\frac{x^n}{n^b}\\&=\zeta_{a+b}(x^2)+\sum_{n=1}^{\infty}\frac{x^n}{n^b}\sum_{m=n+1}^{\infty}\frac{x^m}{m^a}\\&=\zeta_{a+b}(x^2)+\sum_{m=1}^{\infty}\sum_{n=1}^{\infty}\frac{x^{m+2n}}{n^b(m+n)^a}.\end{aligned} \tag{3.89}$$

We determine the coefficients B_k and A_l in the following decomposition:

$$\frac{1}{y^b(m+y)^a}=\sum_{k=1}^{b}\frac{B_k}{y^k}+\sum_{l=1}^{a}\frac{A_l}{(m+y)^l}. \tag{3.90}$$

Relation (3.90) can also be written in the following form:

$$(m+y)^{-a}=\sum_{k=1}^{b}y^{b-k}B_k+\left(\sum_{l=1}^{a}(m+y)^{-l}A_l\right)y^b. \tag{3.91}$$

If

$$f(y) = \sum_{k=1}^{b} y^{b-k} B_k, \tag{3.92}$$

then

$$f^{(s)}(0) = s! B_{b-s} \quad (s = 0, 1, \ldots, b-1), \tag{3.93}$$

$$\left[(m+n)^{-a}\right]^{(s)} = (-1)^s a(a+1)\cdots(a+s-1)(m+y)^{-a-s}. \tag{3.94}$$

From these equations it follows that

$$B_k = (-1)^{b-k} \binom{a+b-1-k}{a-1} \frac{1}{m^{a+b-k}}. \tag{3.95}$$

By analogy, from

$$y^{-b} = \sum_{l=1}^{a} (m+y)^{a-l} A_l + \left(\sum_{k=1}^{b} y^{-k} B_k \right) (m+y)^a, \tag{3.96}$$

$$g(y) = \sum_{l=1}^{a} (m+y)^{a-l} A_l, \tag{3.97}$$

$$g^{(s)}(-m) = s! A_{a-s} \quad (s = 0, 1, \ldots, a-1) \tag{3.98}$$

it follows that

$$A_l = (-1)^b \binom{a+b-1-l}{b-1} \frac{1}{m^{a+b-l}}. \tag{3.99}$$

From the last four relations we obtain

$$\begin{aligned} F_{(a,b)}(x) &= \zeta_{a+b}(x^2) + \sum_{k=1}^{b} (-1)^{b-k} \binom{a+b-1-k}{a-1} \zeta_{a+b-k}(x) \zeta_k(x^2) \\ &\quad + \sum_{l=1}^{a} (-1)^b \binom{a+b-1-l}{b-1} \sum_{m=1}^{\infty} \frac{x^m}{m^{a+b-l}} \sum_{n=1}^{\infty} \frac{x^{2n}}{(m+n)^l} \quad (0 \leqslant x < 1). \end{aligned} \tag{3.100}$$

From formulas

$$\sum_{m=1}^{\infty} \frac{x^m}{m^{a+b-l}} \sum_{n=1}^{\infty} \frac{x^{2n}}{(m+n)^l} = \sum_{k=2}^{\infty} \frac{x^k}{k^l} \sum_{n=1}^{k-1} \frac{x^{k-n}}{n^{a+b-l}}, \tag{3.101}$$

$$\sum_{k=1}^{\infty} \frac{x^k}{k^l} \sum_{n=1}^{k} \frac{x^{k-n}}{n^{a+b-l}} = \zeta_{a+b}(x) + \sum_{k=2}^{\infty} \frac{x^k}{k^l} \sum_{n=1}^{k-1} \frac{x^{k-n}}{n^{a+b-l}} \tag{3.102}$$

and from (3.100) we obtain

$$F_{(a,b)}(x) = \zeta_{a+b}(x^2) + \sum_{k=1}^{b}(-1)^{b-k}\binom{a+b-1-k}{a-1}\zeta_{a+b-k}(x)\zeta_k(x^2)$$
$$+\sum_{l=1}^{a}(-1)^b\binom{a+b-1-l}{b-1}\left(\sum_{m=1}^{\infty}\frac{x^m}{m^l}\sum_{n=1}^{m}\frac{x^{m-n}}{n^{a+b-l}} - \zeta_{a+b}(x)\right) \quad (0 \leqslant x < 1). \quad (3.103)$$

Equality (3.103) can also be written in the form

$$F_{(a,b)}(x) = \zeta_{a+b}(x^2) + \sum_{k=2}^{b}(-1)^{b-k}\binom{a+b-1-k}{a-1}\zeta_{a+b-k}(x)\zeta_k(x^2)$$
$$+ (-1)^{b+1}\zeta_{a+b}(x)\sum_{l=2}^{a}\binom{a+b-1-l}{b-1}$$
$$+ (-1)^b\sum_{l=2}^{a}\binom{a+b-1-l}{b-1}\sum_{m=1}^{\infty}\frac{x^m}{m^l}\sum_{n=1}^{m}\frac{x^{m-n}}{n^{a+b-l}}$$
$$+ (-1)^b\binom{a+b-2}{b-1}\left(\sum_{m=1}^{\infty}\sum_{n=1}^{\infty}\frac{x^{m+2n}}{m^{a+b-1}(m+n)} - \zeta_{a+b-1}(x)\zeta_1(x^2)\right) \quad (0 \leqslant x < 1). \quad (3.104)$$

The expression in the last parentheses of equality (3.104) is equal to the following sum:

$$\sum_{m=1}^{\infty}\sum_{n=1}^{\infty}\left(\frac{x^{m+2n}}{m^{a+b-1}(m+n)} - \frac{x^{m+2n}}{m^{a+b-1}n}\right) = -\sum_{m=1}^{\infty}\sum_{n=1}^{\infty}\frac{x^{m+2n}}{m^{a+b-1}n(m+n)}. \quad (3.105)$$

As $x \to 1$, this expression tends to

$$-\sum_{m=1}^{\infty}\sum_{n=1}^{\infty}\frac{x^{m+2n}}{m^{a+b-2}n(m+n)} = -\sum_{m=1}^{\infty}\frac{1}{m^{a+b-2}}\sum_{n=1}^{\infty}\frac{1}{n(m+n)} = -F(a+b-1,1), \quad (3.106)$$

because

$$\sum_{n=1}^{\infty}\frac{1}{n(m+n)} = \frac{1}{m}\sum_{n=1}^{m}\frac{1}{n}. \quad (3.107)$$

From (3.106) it follows that (3.104) is satisfied for $x = 1$ as well. Let $C_{k,l} = \sum_{j=0}^{l}\binom{k+j}{j}$. Then

$$C_{k,l} = 1 + \sum_{j=0}^{l}\left[\binom{k+j-1}{j-1} + \binom{k+j-1}{j}\right] = C_{k,l} - \binom{k+l}{l} + C_{k-1,l}. \quad (3.108)$$

It follows that

$$\sum_{j=0}^{l}\binom{k+j}{j} = \binom{k+l+1}{k+1}, \quad (3.109)$$

and accordingly,

$$\sum_{l=2}^{a}\binom{a+b-1-l}{b-1}=\sum_{j=0}^{a-2}\binom{b-1+j}{j}=\binom{b+a-2}{a-2}. \tag{3.110}$$

Together with (3.104) and (3.106), this proves equality (3.84). □

Proof. To prove formula (3.85), it is clear from the expression for S_k that

$$S_k=\sum_{r=2}^{\infty}\frac{1}{r^k}\sum_{n=1}^{r-1}\frac{1}{n^k(r-n)^k}. \tag{3.111}$$

In exactly the same way as it was done in (3.90), (3.95), and (3.99), we arrive at the following decomposition:

$$\frac{1}{n^k(r-n)^k}=\sum_{j=1}^{k}\binom{2k-1-j}{k-1}r^{-2k+j}n^{-j}+\sum_{j=1}^{k}\binom{2k-1-j}{k-1}r^{-2k+j}(r-n)^{-j}. \tag{3.112}$$

In this way, we get

$$\begin{aligned}S_k&=\sum_{j=1}^{k}\binom{2k-1-j}{k-1}\sum_{r=2}^{\infty}\frac{1}{r^k}r^{-2k+j}\sum_{n=1}^{r-1}(n^{-j}+(r-n)^{-j})\\&=2\sum_{j=1}^{k}\binom{2k-1-j}{k-1}\sum_{r=2}^{\infty}\frac{1}{r^k}r^{-2k+j}\sum_{n=1}^{r-1}\frac{1}{n^j}\\&=2\sum_{j=1}^{k}\binom{2k-1-j}{k-1}\left(\sum_{r=1}^{\infty}\frac{1}{r^{3k-j}}\sum_{n=1}^{r}\frac{1}{n^j}-\zeta(3k)\right).\end{aligned} \tag{3.113}$$

Since

$$\sum_{j=1}^{k}\binom{2k-1-j}{k-1}=\sum_{l=0}^{k-1}\binom{k-1+l}{l}\overset{(3.109)}{=}\binom{2k-1}{k-1}, \tag{3.114}$$

by (3.83) we obtain (3.85). □

Proof. To prove formula (3.86), from (3.85) it follows that

$$S_3=12F(8,1)+6F(7,2)+2F(6,3)-20\zeta(9). \tag{3.115}$$

Using the relation

$$F(p,q)+F(q,p)=\zeta(p)\zeta(q)+\zeta(p+q), \tag{3.116}$$

we obtain

$$S_3 = 2\zeta(6)\zeta(3) + 6\zeta(7)\zeta(2) - 12\zeta(9) - 2F(3,6) - 6F(2,7) + 12F(8,1). \tag{3.117}$$

Taking $a = 2$ and $b = 7$ in (3.84), it follows that

$$2F(2,7) = 2\zeta(9) - 5\zeta(7)\zeta(2) + 3\zeta(6)\zeta(3) - \zeta(5)\zeta(4) + 7F(8,1). \tag{3.118}$$

From (10.4) it follows that for $a = 5$ and $b = 4$,

$$4F(4,5) = 34\zeta(9) - 15\zeta(7)\zeta(2) + 5\zeta(6)\zeta(3) - \zeta(5)\zeta(4) + 20F(2,7) - 10F(3,6) + 35F(8,1). \tag{3.119}$$

From (3.116) and (3.119) it follows that

$$4F(5,4) = -30\zeta(9) + 15\zeta(7)\zeta(2) - 5\zeta(6)\zeta(3) + 5\zeta(5)\zeta(4) - 20F(2,7) + 10F(3,6) - 35F(8,1). \tag{3.120}$$

From (3.84), for $a = 6$ and $b = 3$, we obtain

$$2F(6,3) = 36\zeta(9) - 6\zeta(7)\zeta(2) + \zeta(6)\zeta(3) - 15F(2,7) - 10F(3,6) - 6F(4,5) - 3F(5,4) + 21F(8,1). \tag{3.121}$$

Substituting the derived values for $F(2,7)$, $F(4,5)$, and $F(5,4)$ into (3.121) and using (3.116) with $p = 3$ and $q = 6$, we get

$$2F(3,6) = 22\zeta(9) + 21\zeta(7)\zeta(2) - 19\zeta(6)\zeta(3) - 9\zeta(5)\zeta(4) - 21F(8,1). \tag{3.122}$$

From the formula

$$2\sum_{r=1}^{\infty} \frac{1}{r^a} \sum_{k=1}^{r} \frac{1}{k} = (a+2)\zeta(a+1) - \sum_{j=1}^{a-2} \zeta(a-j)\zeta(j+1) \quad (a \in \mathbb{N},\ a \geqslant 2), \tag{3.123}$$

and from the results obtained in [96] it follows that

$$F(8,1) = 5\zeta(9) - \zeta(7)\zeta(2) - \zeta(6)\zeta(3) - \zeta(5)\zeta(4). \tag{3.124}$$

Formula (3.86) follows directly from (3.117), (3.122), (3.118), and (3.124), thus proving the theorem. □

The following two formulas can be obtained directly from the Theorem 3.1:

$$2\sum_{m=1}^{\infty} \frac{1}{m^3} \sum_{n=1}^{m} \frac{1}{n^2} = 6\zeta(3)\zeta(2) - 9\zeta(5), \tag{3.125}$$

$$3\sum_{m=1}^{\infty} \frac{1}{m^4} \sum_{n=1}^{m} \frac{1}{n^2} = 3\zeta^2(3) - \zeta(6). \tag{3.126}$$

We further derive formulas similar to those given in Theorem 3.1. Let

$$G_{(a,b)}(x) = \sum_{m=1}^{\infty} \frac{x^m}{m^a} \sum_{n=1}^{m} \frac{1}{n^b} \quad (a, b \in \mathbb{N},\ 0 \leqslant x < 1). \tag{3.127}$$

Then

$$\begin{aligned} G_{(a,b)}(x) &= \zeta_{a+b}(x) + \sum_{k=1}^{b} (-1)^{b-k} \binom{a+b-1-k}{a-1} \zeta_{a+b-k}(x) \zeta_k(x) \\ &\quad + \sum_{l=1}^{a} (-1)^b \binom{a+b-1-l}{b-1} \sum_{m=1}^{\infty} \frac{x^m}{m^{a+b-l}} \sum_{n=1}^{\infty} \frac{x^n}{(m+n)^l}, \end{aligned} \tag{3.128}$$

that is,

$$\begin{aligned} G_{(a,b)}(x) &= \zeta_{a+b}(x) + \sum_{k=1}^{b} (-1)^{b-k} \binom{a+b-1-k}{a-1} \zeta_{a+b-k}(x) \zeta_k(x) \\ &\quad + \left[1 + (-1)^{b+1} \binom{a+b-1}{b} \right] \zeta_{a+b}(x) \\ &\quad + (-1)^b \sum_{l=1}^{a} \binom{a+b-1-l}{b-1} G_{(l,a+b-l)}(x). \end{aligned} \tag{3.129}$$

From this, for $a = 1$ and $b = 2k - 1$, we get

$$\sum_{m=1}^{\infty} \frac{x^m}{m} \sum_{n=1}^{m} \frac{1}{n^{2k-1}} = \zeta_{2k}(x) - \frac{(-1)^k}{2} \zeta_k^2(x) - \sum_{r=1}^{k-1} (-1)^r \zeta_r(x) \zeta_{2k-r}(x) \quad (0 \leqslant x < 1). \tag{3.130}$$

Let us introduce the function

$$H_{(a,b)}(x) = \sum_{m=1}^{\infty} \frac{1}{m^a} \sum_{n=1}^{m} \frac{x^n}{n^b} \quad (a, b \in \mathbb{N},\ 0 \leqslant x < 1). \tag{3.131}$$

From (3.129) it follows that

$$H_{(a,b)}(x) + G_{(b,a)}(x) = \zeta_{a+b}(x) + \zeta(a)\zeta_b(x), \tag{3.132}$$

$$\begin{aligned} H_{(a,b)}(x) &= \zeta(a)\zeta_b(x) + (-1)^{a+1} \sum_{l=1}^{a} (-1)^l \binom{a+b-1-l}{b-1} \zeta_{a+b-l}(x) \zeta_l(x) \\ &\quad + (-1)^{a+1} \sum_{k=1}^{b} \binom{a+b-1-k}{a-1} \zeta(a+b-k) \zeta_k(x) \\ &\quad + (-1^a) \sum_{k=1}^{b} \binom{a+b-1-k}{a-1} H_{(a+b-k,k)}(x) \quad (0 \leqslant x < 1). \end{aligned} \tag{3.133}$$

For $a = 2k - 1$ and $b = 1$, we have

$$\sum_{m=1}^{\infty} \frac{1}{m^{2k-1}} \sum_{n=1}^{m} \frac{x^n}{n} = [\zeta(2k-1) - \zeta_{2k-1}(x)]\zeta_1(x) + \frac{(-1)^k}{2}\zeta_k^2(x) + \sum_{r=2}^{k-1}(-1)^r \zeta_r(x)\zeta_{2k-r}(x)$$
$$(0 \leqslant x < 1). \tag{3.134}$$

It is worth mentioning that

$$\zeta_1(x) = -\log(1-x), \quad \zeta_k'(x) = \frac{\zeta_{k-1}(x)}{x} \quad (k \in \mathbb{N},\ k \geqslant 2). \tag{3.135}$$

Based on the above, we also have

$$\lim_{x \uparrow 1} \zeta_1(x) \cdot [\zeta(2k-1) - \zeta_{2k-1}(x)] = 0. \tag{3.136}$$

From this equation it follows that

$$\sum_{m=1}^{\infty} \frac{1}{m^{2k-1}} \sum_{n=1}^{m} \frac{1}{n} = \frac{(-1)^k}{2}\zeta^2(k) + \sum_{r=2}^{k-1}(-1)^r \zeta_r(x)\zeta(2k-r), \tag{3.137}$$

so, in accordance with (3.116), (3.118), (3.120), (3.122), and (3.124), we obtain the following formulas:

$$2\sum_{m=1}^{\infty} \frac{1}{m^7} \sum_{n=1}^{m} \frac{1}{n^2} = -35\zeta(9) + 14\zeta(7)\zeta(2) + 4\zeta(6)\zeta(3) + 8\zeta(5)\zeta(4), \tag{3.138}$$

$$2\sum_{m=1}^{\infty} \frac{1}{m^6} \sum_{n=1}^{m} \frac{1}{n^3} = 85\zeta(9) - 42\zeta(7)\zeta(2) - 12\zeta(5)\zeta(4), \tag{3.139}$$

$$2\sum_{m=1}^{\infty} \frac{1}{m^5} \sum_{n=1}^{m} \frac{1}{n^4} = -125\zeta(9) + 70\zeta(7)\zeta(2) + 10\zeta(5)\zeta(4). \tag{3.140}$$

It is clear from the last formulas that Theorem 3.1 allows us to recurrently calculate the functions $F(a, b)$. *Subbarao* and *Sitaramachandrarao* [189] obtained the following results:

$$S_{2k} = \frac{(-1)^k \cdot 2^{6k+1} \cdot k \cdot \pi^{6k}}{3 \cdot [(2k)!]^2} \sum_{r=0}^{k} \binom{2k}{2r} \frac{(4k-2r-1)!}{(6k-2r)!} B_{2r}B_{6k-2r}, \tag{3.141}$$

$$S_{2k} = \frac{8k}{3 \cdot [(2k)!]^2} \sum_{r=0}^{k} \binom{2k}{2r} (2r)!(4k-2r-1)!\zeta(2r)\zeta(6k-2r). \tag{3.142}$$

The last equation is a direct consequence of (1.297). In this way the rational coefficients p_{2k} in the expression $S_{2k} = \pi^{6k} p_{2k}$ are determined. At the same time, the authors of the aforementioned paper raised the question of calculating the sums of the forms

$$\sum_{m=1}^{\infty}\sum_{n=1}^{\infty}\frac{(-1)^{m-1}}{m^k n^k (m+n)^k},\quad \sum_{m=1}^{\infty}\sum_{n=1}^{\infty}\frac{(-1)^{m+n}}{m^k n^k (m+n)^k}. \tag{3.143}$$

They themselves calculated both sums for $k = 1$:

$$\sum_{m=1}^{\infty}\sum_{n=1}^{\infty}\frac{(-1)^{m-1}}{mn(m+n)} = \frac{5}{8}\zeta(3), \tag{3.144}$$

$$\sum_{m=1}^{\infty}\sum_{n=1}^{\infty}\frac{(-1)^{m-1}}{mn(m+n)} = \frac{1}{4}\zeta(3). \tag{3.145}$$

Let

$$T_k = \sum_{m=1}^{\infty}\sum_{n=1}^{\infty}\frac{(-1)^{m+n}}{m^k n^k (m+n)^k}. \tag{3.146}$$

In exactly the same way as it was done using (3.111), (3.112) and (3.113) in the derivation of (3.85), we can also derive the formula

$$T_k = \sum_{j=1}^{k}\binom{2k-1-j}{k-1}G_{(3k-j,j)}(-1) - 2\binom{2k-1}{k-1}\zeta_{3k}(-1). \tag{3.147}$$

Since

$$S_1 = 2\zeta(3) = \sum_{m=1}^{\infty}\sum_{n=1}^{\infty}\frac{1}{mn(m+n)} = \sum_{m=1}^{\infty}\frac{1}{m^2}\sum_{n=1}^{\infty}\frac{1}{n}, \tag{3.148}$$

we have that for $a \in \mathbb{N}$, $a \geqslant 2$, and $|x| \leqslant 1$,

$$\begin{aligned}|G_{(a,b)}(x)| &\leqslant \sum_{m=1}^{\infty}\frac{|x|^m}{m^a}\sum_{n=1}^{m}\frac{1}{n^b} \leqslant \sum_{m=1}^{\infty}\frac{|x|^m}{m^a}\sum_{n=1}^{m}\frac{1}{n}\\ &\leqslant \sum_{m=1}^{\infty}\frac{1}{m^2}\sum_{n=1}^{m}\frac{1}{n} = 2\zeta(3),\end{aligned} \tag{3.149}$$

where the series $G_{(a,b)}(x)$ absolutely converges. For $a = 1$ and $|x| < 1$, the following series absolutely converges:

$$|G_{(1,b)}(x)| \leqslant \sum_{m=1}^{\infty}\frac{|x|^m}{m}\sum_{n=1}^{m}\frac{1}{n^b} < \sum_{m=1}^{\infty}|x|^m = \frac{|x|}{1-|x|}. \tag{3.150}$$

Since

$$2\zeta(3) = \sum_{m=1}^{\infty}\frac{1}{m^2}\sum_{n=1}^{m}\frac{1}{n} = \sum_{m=1}^{\infty}\frac{1}{m}\sum_{n=m}^{\infty}\frac{1}{n^2}, \tag{3.151}$$

we have

$$\sum_{m=1}^{\infty}\frac{1}{m}\sum_{n=m}^{m}\frac{1}{n^b}\leqslant 2\zeta(3)\quad(b\geqslant 2). \tag{3.152}$$

Based on

$$\begin{aligned}G_{(1,b)}(x) &= \sum_{m=1}^{\infty}\frac{x^m}{m}\sum_{n=m}^{m}\frac{1}{n^b} = x+\sum_{m=1}^{\infty}\frac{x^m}{m}\sum_{n=m}^{m}\frac{1}{n^b}\\ &= \zeta_{b+1}(x)+\sum_{m=2}^{\infty}\frac{x^m}{m}\sum_{n=m}^{m-1}\frac{1}{n^b}\\ &= \zeta_{b+1}(x)+\sum_{m=2}^{\infty}\frac{x^m}{m}\left(\zeta(b)-\sum_{n=m}^{\infty}\frac{1}{n^b}\right)\\ &= \zeta_{b+1}(x)+\zeta(b)\zeta_1(x)-\sum_{m=1}^{\infty}\frac{x^m}{m}\sum_{n=m}^{\infty}\frac{1}{n^b}\quad(|x|<1)\end{aligned} \tag{3.153}$$

and on

$$\left|\sum_{m=1}^{\infty}\frac{x^m}{m}\sum_{n=m}^{\infty}\frac{1}{n^b}\right|\leqslant 2\zeta(3)\quad(|x|\leqslant 1), \tag{3.154}$$

we can conclude that the following limits exist:

$$\begin{aligned}\lim_{x\uparrow 1}(\zeta(b)\zeta_1(x)-G_{(1,b)}(x)) &= \sum_{m=1}^{\infty}\frac{1}{m}\sum_{n=m+1}^{\infty}\frac{1}{n^b}=\sum_{n=2}^{\infty}\frac{1}{n^b}\sum_{m=1}^{\infty}\frac{1}{m}=\sum_{n=1}^{\infty}\frac{1}{n^b}\sum_{m=1}^{\infty}\frac{1}{m}-\zeta(b+1)\\ &= F(b,1)-\zeta(b+1).\end{aligned} \tag{3.155}$$

The last relation with $\zeta_1(x) = -\log(1-x)$, can be written as

$$\lim_{x\uparrow 1}(\zeta(b)\zeta_1(x)+G_{(1,b)}(x)) = \zeta(b+1)-F(b,1). \tag{3.156}$$

From (3.153) it is clear that $G_{(1,b)}(-1)$ exists and can be expressed in an appropriate form. Since the series in (3.129) absolutely converge for $a\geqslant 2$, $|x|\leqslant 1$ and for $a=1$, $|x|\leqslant b<1$, and since $G_{(1,b)}(-1)$ exists, the following relation (similar to relation (3.84)) follows from (3.129):

$$\begin{aligned}G_{(a,b)}(-1) &= \left[1+(-1)^{b+1}\binom{a+b-1}{b}\right]\zeta_{a+b}(-1)\\ &\quad+(-1)^b\sum_{l=1}^{a}\binom{a+b-1-l}{b-1}G_{(l,a+b-l)}(-1)\\ &\quad+\sum_{k=1}^{b}(-1)^{b-k}\binom{a+b-1-k}{a-1}\zeta_{a+b-k}(-1)\zeta_k(-1).\end{aligned} \tag{3.157}$$

By (3.157) for various values of a, b from (3.147) it follows that

$$\begin{aligned} 3T_2 &= 2\zeta_6(-1) + \zeta_3^2(-1) - 2G_{(3,3)}(-1) \\ &= -\frac{31}{16}\zeta(6) + \frac{9}{16}\zeta^2(3) - 2G_{(3,3)}(-1). \end{aligned} \tag{3.158}$$

3.4 Some results regarding the function $\zeta_k(x)$

3.4.1 Some of the links between the functions $\zeta_k(x)$ and $G_{(a,b)}(x)$ and some sums

Theorem 3.2. *We have the following relations:*

$$\begin{aligned} \zeta_3(1-x) &= \left[\frac{\pi^2}{6} - \zeta_2(x)\right]\log(1-x) - \frac{1}{2}\log x \log^2(1-x) \\ &\quad - \frac{1}{2}\sum_{m=1}^{\infty}\sum_{n=1}^{\infty}\frac{x^{m+n}}{mn(m+n)} + \zeta(3) \quad (0 \leqslant x \leqslant 1), \end{aligned} \tag{3.159}$$

$$\begin{aligned} \zeta_3(1-x) &= \zeta_2(x-1)\log(1-x) + \frac{1}{2}\log^2(1-x)\log(2-x) \\ &\quad + \frac{1}{2}\sum_{m=1}^{\infty}\sum_{n=1}^{\infty}\sum_{p=1}^{\infty}\frac{(\frac{1}{2})^p x^{m+n+p}}{mn(m+n+p)} - \frac{3}{4}\zeta(3) \quad (0 \leqslant x \leqslant 1), \end{aligned} \tag{3.160}$$

$$\begin{aligned} G_{(1,2)}(x) &= \zeta_3(x) + 2\zeta_3(1-x) - 2\zeta(3) \\ &\quad + \left[\log x \cdot \log(1-x) + \zeta_2(x) - \frac{\pi^2}{3}\right]\log(1-x) \quad (0 \leqslant x \leqslant 1), \end{aligned} \tag{3.161}$$

$$\begin{aligned} G_{(1,2)}(x) &= \zeta_3(x) + 2\zeta_3(1-x) - 2\zeta(3) \\ &\quad - \left[\frac{\pi^2}{6} + \zeta_2(1-x)\right]\log(1-x) \quad (0 \leqslant x \leqslant 1), \end{aligned} \tag{3.162}$$

$$\begin{aligned} G_{(2,1)}(x) &= \zeta_3(x) - \zeta_3(1-x) + \left[\frac{\pi^2}{6} - \zeta_2(x)\right]\log(1-x) \\ &\quad + \zeta(3) - \frac{1}{2}\log x \cdot \log^2(1-x) \quad (0 \leqslant x \leqslant 1), \end{aligned} \tag{3.163}$$

$$\sum_{m=1}^{\infty}\frac{(\frac{1}{2})^m}{m^2}\sum_{n=1}^{m}\frac{1}{n} = \zeta(3) - \frac{\pi^2}{12}\log 2, \tag{3.164}$$

$$\sum_{m=1}^{\infty}\frac{(\frac{1}{2})^m}{m}\sum_{n=1}^{m}\frac{1}{n^2} = \frac{5}{8}\zeta(3), \tag{3.165}$$

$$\sum_{m=1}^{\infty}\sum_{n=1}^{\infty}\sum_{p=1}^{\infty}\frac{(\frac{1}{2})^p}{mn(m+n+p)} = \frac{3}{2}\zeta(3). \tag{3.166}$$

Proof. For $0 < x < 1$, we have

$$\begin{aligned} \zeta_1(x) &= -\log(1-x), \quad \zeta_{k+1}'(x) = \frac{1}{x}\zeta_k(x), \\ [\zeta_{k+1}(1-x)]_x' &= -\frac{1}{1-x}\zeta_k(1-x). \end{aligned} \tag{3.167}$$

Let

$$\begin{aligned} Z(x) &= \zeta_3(1-x) - \left[\frac{\pi^2}{6} + \zeta_2(1-x)\right]\log(1-x) + \frac{1}{2}\log x \cdot \log^2(1-x) \\ &+ \frac{1}{2}\sum_{m=1}^{\infty}\sum_{n=1}^{\infty}\frac{x^{m+n}}{mn(m+n)} \quad (0 < x < 1). \end{aligned} \tag{3.168}$$

We have the following equation:

$$\zeta_2(x) + \zeta_2(1-x) = -\log x \cdot \log(1-x) + \frac{\pi^2}{6} \quad (0 < x < 1), \tag{3.169}$$

which can be easily verified by differentiation using (3.167). From (3.168) by differentiation, together with (3.167) and (3.169), it follows that $z'(x) = 0$, i. e., $z(x) = C$. Since $\lim_{x\downarrow 0}\log x \cdot \log^2(1-x) = 0$, we have $C = -\zeta(3)$. So (3.159) follows in the region $0 \leqslant x < 1$. Since

$$\lim_{x\uparrow 1}\left[\frac{\pi^2}{6} - \zeta_2(x)\right]\log(1-x) = 0, \quad \lim_{x\uparrow 1}\log x \cdot \log^2(1-x) = 0, \tag{3.170}$$

$$\sum_{m=1}^{\infty}\sum_{n=1}^{\infty}\frac{1}{mn(m+n)} = 2\zeta(3), \tag{3.171}$$

we conclude that (3.159) is also true for $x = 1$. Since [48, 65]

$$\zeta_2\left(\frac{1}{2}\right) = -\frac{1}{2}\log^2 2 + \frac{\pi^2}{12}, \tag{3.172}$$

$$\zeta_3\left(\frac{1}{2}\right) = \frac{1}{6}\log^3 2 - \frac{\pi^2}{12}\log 2 + \frac{7}{8}\zeta(3), \tag{3.173}$$

taking $x = \frac{1}{2}$ in (3.159), we obtain

$$\sum_{m=1}^{\infty}\sum_{n=1}^{\infty}\frac{(\frac{1}{2})^{m+n}}{mn(m+n)} = \frac{1}{4}\zeta(3) - \frac{1}{3}\log^3 2. \tag{3.174}$$

Now (3.164) follows by the relation

$$\sum_{m=1}^{\infty}\sum_{n=1}^{\infty}\sum_{n=1}^{\infty}\frac{(x)^{m+n}}{mn(m+n)} = 2G_{(2,1)}(x) - 2\zeta_3(x) \tag{3.175}$$

and (3.174), whereas (3.163) is derived directly from (3.175) and (3.159).

Since

$$2G_{(2,1)}(x) + G_{(1,2)}(x) = 3\zeta_3(x) + \zeta_1(x)\zeta_2(x) \quad (0 \leqslant x < 1), \tag{3.176}$$

from (3.163) and (3.169) we obtain formulas (3.161) and (3.162). Based on (3.176) and (3.164) and the expressions for $\zeta_2(\frac{1}{2})$ and $\zeta_3(\frac{1}{2})$, we obtain (3.165). To prove formula (3.160), We need the following formulas for $0 < x < 1$:

$$\begin{aligned} \zeta_k(x) + \zeta_k(-x) &= 2^{1-k}\zeta_k(x^2), \\ \zeta'_{k+1}(x-1) &= \frac{1}{x-1}\zeta_k(x-1), \\ \zeta_1(x-1) &= -\log(2-x), \end{aligned} \tag{3.177}$$

$$\left[\zeta_3(x-1) - \zeta_2(x-1)\log(1-x) - \frac{1}{2}\log^2(1-x)\log(2-x)\right]' = \frac{1}{2}\frac{\log^2(1-x)}{2-x}, \tag{3.178}$$

$$\left[\sum_{m=1}^{\infty}\sum_{n=1}^{\infty}\sum_{p=1}^{\infty}\frac{(\frac{1}{2})^{m+n+p}}{mn(m+n+p)}\right]' = \frac{\log^2(1-x)}{2-x}. \tag{3.179}$$

It follows that

$$\begin{aligned} \zeta_3(x-1) &= \zeta_2(x-1)\log(1-x) + \frac{1}{2}\log^2(1-x)\log(2-x) \\ &\quad + \frac{1}{2}\sum_{m=1}^{\infty}\sum_{n=1}^{\infty}\sum_{p=1}^{\infty}\frac{(\frac{1}{2})^{m+n+p}}{mn(m+n+p)} + C. \end{aligned} \tag{3.180}$$

Since $\zeta_3(-1) = -\frac{3}{4}\zeta(3)$, by taking $x = 0$ in (3.180), we get that $C = -\frac{3}{4}\zeta(3)$, thus proving (3.160) for $0 \leqslant x < 1$. Since

$$\lim_{x\uparrow 1}\zeta_2(x-1)\log(1-x) = 0, \quad \lim_{x\uparrow 1}\log^2(1-x)\log(2-x) = 0, \tag{3.181}$$

we get (3.160) for $x = 1$. Finally, (3.166) follows from (3.160) for $x = 1$. □

We include further information about the function $\zeta_k(x)$ in the following theorem.

Theorem 3.3. *For $a, b \in \mathbb{N}$, $a = b + 2c$, $c \geqslant 0$, $0 \leqslant x < 1$, we have the following equations:*

$$\int_0^x \frac{\zeta_a(t)\zeta_{b-1}(t)}{t}\,\partial t = \sum_{k=0}^{c-1}(-1)^k\zeta_{a-k}(x)\zeta_{b+k}(x) + \frac{(-1)^c}{2}\zeta_{a-c}^2(x), \tag{3.182}$$

$$\begin{aligned} &\sum_{r=1}^{a}\binom{a+b-2-r}{b-2}G_{(a+b-r,r)}(x) + \sum_{s=1}^{b-1}\binom{a+b-2-s}{a-1}G_{(a+b-s,s)}(x) \\ &\quad = \sum_{k=0}^{c-1}(-1)^k\zeta_{a-k}(x)\zeta_{b+k}(x) + \frac{(-1)^c}{2}\zeta_{a-c}^2(x) + \binom{a+b-1}{a}\zeta_{a+b}(x), \end{aligned} \tag{3.183}$$

$$\sum_{r=1}^{b+2c+2}\binom{2b+2c-r}{b-2}G_{(2b+2c-r,r)}(x)+\sum_{s=1}^{b-1}\binom{2b+2c-s}{b+2c+1}G_{(2b+2c-s,s)}(x)$$
$$=\binom{2b+2c+1}{b+2c+2}\zeta_{(2b+2c)}(x)+\zeta_{b+2c+1}(x)\zeta_{b-1}(x)+\zeta_{b+2c+2}(x)\zeta_{b-2}(x), \tag{3.184}$$

$$\sum_{r=1}^{b-1}\binom{2b-1-r}{b-1}G_{(2b-r,r)}(x)+G_{(b,b)}(x)=\binom{2b-1}{b}\zeta_{2b}(x)+\frac{1}{2}\zeta_b^2(x). \tag{3.185}$$

Proof. For $a=b$,

$$\int_0^x \frac{1}{t}\zeta_a(t)\zeta_{b-1}(t)\,\partial t=\int_0^x \zeta_a(t)\zeta_a'(t)\,\partial t=\frac{1}{2}\zeta_a^2(x), \tag{3.186}$$

and for $c\geqslant 1$,

$$\int_0^x \frac{1}{t}\zeta_a(t)\zeta_{b-1}(t)\,\partial t=\int_0^x \zeta_a(t)\zeta_b'(t)\,\partial t=\zeta_a(x)\zeta_b(x)-\int_0^x \zeta_a'(t)\zeta_b(t)\,\partial t$$
$$=-\int_0^x \frac{1}{t}\zeta_{a-1}(t)\zeta_{b+1}'(t)\,\partial t+\zeta_a(x)\zeta_b(x). \tag{3.187}$$

Going further, we get (3.182). On the other hand, we have

$$\int_0^x \zeta_a(t)\zeta_b'(t)\,\partial t=\sum_{m=1}^{\infty}\sum_{n=1}^{\infty}\frac{x^{m+n}}{m^a n^{b-1}(m+n)}=\sum_{k=2}^{\infty}\frac{x^k}{k}\sum_{m=1}^{k-1}\frac{1}{m^a(k-m)^{b-1}}. \tag{3.188}$$

In the same way as we did for (3.90), we can prove that

$$\frac{1}{m^a(k-m)^{b-1}}=\sum_{r=1}^{a}\frac{1}{m^r}\binom{a+b-2-r}{b-2}\frac{1}{k^{a+b-1-r}}$$
$$+\sum_{s=1}^{b-1}\frac{1}{(k-m)^s}\binom{a+b-2-s}{a-1}\frac{1}{k^{a+b-1-s}}. \tag{3.189}$$

By (3.188) and (3.189) and the formula

$$\sum_{k=2}^{\infty}\frac{x^k}{k^{a+b-r}}\sum_{m=1}^{k-1}\frac{1}{m^r}=G_{(a+b-r,r)}(x)-\zeta_{a+b}(x) \tag{3.190}$$

(3.183) follows. By differentiating (3.183) and using the equality

$$G'_{(a+b-r,r)}(x)=\frac{1}{x}G_{(a+b-1-r,r)}(x), \tag{3.191}$$

after rearranging, we obtain (3.184). Taking $a=b$ and $c=0$ in (3.183), we get (3.185). □

Proposition 3.1. *Relations* (3.183), (3.184), *and* (3.185) *are also valid for* $x = -1$.

From (3.130), which also holds for $x = -1$, it follows that

$$G_{(1,2k-1)}(-1) = \zeta_{2k}(-1) - \frac{(-1)^k}{2}\zeta_k^2(-1) - \sum_{r=1}^{k-1}(-1)^r\zeta_r(-1)\zeta_{2k-r}(-1), \tag{3.192}$$

whereas from (3.134) (valid for $x = -1$) it follows that

$$H_{(2k-1,1)}(-1) = [\zeta_{2k-1}(-1)-\zeta(2k-1)]\log 2++\frac{(-1)^k}{2}\zeta_k^2(-1)+\sum_{r=1}^{k-1}(-1)^r\zeta_r(-1)\zeta_{2k-r}(-1). \tag{3.193}$$

For $k = 1$,

$$H_{(0,1)}(-1) = \sum_{m=1}^{\infty}\frac{1}{m^b}\sum_{n=1}^{m}\frac{(-1)^n}{n} = \frac{\pi^4}{288} - \frac{7}{4}\zeta(3)\log 2. \tag{3.194}$$

3.4.2 Some formulas for $\zeta(3)$

We will present some formulas in which $\zeta(3)$ and *Catalan*'s constant

$$G = \sum_{k=0}^{\infty}\frac{(-1)^k}{(2k+1)^2} \tag{3.195}$$

appear. The formulas are given by the following theorem.

Theorem 3.4. *We have*

$$(\pi-\alpha)\sum_{m=1}^{\infty}\frac{\sin m\alpha}{m^2} - 3\sum_{m=1}^{\infty}\frac{\cos m\alpha}{m^3} + 2\sum_{m=1}^{\infty}\frac{\cos m\alpha}{m^2}\sum_{n=1}^{m}\frac{1}{n} = \zeta(3) \quad (0 \leqslant \alpha \leqslant \pi), \tag{3.196}$$

$$\sum_{m=1}^{\infty}\frac{\cos\frac{m\pi}{2}}{m^2}\sum_{n=1}^{m}\frac{1}{n} = \frac{23}{64}\zeta(3) - \frac{\pi}{4}G, \tag{3.197}$$

$$\sum_{m=1}^{\infty}\frac{\cos\frac{m\pi}{3}}{m^2}\sum_{n=1}^{m}\frac{1}{n} = \zeta(3) + \frac{\pi^3\sqrt{3}}{27} - \frac{\pi\sqrt{3}}{18}\Psi'\left(\frac{1}{3}\right) \tag{3.198}$$

(where $\Psi(x) = \frac{\Gamma'(x)}{\Gamma(x)}$, $x > 0$*)*,

$$\sum_{m=1}^{\infty}\frac{\cos\frac{m\pi}{3}}{m^3} = \frac{\zeta(3)}{3}, \tag{3.199}$$

$$\sum_{m=1}^{\infty}\frac{(-1)^m}{m^2}\sum_{n=1}^{2m}\frac{1}{n} = \frac{23}{16}\zeta(3) - \pi G, \tag{3.200}$$

$$\sum_{m=1}^{\infty}\frac{1}{m^2}\sum_{n=1}^{4m}\frac{1}{n}=\frac{67}{8}\zeta(3)-2\pi G. \tag{3.201}$$

Proof. Let $f(s)=\frac{\log^2(1-s)}{s}$. We will integrate this function along the contour given in Figure 3.1 (where $0<\alpha<\pi$).

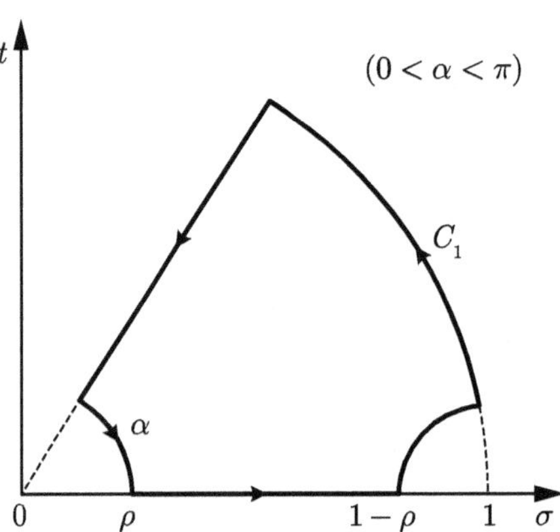

Figure 3.1: Integration contour for the function $f(s)=\frac{\log^2(1-s)}{s}$.

By *Cauchy*'s formula

$$\int_{C_1} f(s)\,\partial s=0, \tag{3.202}$$

and thus the integral

$$I_\rho^{(0)}=\int_{\alpha}^{0}\frac{\log^2(1-\rho e^{\varphi i})}{\rho e^{\varphi i}}i\rho e^{\varphi i}\,\partial\varphi\longrightarrow 0\quad(\rho\longrightarrow 0), \tag{3.203}$$

For the similar integral $I_\rho^{(1)}$, after replacing $s=1-\rho e^{\varphi i}$, we get

$$I_\rho^{(1)}=-\int_{\pi}^{\theta}\frac{\log^2(\rho e^{\varphi i})}{1-\rho e^{\varphi i}}\rho i e^{\varphi i}\,\partial\varphi\longrightarrow 0\quad(\rho\longrightarrow 0), \tag{3.204}$$

since

$$\lim_{\rho\downarrow 0}\rho\log\rho=\lim_{\rho\downarrow 0}\rho\log^2\rho=0. \tag{3.205}$$

Passing to the limits as $\rho \to 0$, we get

$$0 = \int_0^1 \frac{\log^2(1-x)}{x}\,\partial x + \int_0^a \left(\log\left(2\sin\frac{x}{2}\right) + \frac{x-\pi}{2}\mathrm{i}\right)^2 \partial x + \int_1^0 \frac{\log^2(1-xe^{a\mathrm{i}})}{x}\,\partial x = I_1 + I_2 + I_3. \tag{3.206}$$

In this case, we have

$$I_1 = \int_0^1 \frac{\log^2(1-x)}{x}\,\partial x = \int_0^1 \frac{\log^2 x}{1-x}\,\partial x = \sum_{m=1}^{\infty} \int_0^1 t^{m-1}\log^2 t\,\partial t = 2\zeta(3), \tag{3.207}$$

$$\mathfrak{Re}(I_2) = \int_0^a (\pi - x)\log\left(2\sin\frac{x}{2}\right)\partial x. \tag{3.208}$$

Since [46]

$$\log\left(2\sin\frac{x}{2}\right) = -\sum_{m=1}^{\infty} \frac{\cos mx}{m} \quad (0 < x < 2\pi), \tag{3.209}$$

from (3.208) it follows that

$$\begin{aligned}\mathfrak{Re}(I_2) &= -\sum_{m=1}^{\infty} \frac{1}{m}\int_0^a (\pi - x)\cos mx\,\partial x \\ &= -(\pi - a)\sum_{m=1}^{\infty}\frac{\sin ma}{m^2} - \zeta(3) + \sum_{m=1}^{\infty}\frac{\cos ma}{m^3}.\end{aligned} \tag{3.210}$$

The summation and integration signs in this relation can be interchanged, since the corresponding *Fourier* series is multiplied by $\pi - x$ [149]. Since

$$\log(1 - xe^{a\mathrm{i}}) = \frac{1}{2}\log(1 - 2x\cos a + x^2) - \mathrm{i}\arctan\frac{x\sin a}{1 - x\cos a}, \tag{3.211}$$

from (3.206) it follows that

$$\mathfrak{Re}(I_3) = -\frac{1}{4}\int_0^1 \frac{1}{x}\log^2(1 - 2x\cos a + x^2)\,\partial x + \int_0^1 \frac{1}{x}\left(\arctan\frac{x\sin a}{1 - x\cos a}\right)^2 \partial x. \tag{3.212}$$

Since [46]

$$\log(1 - 2x\cos a + x^2) = -2\sum_{m=1}^{\infty}\frac{x^m}{m}\cos ma \quad (-1 < x < 1, 0 < a < 2\pi) \tag{3.213}$$

and

$$\arctan\frac{x\sin\alpha}{1-x\cos\alpha}=\sum_{m=1}^{\infty}\frac{x^m}{m}\sin m\alpha\quad(-1<x<1,0<\alpha<2\pi),\tag{3.214}$$

after the integration, from (3.212) we get

$$\begin{aligned}\mathfrak{Re}(I_2)&=-\sum_{m=1}^{\infty}\sum_{n=1}^{\infty}\frac{\cos(m+n)\alpha}{mn(m+n)}=-\sum_{k=2}^{\infty}\frac{\cos k\alpha}{k}\sum_{n=1}^{k-1}\frac{1}{n(k-n)}\\&=-2\sum_{k=2}^{\infty}\frac{\cos k\alpha}{k^2}\sum_{n=1}^{k-1}\frac{1}{n}.\end{aligned}\tag{3.215}$$

This, together with

$$\sum_{k=1}^{\infty}\frac{\cos k\alpha}{k^2}\sum_{n=1}^{k}\frac{1}{n}=\sum_{k=1}^{\infty}\frac{\cos k\alpha}{k^3}+\sum_{k=2}^{\infty}\frac{\cos k\alpha}{k^2}\sum_{n=1}^{k-1}\frac{1}{n},\tag{3.216}$$

(3.206), (3.207), and (3.210), gives (3.196). Taking $\alpha=\frac{\pi}{2}$ in (3.196), from

$$\sum_{m=1}^{\infty}\frac{\sin\frac{m\pi}{2}}{m^2}=\sum_{m=1}^{\infty}\frac{(-1)^m}{(2m+1)^2}=G\tag{3.217}$$

and:

$$\sum_{m=1}^{\infty}\frac{\cos\frac{m\pi}{2}}{m^3}=\frac{1}{8}\sum_{m=1}^{\infty}\frac{(-1)^m}{m^3}=-\frac{3}{32}\zeta(3)\tag{3.218}$$

we get (3.197), which is a different expression for relation (3.200). To prove relation (3.199), we will observe the function $g(s)=\frac{\log^2(1-s^2)}{s}$ and the contour C_2 given in Figure 3.2 (the way how the contour C_2 should be used is described in detail in [71]), which consists of the elements $s=\sigma$ ($\rho\leqslant\sigma\leqslant1-\rho$), $s=\mathfrak{e}^{\varphi\mathrm{i}}$ ($\alpha_0\leqslant\varphi\leqslant\frac{\pi}{6}$), and $s=\mathfrak{e}^{\varphi\mathrm{i}}\sqrt{2\cos2\varphi}$ ($\frac{\pi}{6}\leqslant\varphi\leqslant\frac{\pi}{4}$) and two small circular arcs with respect to $s=0$ and $s=1$.

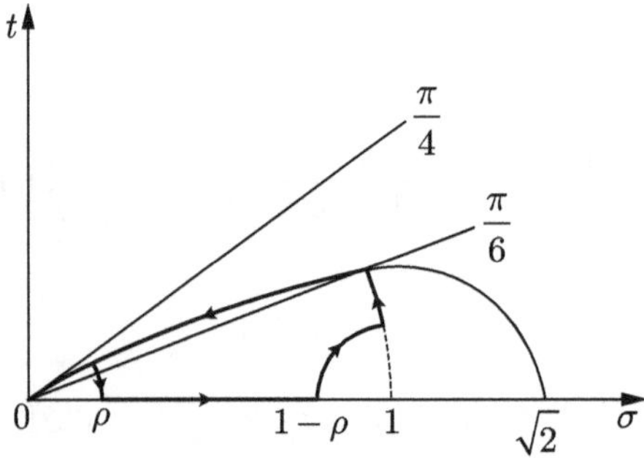

Figure 3.2: Integration contour for the function $g(s)$.

As in the previous transformations and calculations, we get

$$0 = \Re\mathfrak{e} \int_{C_2} g(s)\,\partial s = \int_0^1 \frac{\log^2(1-x^2)}{x}\,\partial x - 2\int_0^{\frac{\pi}{6}} \left(x - \frac{\pi}{2}\right)\log(2\sin x)\,\partial x$$

$$+ \frac{1}{2}\int_{\frac{\pi}{3}}^{\frac{\pi}{2}} (2x-\pi)^2 \tan x\,\partial x = J_1 + J_2 + J_3, \tag{3.219}$$

$$J_1 = \int_0^1 \frac{\log^2(1-x^2)}{x} \iff \partial x = \frac{1}{2}\int_0^1 \frac{\log^2 x}{1-x}dx \overset{(3.207)}{=} \zeta(3), \tag{3.220}$$

$$J_3 = 2\int_0^{\frac{\pi}{6}} x^2 \operatorname{ctan} x\,\partial x = 2(x^2 \log\sin x)\Big|_0^{\frac{\pi}{6}} - 4\int_0^{\frac{\pi}{6}} x\sin x\,\partial x, \tag{3.221}$$

and from (3.219), after rearranging, it follows that

$$\zeta(3) = 6\int_0^{\frac{\pi}{6}} \left(x - \frac{\pi}{6}\right)\log(2\sin x)\,\partial x. \tag{3.222}$$

From (3.222), (3.209), and:

$$\int_0^{\frac{\pi}{6}} \left(x - \frac{\pi}{6}\right)\cos 2mx\,\partial x = \frac{1}{4m^2}\left(\cos\frac{m\pi}{3} - 1\right) \tag{3.223}$$

we obtain (3.199). Taking $a = \frac{\pi}{3}$ in (3.196), from (3.199) and the relations

$$\sum_{m=1}^{\infty} \frac{\sin\frac{m\pi}{3}}{m^2} = \frac{\sqrt{3}}{2}\sum_{m=1}^{\infty}(-1)^m\left(\frac{1}{(3m+1)^2} + \frac{1}{(3m+1)^2}\right) \tag{3.224}$$

and

$$\sum_{m=1}^{\infty}(-1)^m\left(\frac{1}{(3m+1)^2} + \frac{1}{(3m+1)^2}\right) = \frac{1}{3}\Psi'\left(\frac{1}{3}\right) - \frac{2\pi^2}{9} \tag{3.225}$$

(the last formula was the subject of research in [78]) we obtain (3.198).

Since [31]

$$\sum_{m=1}^{\infty}\frac{1}{m^2}\sum_{n=1}^{2m}\frac{1}{n} = \frac{11}{4}\zeta(3), \tag{3.226}$$

by adding the expressions in relations (3.200) and (3.226) we get (3.201). □

Proposition 3.2. *Relation* (3.196) *for* $a = 0$ *gives*

$$\sum_{m=1}^{\infty}\sum_{n=1}^{\infty}\frac{1}{mn(m+n)} = 2\zeta(3), \tag{3.227}$$

whereas for $a = \pi$, *we get*

$$\sum_{m=1}^{\infty}\frac{(-1)^m}{m^2}\sum_{n=1}^{m}\frac{1}{n} = -\frac{5}{8}\zeta(3), \tag{3.228}$$

which is also equality (3.144).

3.4.3 $\zeta_k(x)$ and *Apéry*'s sums

When proving that $\zeta(3)$ is an irrational number, *Apéry* used the sums of the form

$$A_k = \sum_{n=1}^{\infty}\frac{1}{n^k\binom{2n}{n}} \quad (k \text{ nonnegative integer}), \tag{3.229}$$

$$B_k = \sum_{n=1}^{\infty}\frac{(-1)^{n+1}}{n^k\binom{2n}{n}} \quad (k \text{ nonnegative integer}). \tag{3.230}$$

Denoting

$$C_k = \int_0^{\frac{1}{4}}\frac{\zeta_k(t)}{\sqrt{1-4t}}\,\partial t, \quad D_k = \int_0^{\frac{1}{4}}\frac{\zeta_k(-t)}{\sqrt{1-4t}}\,\partial t, \tag{3.231}$$

and

$$\zeta_0(t) = \frac{t}{1-t}, \tag{3.232}$$

we obtain the following results.

Theorem 3.5. *We have*

$$\begin{gathered} A_0 = \frac{1}{3} + \frac{2\sqrt{3}}{27}\pi, \quad A_1 = \frac{\sqrt{3}}{9}\pi, \quad A_2 = \frac{1}{3}\zeta(2), \\ A_3 = -\frac{4}{3}\zeta(3) + \frac{\sqrt{3}}{9}\pi\Psi'\left(\frac{1}{3}\right) - \frac{2\sqrt{3}}{27}\pi^3, \\ B_0 = \frac{1}{5} + \frac{4\sqrt{5}}{25}\log a, \quad B_1 = \frac{2\sqrt{5}}{5}\log a, \quad B_2 = 2\log^2 a, \quad B_3 = \frac{2}{5}\zeta(3), \\ C_0 = -\frac{1}{2} + \frac{\sqrt{3}}{9}\pi, \quad C_1 = 1 - \frac{\sqrt{3}}{6}\pi, \quad C_2 = -2 + \frac{\sqrt{3}}{3}\pi + \frac{1}{6}\zeta(2), \end{gathered} \tag{3.233}$$

$$C_3 = 4 - \frac{2\sqrt{3}}{3}\pi - \zeta(3) + \frac{\sqrt{3}}{18}\pi\Psi'\left(\frac{1}{3}\right) - \frac{2\sqrt{3}}{27}\pi^3,$$

$$D_0 = -\frac{1}{2} + \frac{2\sqrt{5}}{5}\log\alpha, \quad D_1 = 1 - \sqrt{5}\log\alpha, \quad D_2 = -\log^2\alpha + 2\sqrt{5}\log\alpha - 2,$$

$$D_3 = 2\log^2\alpha - 4\sqrt{5}\log\alpha + 4 - \frac{1}{5}\zeta(3),$$

where $\alpha = \frac{\sqrt{5}+1}{2}$.

Theorem 3.6. *We have*

$$\sum_{m=1}^{\infty}\frac{(-1)^m}{m^2}\sum_{n=1}^{3m}\frac{1}{n} = \frac{33}{8}\zeta(3) - \frac{\sqrt{3}}{3}\pi\Psi'\left(\frac{1}{3}\right) + \frac{2\sqrt{3}}{9}\pi^3, \tag{3.234}$$

$$G_{(2,1)}(\beta) = \zeta_3(\beta) + \frac{2}{3}\log^3\beta + \frac{1}{5}\zeta(3), \tag{3.235}$$

$$G_{(2,1)}(\beta) = \zeta_3(\beta) + \frac{2}{3}\log^3\beta - \frac{\pi^2}{5}\log\beta - \frac{2}{3}\zeta(3), \tag{3.236}$$

where $\beta = \frac{\sqrt{5}-1}{2}$.

Remark 3.1. A simpler formula for $\zeta_3(\beta)$ has not yet been found.

Proof. To prove Theorems 3.5 and 3.6, we will start with the assumption that

$$a_{n,k} = \frac{1}{n^k\binom{2n}{n}} \quad \Longrightarrow \quad A_k = \sum_{n=1}^{\infty} a_{n,k}. \tag{3.237}$$

Since

$$\int_0^1 t^{n-1}(1-t)^{-\frac{1}{2}}\,\partial t = B\left(n, \frac{1}{2}\right) = \frac{\Gamma(n)\Gamma(\frac{1}{2})}{\Gamma(n+\frac{1}{2})} = \frac{1}{n}\frac{(2n)!!}{(2n-1)!!} \quad (n \geqslant 1), \tag{3.238}$$

we have

$$a_{n,k} = n^{1-k}2^{-2n}\int_0^1 t^{n-1}(1-t)^{-\frac{1}{2}}\,\partial t = n^{1-k}\int_0^{\frac{1}{4}} t^{n-1}(1-4t)^{-\frac{1}{2}}\,\partial t \quad (n \geqslant 1, k \geqslant 0). \tag{3.239}$$

From the equations above it follows that

$$A_k = \sum_{n=1}^{\infty} n^{1-k}\int_0^{\frac{1}{4}} t^{n-1}(1-4t)^{-\frac{1}{2}}\,\partial t = \int_0^{\frac{1}{4}}(1-4t)^{-\frac{1}{2}}\sum_{n=1}^{\infty} t^{n-1}n^{1-k}\,\partial t = \int_0^{\frac{1}{4}}\frac{\zeta_k'(t)}{\sqrt{1-4t}}\,\partial t \quad (k \geqslant 0). \tag{3.240}$$

However,

$$\begin{aligned}A_k &= \int_0^{\frac{1}{4}} \frac{\sqrt{1-4t}}{t(1-4t)} \zeta_k(t)\,\partial t = \int_0^{\frac{1}{4}} \sqrt{1-4t}\left(\frac{1}{t} + \frac{4}{1-4t}\right)\zeta_{k-1}(t)\,\partial t \\ &= 4\int_0^{\frac{1}{4}} \frac{\zeta_{k-1}(t)}{\sqrt{1-4t}}\,\partial t + \int_0^{\frac{1}{4}} \frac{\sqrt{1-4t}}{t}\zeta_{k-1}(t)\,\partial t,\end{aligned} \tag{3.241}$$

and

$$\int_0^{\frac{1}{4}} \frac{\sqrt{1-4t}}{t}\zeta_{k-1}(t)\,\partial t = \int_0^{\frac{1}{4}} \sqrt{1-4t}\zeta_k'(t)\,\partial t = 2\int_0^{\frac{1}{4}} \frac{\zeta_k(t)}{\sqrt{1-4t}}\,\partial t. \tag{3.242}$$

Therefore

$$A_k = 2C_k + 4C_{k-1} \quad (k \geqslant 1). \tag{3.243}$$

From (3.231) it follows that

$$C_1 = \int_0^{\frac{1}{4}} \frac{\log(1-t)}{\sqrt{1-4t}}\,\partial t = \frac{1}{2}\int_0^{\frac{1}{4}} \frac{\sqrt{1-4t}}{1-t}\,\partial t = 1 - \frac{\sqrt{3}}{6}\pi. \tag{3.244}$$

For C_k, we have

$$C_k = 2\int_0^{\frac{1}{2}} \frac{t\zeta_k(t^2)}{\sqrt{1-4t^2\,\partial t}} = \left|\begin{matrix}\zeta_k(t^2) = u, \ u' = \frac{2}{t}\zeta_{k-1}(t^2) \\ v = -\frac{1}{4}\sqrt{1-4t^2}\end{matrix}\right| = \int_0^{\frac{1}{2}} \frac{\sqrt{1-4t^2}}{t}\zeta_{k-1}(t^2)\,\partial t \tag{3.245}$$

and, accordingly,

$$C_k = \int_0^{\frac{1}{2}} \frac{\sqrt{1-4t^2}}{t}\zeta_{k-1}(t^2)\,\partial t. \tag{3.246}$$

From the last relation it follows that (in the particular case $k = 2$)

$$\begin{aligned}C_2 &= -\int_0^{\frac{1}{2}} \frac{\sqrt{1-4t^2}}{t}\log(1-t^2)\,\partial t = \left|t = \frac{u}{1+u^2}\right| \\ &= -\int_0^{1} \frac{1}{u}\frac{(1-u^2)^2}{(1+u^2)^2}\log\frac{u^4+u^2+1}{(1+u^2)^2}\,\partial u\end{aligned}$$

$$= -\frac{1}{2}\int_0^1 \frac{(1-t)^2}{t(1+t)^2}\log\frac{1+t+t^2}{(1+t)^2}\,\partial t = \left|1+t=\frac{1}{u}\right|$$

$$= -\frac{1}{2}\int_{\frac{1}{2}}^1 \frac{(2u-1)^2}{u(1-u)}\log(u^2-u+1)\,\partial u$$

$$= 2\int_{\frac{1}{2}}^1 \log(t^2-t+1)\,\partial t - \frac{1}{2}\int_{\frac{1}{2}}^1 \frac{\log(t^2-t+1)}{t(1-t)}\,\partial t$$

$$= \int_0^1 \log(t^2-t+1)\,\partial t - \frac{1}{4}\int_0^1 \frac{\log(t^2-t+1)}{t(1-t)}\,\partial t. \tag{3.247}$$

From the equality

$$\int_0^1 \log(t^2-t+1)\,\partial t = -2 + \frac{\sqrt{3}}{3}\pi \tag{3.248}$$

it follows that

$$\int_0^1 \frac{\log(t^2-t+1)}{t(1-t)}\,\partial t = \int_0^1 \frac{1}{t}\log(t^2-t+1)\,\partial t + \int_0^1 \frac{\log(t^2-t+1)}{t(1-t)}\,\partial t$$

$$= 2\int_0^1 \frac{1}{t}\log(t^2-t+1)\,\partial t$$

$$= 2\int_0^1 \frac{1}{t}\{\log(t^3+1) - \log(t+1)\}\,\partial t$$

$$= -\frac{4}{3}\int_0^1 \frac{1}{t}\log(1+t)\,\partial t = \frac{4}{3}\int_0^1 \frac{\log t}{1+t}\,\partial t = \frac{4}{3}\sum_{n=0}^{\infty}(-1)^n\int_0^1 t^n\log t\,\partial t$$

$$= -\frac{2}{3}\zeta(2), \tag{3.249}$$

so the given expression for C_2 follows from (3.247). From (3.243) and the formulas for C_1 and C_2 we get A_2. By the relation for $\zeta_0(t)$ and (3.231) we obtain the equality that determines the value for C_0. The formula for A_1 follows from the formulas for C_0 and C_1 and from (3.243). From $\zeta_0(t)$ and from (3.240) it follows that

$$A_0 = \int_0^{\frac{1}{4}} \frac{\partial t}{(1-t)^2\sqrt{1-4t}} = 8\int_0^{\frac{1}{4}} \frac{\partial t}{(3+t^2)^2} = \frac{1}{3} + \frac{2\sqrt{3}}{27}\pi. \tag{3.250}$$

For A_3, we have

$$A_3 = \int_0^{\frac{1}{4}} \frac{\zeta_2(t)}{t\sqrt{1-4t}}\,\partial t = \left| \begin{matrix} u = \zeta_2(t) \\ v = \log\frac{1-\sqrt{1-4t}}{1+\sqrt{1-4t}} \end{matrix} \right|$$

$$= -\int_0^{\frac{1}{4}} \frac{1}{t}\log(1-t)\log\frac{1-\sqrt{1-4t}}{1+\sqrt{1-4t}}\,\partial t = |\sqrt{1-4t} = 1-2u|$$

$$= \int_0^{\frac{1}{2}} \frac{1-2t}{t-t^2}\log(1-t+t^2)\log\frac{t}{1-t}\,\partial t$$

$$= \frac{1}{2}\int_0^1 \frac{1-2t}{t-t^2}\log(1-t+t^2)\log\frac{t}{1-t}\,\partial t$$

$$= \int_0^1 \frac{1}{t}\log(1-t+t^2)\log\frac{t}{1-t}\,\partial t$$

$$= \int_0^1 \frac{1}{t}\log(1-t+t^2)\log t\,\partial t - \int_0^1 \frac{1}{t}\log(1-t+t^2)\log(1-t)\,\partial t = P - Q, \tag{3.251}$$

where

$$P = -\frac{1}{2}\int_0^1 \frac{2t-1}{1-t+t^2}\log^2 t\,\partial t. \tag{3.252}$$

Taking $\alpha_1 = e^{-\frac{\pi}{3}i}$ and $\alpha_2 = e^{-\frac{\pi}{3}i}$, we have

$$\frac{2t-1}{1-t+t^2} = \frac{1}{t-\alpha_1} + \frac{1}{t-\alpha_2} \tag{3.253}$$

and

$$P = -\frac{1}{2}\int_0^1 \left(\frac{1}{t-\alpha_1} + \frac{1}{t-\alpha_2}\right)\log^2 t\,\partial t = -\frac{1}{2}(P_1+P_2), \tag{3.254}$$

where

$$P_1 = -\frac{1}{\alpha_1}\int_0^1 \frac{1}{t-\frac{t}{\alpha_1}}\log^2 t\,\partial t = -2\sum_{n=1}^{\infty}\frac{\alpha_2^n}{n^3}, \tag{3.255}$$

and, accordingly,

$$P_2 = -2\sum_{n=1}^{\infty}\frac{\alpha_1^n}{n^3}, \tag{3.256}$$

which finally leads to the value

$$P = \sum_{n=1}^{\infty}\frac{\alpha_1^n + \alpha_2^n}{n^3} = 2\sum_{n=1}^{\infty}\frac{\cos\frac{n\pi}{3}}{n^3} \overset{(3.199)}{=} \frac{2}{3}\zeta(3). \tag{3.257}$$

In addition,

$$\begin{aligned} Q &= \left|\begin{matrix} u = \log(1-t+t^2) \\ v = \zeta_2(t) \end{matrix}\right| = -\int_0^1 \frac{2t-1}{1-t+t^2}\zeta_2(t)\,\partial t \\ &= -\int_0^1 \left(\frac{1}{t-\alpha_1} + \frac{1}{t-\alpha_2}\right)\zeta_2(t)\,\partial t = -(Q_1 + Q_2), \end{aligned} \tag{3.258}$$

where

$$Q_1 = \int_0^1 \frac{\zeta_2(t)}{\alpha_1 - t}\,\partial t = \frac{1}{\alpha_1}\sum_{m=1}^{\infty}\left(\frac{1}{\alpha_1}\right)^{m-1}\sum_{n=1}^{\infty}\frac{1}{n^2}\int_0^1 t^{m+n-1}\,\partial t = \sum_{m=1}^{\infty}\alpha_2^m\sum_{n=1}^{\infty}\frac{1}{n^2(m+n)}. \tag{3.259}$$

Since

$$\frac{1}{n^2(m+n)} = \frac{1}{m}\frac{1}{n^2} + \frac{1}{m^2}\left(\frac{1}{m+n} - \frac{1}{n}\right), \tag{3.260}$$

we get

$$Q_1 = \frac{\pi^2}{6}\zeta_1(\alpha_2) - \sum_{m=1}^{\infty}\frac{\alpha_2^m}{m^2}\sum_{n=1}^{m}\frac{1}{n}. \tag{3.261}$$

From the analogous expression for Q_2 and from $\zeta_1(\alpha_2) = -\log\alpha_1$ it follows that

$$\begin{aligned} Q &= -\sum_{m=1}^{\infty}\frac{\alpha_1^m}{m^2}\sum_{n=1}^{m}\frac{1}{n} - \sum_{m=1}^{\infty}\frac{\alpha_2^m}{m^2}\sum_{n=1}^{m}\frac{1}{n} = -2\sum_{m=1}^{\infty}\frac{\cos\frac{m\pi}{3}}{m^2}\sum_{n=1}^{\infty}\frac{1}{n} \\ &\overset{(3.198)}{=} -2\zeta(3) + \frac{\sqrt{3}}{9}\pi\Psi'\left(\frac{1}{3}\right) - \frac{2\sqrt{3}}{27}\pi^3. \end{aligned} \tag{3.262}$$

In this way, based on (3.251), the formula for A_3 is derived. The formula for C_3 follows from (3.243) and the expressions for A_3 and C_2. Let us express Q in another way:

$$Q = -\int_0^1 \frac{1}{t}\log(1-t)\log(1-t+t^2)\,\partial t$$

$$= -\int_0^1 \frac{1}{t}\log(1-t)\{\log(1+t^3) - \log(1+t)\}\,\partial t$$

$$= \sum_{m=1}^{\infty} \frac{(-1)^{m-1}}{m} \sum_{n=1}^{\infty} \frac{1}{n} \int_0^1 \left(t^{3m+n-1} - t^{m+n-1}\right) \partial t$$

$$= \sum_{m=1}^{\infty} \frac{(-1)^{m-1}}{m} \sum_{n=1}^{\infty} \frac{1}{n}\left(\frac{1}{3m+n} - \frac{1}{m+n}\right). \tag{3.263}$$

Since

$$\sum_{n=1}^{\infty} \frac{1}{n(3m+n)} = \frac{1}{3m}\sum_{n=1}^{3m}\frac{1}{n}, \quad \sum_{m=1}^{\infty} \frac{1}{n(m+n)} = \frac{1}{m}\sum_{n=1}^{m}\frac{1}{n}, \tag{3.264}$$

we get

$$Q = -\frac{1}{3}\sum_{m=1}^{\infty} \frac{(-1)^m}{m^2} \sum_{n=1}^{3m} \frac{1}{n} + \sum_{m=1}^{\infty} \frac{(-1)^m}{m^2} \sum_{n=1}^{m} \frac{1}{n}. \tag{3.265}$$

Formula (3.234) follows from (3.262) and the fact that $G_{(2,1)}(-1) = -\frac{5}{8}\zeta(3)$ (see (3.144). Further, denoting

$$B_k = \int_0^{\frac14} (1-4t)^{-\frac12} \sum_{n=1}^{\infty} (-t)^{n-1} n^{1-k}\,\partial t = -\int_0^{\frac14} \frac{\zeta_{k-1}(-t)}{t\sqrt{1-4t}}\,\partial t = -\int_0^{\frac14} \frac{[\zeta_k(-t)]_t'}{\sqrt{1-4t}}\,\partial t, \tag{3.266}$$

by the way used for deriving (3.243) we obtain

$$B_k = -2D_k - 4D_{k-1} \quad (k \geqslant 1). \tag{3.267}$$

For $k = 0$,

$$D_0 = \int_0^{\frac14} \frac{\zeta_0(-t)}{\sqrt{1-4t}}\,\partial t = -\int_0^{\frac14} \frac{t}{(1+t)\sqrt{1-4t}}\,\partial t = -\frac{1}{2}\int_0^1 \frac{1-t^2}{5-t^2}\,\partial t = -\frac{1}{2} + \frac{2\sqrt{5}}{5}\log\alpha, \tag{3.268}$$

whereas

$$B_0 = -\int_0^{\frac14} \frac{[\zeta_k(-t)]_t'}{\sqrt{1-4t}}\,\partial t = \int_0^{\frac14} \frac{\partial t}{(1+t)^2\sqrt{1-4t}} = 8\int_0^1 \frac{\partial t}{(5-t^2)^2} = \frac{1}{5} + \frac{4\sqrt{5}}{25}\log\alpha, \tag{3.269}$$

$$D_1 = \int_0^{\frac14} \frac{\zeta_1(-t)}{\sqrt{1-4t}}\,\partial t = -\int_0^{\frac14} \frac{\log(1+t)}{\sqrt{1-4t}}\,\partial t = -\frac{1}{2}\int_0^{\frac14} \frac{\sqrt{1-4t}}{1+t}\,\partial t = -\int_0^1 \frac{t^2}{5-t^2}\,\partial t = 1 - \sqrt{5}\log\alpha. \tag{3.270}$$

We are now in the position to determine B_1 from (3.267). For B_2, we have

$$
\begin{aligned}
B_2 &= -\int_0^{\frac{1}{4}} \frac{\zeta_1(-t)}{t\sqrt{1-4t}}\,\partial t = \int_0^{\frac{1}{4}} \frac{\log(1+t)}{t\sqrt{1-4t}}\,\partial t = |1-4t=u^2| \\
&= 2\int_0^1 \frac{\log\frac{5-t^2}{4}}{1-t^2}\,\partial t = |t=2u-1| = \int_{\frac{1}{2}}^1 \frac{\log(1+t-t^2)}{t-t^2}\,\partial t \\
&= \frac{1}{2}\int_0^1 \frac{\log(1+t-t^2)}{t-t^2}\,\partial t = \frac{1}{2}\int_0^1 \left(\frac{1}{t}+\frac{1}{1-t}\right)\log(1+t-t^2)\,\partial t \\
&= \int_0^1 \frac{\log(1+t-t^2)}{t}\,\partial t = -\int_0^1 \frac{1-2t}{1+t-t^2}\log t\,\partial t.
\end{aligned}
\tag{3.271}
$$

Since $1+t-t^2=(t+\beta)(\alpha-t)$, $\frac{1-2t}{1+t-t^2}=\frac{1}{t+\beta}-\frac{1}{\alpha-t}$, and from relation (3.271) it follows that

$$
B_2 = \int_0^1 \left(\frac{1}{\alpha-t}-\frac{1}{t+\beta}\right)\log t\,\partial t. \tag{3.272}
$$

From equalities

$$
\int_0^1 \frac{1}{\alpha-t}\log t\,\partial t = \int_0^1 \sum_{n=0}^{\infty}\frac{1}{\alpha}\left(\frac{t}{\alpha}\right)^n \log t\,\partial t = -\zeta_2(\beta), \tag{3.273}
$$

$$
\int_0^1 \frac{\log t}{t+\beta}\,\partial t = \int_0^{\beta} \frac{\log t}{t+\beta}\,\partial t + \int_{\beta}^1 \frac{\log t}{t+\beta}\,\partial t, \tag{3.274}
$$

$$
\begin{aligned}
\int_{\beta}^1 \frac{\log t}{t+\beta}\,\partial t &= \left|t=\frac{1}{u}\right| = -\int_1^{\alpha} \frac{\log t}{t(1+\beta t)}\,\partial t = -\int_1^{\alpha} \sum_{n=0}^{\infty}(-\beta t)^n \frac{\log t}{t}\,\partial t \\
&= -\int_1^{\alpha} \frac{\log t}{t}\,\partial t - \sum_{n=0}^{\infty}(-\beta)^n \int_1^{\alpha} t^{n-1}\log t\,\partial t,
\end{aligned}
\tag{3.275}
$$

and

$$
\int_1^{\alpha} t^{n-1}\log t\,\partial t = \frac{\alpha^n}{n}\log\alpha - \frac{\alpha^n}{n^2} + \frac{1}{n^2} \tag{3.276}
$$

it follows that:

$$\int_{\beta}^{1} \frac{\log t}{t+\beta}\,\partial t = -\frac{1}{2}\log^2\alpha + (\log 2)\log\alpha - \frac{\pi^2}{12} - \zeta_2(-\beta), \tag{3.277}$$

$$\int_{0}^{\beta} \frac{\log t}{t+\beta}\,\partial t = \int_{0}^{1} \frac{\log t}{1+t}\,\partial t + (\log\beta)\int_{0}^{1} \frac{\partial t}{1+t} = (\log 2)\log\beta - \frac{\pi^2}{12}. \tag{3.278}$$

The formula for B_2 follows from (3.272), (3.273), (3.274), (3.277), and (3.278):

$$B_2 = \zeta_2(-\beta) - \zeta_2(\beta) + \frac{\pi^2}{6} + \frac{1}{2}\log^2\alpha. \tag{3.279}$$

Since

$$\zeta_k(t) + \zeta_k(-t) = 2^{1-k}\zeta_k(t^2), \tag{3.280}$$

we get

$$B_2 = \frac{1}{2}\zeta_2(\beta^2) - 2\zeta_2(\beta) + \frac{\pi^2}{6} + \frac{1}{2}\log^2\alpha. \tag{3.281}$$

From the last equality and from

$$1 - \beta - \beta^2 = 0, \quad \zeta_2(t) + \zeta_2(1-t) = -\log(1-t)\log t + \frac{\pi^2}{6} \tag{3.282}$$

it follows that

$$B_2 = -\frac{5}{2}\zeta_2(\beta) + \frac{1}{2}\log^2\alpha - \log^2\beta + \frac{\pi^2}{4}. \tag{3.283}$$

However, since [180]

$$\zeta_2(\beta) = \frac{\pi^2}{10} - \log^2\beta, \tag{3.284}$$

we get $B_2 = 2\log^2\alpha$. The formula for D_2 follows from (3.267) and the expressions for D_1 and B_2. Further,

$$B_3 = -\int_{0}^{\frac{1}{4}} \frac{\zeta_2(-t)}{t\sqrt{1-4t}}\,\partial t = \left| \begin{matrix} -\zeta_2(-t) = u \\ \frac{\partial t}{t\sqrt{1-4t}} = \partial v \end{matrix} \right|$$

$$= \int_{0}^{\frac{1}{4}} \frac{1}{t}\log(1+t)\log\frac{1+\sqrt{1-4t}}{1-\sqrt{1-4t}}\,\partial t = |\sqrt{1-4t} = 1-2u|$$

$$
\begin{aligned}
&= \int_0^{\frac{1}{2}} \frac{1-2t}{t-t^2} \log(1+t-t^2) \log\frac{1-t}{t} \, \eth t \\
&= \frac{1}{2} \int_0^1 \frac{1-2t}{t-t^2} \log(1+t-t^2) \log\frac{1-t}{t} \, \eth t \\
&= \int_0^1 \frac{1}{t} \log(1+t-t^2) \log\frac{1-t}{t} \, \eth t \\
&= -\int_0^1 \frac{1}{t} \log(1+t-t^2) \log t \, \eth t - \int_0^1 \zeta_2'(t) \log(1+t-t^2) \, \eth t \\
&= B_{3,1} + B_{3,2},
\end{aligned}
\tag{3.285}
$$

where

$$
B_{3,1} = \begin{vmatrix} u = \log(1+t-t^2) \\ v = \frac{1}{2}\log^2 t \end{vmatrix} = \frac{1}{2} \int_0^1 \frac{1-2t}{1+t-t^2} \log^2 t \, \eth t, \tag{3.286}
$$

$$
B_{3,2} = \int_0^1 \frac{1-2t}{1+t-t^2} \zeta_2(t) \, \eth t. \tag{3.287}
$$

Let us determine $B_{3,1}$:

$$
\begin{aligned}
B_{3,1} &= \frac{1}{2} \int_0^1 \left(\frac{1}{t+\beta} + \frac{1}{t-\alpha} \right) \log^2 t \, \eth t \\
&= \frac{1}{2} \int_0^{\beta} \frac{\log^2 t}{t+\beta} \eth t + \frac{1}{2} \int_{\beta}^1 \frac{\log^2 t}{t+\beta} \eth t + \frac{1}{2} \int_0^1 \frac{\log^2 t}{t-\alpha} \eth t = X_1 + X_2 + X_3.
\end{aligned}
\tag{3.288}
$$

Of course, the values of the expressions X_1, X_2, and X_3 must also be specified. We have

$$
X_1 = \frac{1}{2} \int_0^{\beta} \frac{1}{\beta} \frac{1}{1+\frac{t}{\beta}} \log^2 t \, \eth t = \frac{1}{2} \sum_{n=0}^{\infty} (-1)^n \beta^{-n-1} \int_0^{\beta} t^n \log^2 t \, \eth t. \tag{3.289}
$$

Since

$$
\int_0^{\beta} t^n \log^2 t \, \eth t = \beta^{n+1} \left(\frac{\log^2 \beta}{n+1} - 2\frac{\log \beta}{(n+1)^2} - \frac{2}{(n+1)^3} \right), \tag{3.290}
$$

we get

$$X_1 = -\frac{1}{2}\zeta_1(-1)\log^2\beta + \zeta_2(-1)\log\beta - \zeta_3(-1). \tag{3.291}$$

From the expression

$$X_2 = \frac{1}{2}\int_\beta^1 \frac{\log^2 t}{t(1+\frac{\beta}{t})}\,\partial t = \frac{1}{2}\sum_{n=0}^{\infty}(-1)^n\beta^n\int_\beta^1 t^{-n-1}\log^2 t\,\partial t, \tag{3.292}$$

since

$$\int_\beta^1 t^{-n-1}\log^2 t\,\partial t = -\frac{2}{n^3} + \beta^{-n}\left(\frac{\log^2\beta}{n} + 2\frac{\log\beta}{n^2} + \frac{2}{n^3}\right) \quad (n \geqslant 1), \tag{3.293}$$

we obtain that

$$X_2 = -\frac{1}{6}\log^3\beta - \zeta_3(-\beta) + \frac{1}{2}\zeta_1(-1)\log^2\beta + \zeta_2(-1)\log\beta + \zeta_3(-1). \tag{3.294}$$

Finally, for X_3, we have

$$\begin{aligned} X_3 &= -\frac{1}{2}\int_0^1 \frac{1}{\alpha}\frac{\log^2 t}{1-\frac{t}{\alpha}}\,\partial t \\ &= -\frac{1}{2}\sum_{n=0}^{\infty}\beta^{n+1}\left[t^{n+1}\left(\frac{\log^2 t}{n+1} - 2\frac{\log t}{(n+1)^2} + \frac{2}{(n+1)^3}\right)\right]_0^1 = -\zeta_3(\beta) \end{aligned} \tag{3.295}$$

and, accordingly,

$$B_{3,1} = -\frac{1}{6}\log^3\beta + 2\zeta_2(-1)\log\beta - [\zeta_3(\beta) + \zeta_3(-\beta)]. \tag{3.296}$$

Finally, because of $\zeta_2(-1) = -\frac{\pi^2}{12}$ and $\zeta_3(\beta) + \zeta_3(-\beta) = \frac{1}{4}\zeta_3(\beta^2)$, it follows that

$$B_{3,1} = -\frac{1}{6}\log^3\beta - \frac{\pi^2}{6}\log\beta - \frac{1}{4}\zeta_3(\beta^2). \tag{3.297}$$

The expression $\zeta_3(\beta^2)$ can be calculated using the *Spence* relation (proof of this relation can be found in *Watson* [41])

$$\begin{aligned} \zeta_3\left(\frac{x}{x-1}\right) &= -\zeta_3(x) - \zeta_3(1-x) + \zeta(3) + \frac{\pi^2}{6}\log(1-x) \\ &\quad + \frac{1}{6}\log^3(1-x)\log x \quad (0 < x < 1). \end{aligned} \tag{3.298}$$

Replacing $x = \beta^2$ and using formula (3.280) with $k = 3$ and the relation $1-\beta-\beta^2 = 0$, we conclude that

$$\zeta_3(\beta^2) = \frac{4}{5}\zeta(3) + \frac{2\pi^2}{15}\log\beta - \frac{2}{3}\log^3\beta. \tag{3.299}$$

From (3.299) and the derived expression for $B_{3,1}$ we get

$$B_{3,1} = -\frac{1}{5}\zeta(3) - \frac{\pi^2}{5}\log\beta. \tag{3.300}$$

We are now in the position to determine the value for $B_{3,2}$:

$$B_{3,2} = \int_0^1 \left(\frac{1}{t+\beta} + \frac{1}{t-\alpha}\right)\zeta_2(t)\,\partial t = \int_0^1 \frac{\zeta_2(t)}{t+\beta}\,\partial t + \int_0^1 \frac{\zeta_2(t)}{t-\alpha}\,\partial t = Y_1 + Y_2, \tag{3.301}$$

where

$$\begin{aligned}
Y_1 &= |t = 1-u| = -\int_0^1 \frac{1}{t-\alpha}\zeta_2(1-t)\,\partial t \\
&= -\int_0^1 \frac{1}{t-\alpha}\left[\frac{\pi^2}{6} - \log(1-t)\log t - \zeta_2(t)\right]\partial t \\
&= \int_0^1 \frac{\log(1-t)\log t}{t-\alpha}\,\partial t - \frac{\pi^2}{3}\log\beta + \int_0^1 \frac{\zeta_2(t)}{t-\alpha}\,\partial t = -\frac{\pi^2}{3}\log\beta + Z_1 + Y_2,
\end{aligned} \tag{3.302}$$

$$Z_1 = -\int_0^1 \left(\frac{\log(1-t)}{t} - \int_0^1 \frac{\log t}{1-t}\right)\log(\alpha - t)\,\partial t, \tag{3.303}$$

$$Y_2 = \frac{\pi^2}{6}\log\beta + \int_0^1 \frac{\log(1-t)}{t}\log(\alpha - t)\,\partial t. \tag{3.304}$$

It follows that

$$B_{3,2} = \int_0^1 \left(\frac{\log(1-t)}{t} + \frac{\log t}{1-t}\right)\log(\alpha - t)\,\partial t = T_1 + T_2, \tag{3.305}$$

where

$$\begin{aligned}
T_1 &= \int_0^1 \log\left[\alpha\left(1 - \frac{t}{\alpha}\right)\right]\frac{\log(1-t)}{t}\,\partial t \\
&= (\log\alpha)\int_0^1 \frac{\log(1-t)}{t}\,\partial t + \int_0^1 \log\left(1 - \frac{t}{\alpha}\right)\frac{\log(1-t)}{t}\,\partial t
\end{aligned}$$

$$
\begin{aligned}
&= \frac{\pi^2}{6}\log\beta + \int_0^1 \left(\sum_{m=1}^{\infty} \frac{(\frac{t}{\alpha})^m}{m} \sum_{n=1}^{\infty} \frac{t^{n-1}}{n} \right) \partial t \\
&= \frac{\pi^2}{6}\log\beta + \sum_{m=1}^{\infty} \frac{\beta^m}{m} \sum_{n=1}^{\infty} \frac{t^{n-1}}{n(m+n)} = \frac{\pi^2}{6}\log\beta + G_{(2,1)}(\beta),
\end{aligned} \tag{3.306}
$$

and

$$
\begin{aligned}
T_2 &= \int_0^1 \log\left[\alpha\left(1 - \frac{t}{\alpha}\right)\right] \frac{\log t}{1-t}\, \partial t \\
&= \frac{\pi^2}{6}\log\beta - \int_0^1 \left(\sum_{m=1}^{\infty} \frac{(\frac{t}{\alpha})^m}{m} \sum_{n=1}^{\infty} \frac{t^{n-1}}{n} \right) \log t\, \partial t \\
&= \frac{\pi^2}{6}\log\beta - \sum_{m=1}^{\infty} \frac{\beta^m}{m} \sum_{n=1}^{\infty} \int_0^1 t^{m+n-1} \log t\, \partial t \\
&= \sum_{m=1}^{\infty} \frac{\beta^m}{m} \left(\frac{\pi^2}{6} - \sum_{n=1}^{m} \frac{1}{n^2} \right) + \frac{\pi^2}{6}\log\beta = -\frac{\pi^2}{6}\log\beta - G_{(1,2)}(\beta).
\end{aligned} \tag{3.307}
$$

Finally, we have

$$
B_{3,2} = G_{(2,1)}(\beta) - G_{(1,2)}(\beta). \tag{3.308}
$$

Relations (3.235) and (3.236) and the expression for $B_{3,2}$ follow from (3.162), (3.163) for $x = \beta$, and (3.299):

$$
B_{3,2} = \frac{3}{5}\zeta(3) + \frac{\pi^2}{5}\log\beta. \tag{3.309}
$$

By (3.300) and (3.309) it finally follows that

$$
B_3 = \frac{2}{5}\zeta(3). \tag{3.310}
$$

The formula for D_3 is obtained from the derived relation (3.267) and the expressions for D_2 and B_3. So Theorems 3.5 and 3.6 are proved. □

Remark 3.2. $\zeta_3(x)$ in the region $0 \leqslant x \leqslant 1$ is represented in a simpler form only for $x = 0, 1, \frac{1}{2}, \beta^2$ [190].

Remark 3.3. In 1979, *van der Poorten* [141] lists the formulas for A_0, A_1, A_2, A_4 without providing any proof. He only proves the formula for B_3, assuming that the formula for B_3 appears in [57].

Cohen [153] obtained formulas for A_2, A_4, and B_3, where

$$A_4 = \frac{17}{36}\zeta(4), \tag{3.311}$$

and derived the formula

$$\zeta(5) = \frac{5}{2}\sum_{n=1}^{\infty}\frac{(-1)^n}{n^3\binom{2n}{n}}\left(\sum_{k=1}^{n-1}\frac{1}{k^2} - \frac{4}{5n^2}\right). \tag{3.312}$$

Zucker [190] stated a formula similar to that for A_3 and proved formulas for A_4, A_5, and A_3. It is important to point out that the method used to prove the theorems in this chapter was not used in the aforementioned papers. *Leshchiner* [157] stated the following result.

Theorem 3.7. *We have*

$$\zeta(2k+1) = \sum_{n=1}^{\infty}\frac{2(-1)^n}{n^3\binom{2n}{n}}\sum_{\varepsilon=1}^{k-1}\frac{(-1)^{k-1-\varepsilon}B_\varepsilon}{n^{2\varepsilon}}f_n^{k-1-\varepsilon}, \tag{3.313}$$

where

$$f_k^r = \sum_{0<l_1<l_2<\cdots<l_r<k}\prod_{j=1}^{r} l_j^{-2}, \quad f_k^r = 0\ (r<0,\ r \geqslant k),\ f_k^0 = 1, \tag{3.314}$$

$$B_\varepsilon = \frac{5}{4} \quad (\varepsilon = 0), \quad B_\varepsilon = 1 \quad (\varepsilon > 0). \tag{3.315}$$

Remark 3.4. Concerning the numbers $\zeta(2k+1)$ $(k \in \mathbb{N})$, it has only been proven that $k = 1$ is an irrational number. Neither the irrationality nor the rationality of the numbers $\zeta'(2n)$, $(n \in \mathbb{N})$, *Euler*'s constant γ, and *Catalan*'s constant G have been proven.

4 The problem of basic double sums

4.1 About the problem of basic double sums

Euler [1, 2] and *Nielsen* [24] brought up the problem of determining the values of the coefficients $C_{p,q}$, $\gamma_{p,q}$, $d_{p,q}$, $\delta_{p,q}$ defined as follows (where p and q are positive integers):

$$C_{p,q} = \sum_{r=1}^{\infty} \frac{1}{(r+1)^p}\left(\frac{1}{1^q} + \frac{1}{2^q} + \frac{1}{3^q} + \cdots + \frac{1}{r^q}\right) = \sum_{r=2}^{\infty} \frac{1}{r^p} \sum_{s=1}^{r-1} \frac{1}{s^q}, \tag{4.1}$$

$$\gamma_{p,q} = \sum_{r=1}^{\infty} \frac{(-1)^{r+1}}{(r+1)^p}\left(\frac{1}{1^q} + \frac{1}{2^q} + \frac{1}{3^q} + \cdots + \frac{1}{r^q}\right) = \sum_{r=2}^{\infty} \frac{(-1)^r}{r^p} \sum_{s=1}^{r-1} \frac{1}{s^q}, \tag{4.2}$$

$$d_{p,q} = \sum_{r=1}^{\infty} \frac{1}{(r+1)^p}\left(\frac{1}{1^q} - \frac{1}{2^q} + \frac{1}{3^q} - \cdots + \frac{(-1)^{r-1}}{r^q}\right) = \sum_{r=2}^{\infty} \frac{1}{r^p} \sum_{s=1}^{r-1} \frac{(-1)^{s-1}}{s^q}, \tag{4.3}$$

$$\delta_{p,q} = \sum_{r=1}^{\infty} \frac{(-1)^{r+1}}{(r+1)^p}\left(\frac{1}{1^q} - \frac{1}{2^q} + \frac{1}{3^q} - \cdots + \frac{(-1)^{r-1}}{r^q}\right) = \sum_{r=2}^{\infty} \frac{(-1)^r}{r^p} \sum_{s=1}^{r-1} \frac{(-1)^{s-1}}{s^q}. \tag{4.4}$$

In this chapter, the case where $p + q = 2k + 1$ is solved, where k is a positive integer, i. e., some special combinations of the numbers p and q are found for which the problem can be solved. Appropriate formulas are indicated in all the cases that can be solved. Based on the relations already established in the previous chapters and relations (4.1)–(4.4), it is possible to introduce the following connections:

$$C_{p,q} = G^{(1)}_{(p,q)}(1) - \zeta_{p+q}(1); \quad \zeta_r(1) = \zeta(r), \quad r \geqslant 2, \tag{4.5}$$

$$\gamma_{p,q} = G_{(p,q)}(-1) - \zeta_{p+q}(-1); \quad \zeta_r(-1) = -\left(1 - \frac{1}{2^{r-1}}\right)\zeta(r), \quad r \geqslant 2, \tag{4.6}$$

$$d_{p,q} = \zeta_{p+q}(-1) - H_{(p,q)}(-1); \quad \zeta_1(-1) = -\log 2, \tag{4.7}$$

$$\delta_{p,q} = \zeta_{p+q}(1) - F_{(p,q)}(-1). \tag{4.8}$$

Some simpler facts about the functions introduced in this chapter will be given at the end in a separate Appendix. When calculating the aforementioned coefficients, various functional relations are obtained for certain continuous functions (given in the form of double series) for $|z| \leqslant 1$ or $|z| = 1$. It is shown that some of these relations are inverses of each other in terms of the *Jacobi* inversion for combinatorial relations. In addition, formulas are obtained for all the expressions of the form $\Omega_j(p,q)$, where

$$\Omega_j(p,q) = \begin{cases} \omega_j(p,q), & p+q = 2k+1, \\ 0, & p+q = 2k, \end{cases} \quad j = 1,2,3,5,6,7,9,10,11,16, \tag{4.9}$$

$$\Omega_j(p,q) = \begin{cases} 0, & p+q = 2k+1, \\ \omega_j(p,q), & p+q = 2k, \end{cases} \quad j = 4,8,12,13,14,15, \tag{4.10}$$

https://doi.org/10.1515/9783112233276-004

with

$$\omega_1(p,q)=\sum_{m=1}^{\infty}\frac{1}{m^p}\sum_{n=1}^{m}\frac{1}{n^q}=G_{(p,q)}(1),\ \omega_2(p,q)=\sum_{m=1}^{\infty}\frac{1}{m^p}\sum_{n=1}^{m}\frac{(-1)^n}{n^q}=H_{(p,q)}(-1), \tag{4.11}$$

$$\omega_3(p,q)=\sum_{m=1}^{\infty}\frac{1}{m^p}\sum_{n=1}^{m}\frac{1}{(2n-1)^q},\quad \omega_4(p,q)=\sum_{m=1}^{\infty}\frac{1}{m^p}\sum_{n=1}^{m}\frac{(-1)^n}{(2n-1)^q}, \tag{4.12}$$

$$\omega_5(p,q)=\sum_{m=1}^{\infty}\frac{(-1)^m}{m^p}\sum_{n=1}^{m}\frac{1}{n^q}=G_{(p,q)}(-1),\quad \omega_6(p,q)=\sum_{m=1}^{\infty}\frac{(-1)^m}{m^p}\sum_{n=1}^{m}\frac{(-1)^n}{n^q}=F_{(p,q)}(-1), \tag{4.13}$$

$$\omega_7(p,q)=\sum_{m=1}^{\infty}\frac{(-1)^m}{m^p}\sum_{n=1}^{m}\frac{1}{(2n-1)^q},\quad \omega_8(p,q)=\sum_{m=1}^{\infty}\frac{(-1)^m}{m^p}\sum_{n=1}^{m}\frac{(-1)^n}{(2n-1)^q}, \tag{4.14}$$

$$\omega_9(p,q)=\sum_{m=1}^{\infty}\frac{1}{(2m-1)^p}\sum_{n=1}^{m}\frac{1}{n^q},\quad \omega_{10}(p,q)=\sum_{m=1}^{\infty}\frac{1}{(2m-1)^p}\sum_{n=1}^{m}\frac{(-1)^n}{n^q}, \tag{4.15}$$

$$\omega_{11}(p,q)=\sum_{m=1}^{\infty}\frac{1}{(2m-1)^p}\sum_{n=1}^{m}\frac{1}{(2n-1)^q},\quad \omega_{12}(p,q)=\sum_{m=1}^{\infty}\frac{1}{(2m-1)^p}\sum_{n=1}^{m}\frac{(-1)^n}{(2n-1)^q}, \tag{4.16}$$

$$\omega_{13}(p,q)=\sum_{m=1}^{\infty}\frac{(-1)^m}{(2m-1)^p}\sum_{n=1}^{m}\frac{1}{n^q},\quad \omega_{14}(p,q)=\sum_{m=1}^{\infty}\frac{(-1)^m}{(2m-1)^p}\sum_{n=1}^{m}\frac{(-1)^n}{n^q}, \tag{4.17}$$

$$\omega_{15}(p,q)=\sum_{m=1}^{\infty}\frac{(-1)^m}{(2m-1)^p}\sum_{n=1}^{m}\frac{1}{(2n-1)^q},\quad \omega_{16}(p,q)=\sum_{m=1}^{\infty}\frac{(-1)^m}{(2m-1)^p}\sum_{n=1}^{m}\frac{(-1)^n}{(2n-1)^q}. \tag{4.18}$$

Sitaramachandrarao [196] introduced the series $H_j(s)$, $j=1,2,\dots,12$, which can be expressed in terms of $\omega_j(p,q)$ as follows:

$$H_1(s)=\omega_1(s,1), \tag{4.19}$$

$$H_2(s)=\omega_3(s,1)-\frac{1}{2}\omega_1(s,1), \tag{4.20}$$

$$H_3(s)=\omega_3(s,1)+\frac{1}{2}\omega_1(s,1)-\frac{1}{2}\zeta_{s+1}(1), \tag{4.21}$$

$$H_4(s)=-\omega_2(s,1), \tag{4.22}$$

$$H_5(s)=-\frac{1}{2}\omega_5(s,1)+\omega_6(s,1)+\omega_7(s,1), \tag{4.23}$$

$$H_6(s)=\omega_3(s,1)-\frac{1}{2}\omega_1(s,1)+\frac{1}{2}\zeta_{s+1}(1), \tag{4.24}$$

$$H_7(s)=-\omega_5(s,1), \tag{4.25}$$

$$H_8(s)=\frac{1}{2}\omega_5(s,1)-\omega_7(s,1), \tag{4.26}$$

$$H_9(s)=-\frac{1}{2}\omega_5(s,1)-\omega_7(s,1)+\frac{1}{2}\zeta_{s+1}(-1), \tag{4.27}$$

$$H_{10}(s)=\omega_6(s,1), \tag{4.28}$$

$$H_{11}(s) = -\omega_2(s,1) - \omega_3(s,1) + \frac{1}{2}\omega_1(s,1), \tag{4.29}$$

$$H_{12}(s) = \frac{1}{2}\omega_5(s,1) - \omega_7(s,1) - \frac{1}{2}\zeta_{s+1}(-1). \tag{4.30}$$

Regarding the expression $\omega_1(s,1)$ for integer $s = a \geqslant 2$, according to [196], *Euler* was the first to obtain the result for $\omega_1(2,1)$ [2, 4]. *Nielsen* [24] developed a method to derive the formula for $\omega_1(a,1)$ and to obtain some similar results. For $a = 2$, the formula for $\omega_1(a,1)$ was rediscovered by *Ramanujan* [65], *Briggs* etc. [60], *Klamkin* [52], and *Bruckman* [65]. In the case $a = 3$, and expression for $\omega_1(3,1)$ in a somewhat different form was obtained by *Rutledge* and *Douglass* [43], who attributed this *Morley*. This expression also occurs in the problem suggested by *Klamkin* [62]. *Williams* [56] in 1953 and *Georghiou* and *Philippou* [167] in 1983 rediscovered the formula for $\omega_1(a,1)$, and in 1979, *Sitaramachandrarao* and *Sivaramasarma* [143] provided a new proof of this formula based on *Apostol's* extension of the transformation formula of *Lehner* and *Newman* [126]. *Matsuoka* [164], and *Apostol. Vu* [175] independently discussed the nature of the analytic extension for $H_1(s)$. *Jordan* was the first to obtain the formula for $\omega_3(2b,1)$ [107] when b is a positive integer. *Gupta* [79] obtained a result for $\omega_3(2,1)$, whereas a formula for $\omega_3(2,1)$ appears in the paper by *Ramanujan* [65]. *Sitaramachandrarao* [196] obtained expressions for $\omega_2(a,1)$, $\omega_5(2b,1)$, and $\omega_6(2b,1)$. All these results are only particular cases of the results obtained in this chapter.

Regarding the term multiple sum, the following should be noted: it is intuitively obvious that in the most general case, the sum of multiplicity r should be the sum in which r signs $\sum$ appear. One such example of a sufficiently broad class of sums of multiplicity r and series S contains the elements

$$\sum_{k_1=1}^{\infty} \frac{z^{a_1 k_1}}{P_1^{\lambda_1}} \sum_{k_2=1}^{Q_1} \frac{z^{a_2 k_2}}{P_2^{\lambda_2}} \sum_{k_3=1}^{Q_2} \frac{z^{a_3 k_3}}{P_3^{\lambda_3}} \cdots \sum_{k_r=1}^{Q_{r-1}} \frac{z^{a_r k_r}}{P_r^{\lambda_r}} \tag{4.31}$$

satisfying the following conditions:

(1) $a_j = 0$ or $a_j \in \mathbb{N}$ $(j = 1,2,\ldots,r)$.

(2) $\lambda_j \in \mathbb{N}$ $(j = 1,2,\ldots,r)$.

(3) $S = \sum_{j=1}^{r} \lambda_j$.

(4)

$$P_j = P_j(k_1,k_2,\ldots,k_j) = \sum_{l=1}^{j} m_j^{(l)} k_l + n_j \quad (j = 1,2,\ldots,r), \tag{4.32}$$

$$m_j^{(l)} \in \mathbb{Z}, \quad n_j \in \mathbb{Z}, \quad P_j \in \mathbb{N} \quad (j = 1,2,\ldots,r;\ l = 1,2,\ldots,j). \tag{4.33}$$

(5)

$$Q_j = Q_j(k_1,k_2,\ldots,k_j) = \sum_{l=1}^{j} p_j^{(l)} k_l + q_j \quad (j = 1,2,\ldots,r-1), \tag{4.34}$$

$$p_j^{(l)} \in \mathbb{Z}, \quad q_j \in \mathbb{Z}, \quad P_j \in \mathbb{N} \quad \text{or} \quad P_j = \infty \quad (j = 1,2,\ldots,r-1;\ l = 1,2,\ldots,j). \tag{4.35}$$

First of all, let us introduce the following functions:

$$\zeta_k(z)=\sum_{n=1}^{\infty}\frac{z^n}{n^k},\quad F_{(p,q)}(z)=\sum_{m=1}^{\infty}\frac{z^m}{m^p}\sum_{n=1}^{m}\frac{z^n}{n^q},\quad G_{(p,q)}(z)=\sum_{m=1}^{\infty}\frac{z^m}{m^p}\sum_{n=1}^{m}\frac{1}{n^q},\tag{4.36}$$

$$H_{(p,q)}(z)=\sum_{m=1}^{\infty}\frac{1}{m^p}\sum_{n=1}^{m}\frac{z^n}{n^q},\quad I_{(p,q)}(z)=\sum_{m=1}^{\infty}\frac{z^m}{m^p}\sum_{n=1}^{m}\frac{(\overline{z})^n}{n^q},\quad J_{(p,q)}(z)=\sum_{m=1}^{\infty}\frac{z^m}{m^p}\sum_{n=1}^{m}\frac{z^{2n}}{n^q}.\tag{4.37}$$

The function $I_{(p,q)}(z)$ is defined for $|z|=1$, whereas the other functions are defined for $|z|\leqslant 1$, $z\in\mathbb{C}$. In addition to these functions, we will also use the following decompositions in the derivation process (for the proof, see (4.A29) in the Appendix to this chapter) [102, 119]:

$$\frac{1}{n^p(m+n)^q}=(-1)^p\sum_{l=1}^{p}(-1)^l\binom{p+q-1-l}{q-1}\frac{1}{m^{p+q-l}n^l}+(-1)^p\sum_{l=1}^{q}\binom{p+q-1-l}{p-1}\frac{1}{m^{p+q-l}(m+n)^l},\tag{4.38}$$

$$\frac{1}{n^p(m-n)^q}=\sum_{l=1}^{p}\binom{p+q-1-l}{q-1}\frac{1}{m^{p+q-l}n^l}+\sum_{l=1}^{q}\binom{p+q-1-l}{p-1}\frac{1}{m^{p+q-l}(m-n)^l}.\tag{4.39}$$

Relations (4.38) and (4.39) can be written as follows:

$$\begin{aligned}\frac{1}{n^p}\left(\frac{1}{(m+n)^q}-\frac{1}{m^q}\right)&=(-1)^p\sum_{l=2}^{p-1}(-1)^l\binom{p+q-1-l}{q-1}\frac{1}{m^{p+q-l}n^l}\\&\quad+(-1)^p\sum_{l=2}^{q}\binom{p+q-1-l}{p-1}\frac{1}{m^{p+q-l}(m+n)^l}\\&\quad+(-1)^p\binom{p+q-2}{p-1}\frac{1}{m^{p+q-1}}\left(\frac{1}{m+n}-\frac{1}{n}\right),\end{aligned}\tag{4.40}$$

$$\begin{aligned}\frac{1}{n^p}\left(\frac{1}{(m-n)^q}-\frac{1}{m^q}\right)&=\sum_{l=2}^{p-1}\binom{p+q-1-l}{q-1}\frac{1}{m^{p+q-l}n^l}\\&\quad+\sum_{l=2}^{q}\binom{p+q-1-l}{p-1}\frac{1}{m^{p+q-l}(m-n)^l}\\&\quad+\binom{p+q-2}{p-1}\frac{1}{m^{p+q-1}}\left(\frac{1}{m-n}+\frac{1}{n}\right).\end{aligned}\tag{4.41}$$

We will multiply equations (4.40) and (4.41) by z^{am+bn} $(m,n\in\mathbb{N},\ a,b\in\mathbb{Z})$ and then sum by the arranged pairs (m,n) on some subset of the set $\mathbb{N}\times\mathbb{N}$. Where z appears with a negative integer exponent in the sums, we will take $\overline{z}$ instead of z^{-1}, and thus the summation will be valid for $z\in\mathbb{C}$, $|z|=1$. First, we multiply relation (4.40) by z^{am+bn} and sum by $\sum_{n=1}^{\infty}\sum_{m=n+1}^{\infty}$:

$$
\begin{aligned}
L_1 &= \sum_{n=1}^{\infty}\sum_{m=n+1}^{\infty}\frac{z^{am+bn}}{n^p}\left(\frac{1}{(m+n)^q}-\frac{1}{m^q}\right)\\
&= \sum_{n=1}^{\infty}\frac{z^{(b-a)n}}{n^p}\sum_{m=n+1}^{\infty}\frac{z^{a(m+n)}}{(m+n)^q}-\sum_{n=1}^{\infty}\frac{z^{bn}}{n^p}\sum_{m=n+1}^{\infty}\frac{z^{am}}{m^q}\\
&= \sum_{n=1}^{\infty}\frac{z^{(b-a)n}}{n^p}\sum_{m=2n+1}^{\infty}\frac{z^{am}}{m^q}-\left(\zeta_q(z^a)\zeta_p(z^b)-\sum_{n=1}^{\infty}\frac{z^{bn}}{n^p}\sum_{m=1}^{\infty}\frac{z^{am}}{m^q}\right)\\
&= \sum_{m=1}^{\infty}\frac{z^{bm}}{m^p}\sum_{n=1}^{m}\frac{z^{an}}{n^q}+\zeta_q(z^a)[\zeta_p(z^{b-a})-\zeta_p(z^b)]-\sum_{m=1}^{\infty}\frac{z^{(b-a)m}}{m^p}\sum_{n=1}^{2m}\frac{z^{an}}{n^q},
\end{aligned}
\tag{4.42}
$$

$$
\begin{aligned}
D_{11} &= (-1)^p\sum_{l=2}^{p-1}(-1)^l\binom{p+q-1-l}{q-1}\sum_{m=2}^{\infty}\frac{z^{am}}{m^{p+q-l}}\sum_{n=1}^{m-1}\frac{z^{bn}}{n^l}\\
&= (-1)^p\sum_{l=2}^{p-1}(-1)^l\binom{p+q-1-l}{q-1}\cdot\left(\sum_{m=1}^{\infty}\frac{z^{am}}{m^{p+q-l}}\sum_{n=1}^{m}\frac{z^{bn}}{n^l}-\zeta_{p+q}(z^{a+b})\right),
\end{aligned}
\tag{4.43}
$$

$$
\begin{aligned}
D_{21} &= (-1)^p\sum_{l=2}^{q}\binom{p+q-1-l}{p-1}\sum_{m=2}^{\infty}\frac{z^{(a-b)m}}{m^{p+q-l}}\sum_{n=1}^{m-1}\frac{z^{b(m+n)}}{(m+n)^l}\\
&= (-1)^p\sum_{l=2}^{q}\binom{p+q-1-l}{p-1}\sum_{m=2}^{\infty}\frac{z^{(a-b)m}}{m^{p+q-l}}\sum_{n=m+1}^{2m-1}\frac{z^{bn}}{n^l}\\
&= (-1)^p\sum_{l=2}^{q}\binom{p+q-1-l}{p-1}\Big(\sum_{m=1}^{\infty}\frac{z^{(a-b)m}}{m^{p+q-l}}\sum_{n=1}^{2m}\frac{z^{bn}}{n^l}-\sum_{m=1}^{\infty}\frac{z^{(a-b)m}}{m^{p+q-l}}\sum_{n=1}^{m}\frac{z^{bn}}{n^l}-\frac{1}{2^l}\zeta_{(p+q)}(z^{a+b})\Big),
\end{aligned}
\tag{4.44}
$$

$$
\begin{aligned}
D_{31} &= (-1)^p\binom{p+q-2}{p-1}\left(\sum_{m=2}^{\infty}\frac{z^{(a-b)m}}{m^{p+q-1}}\sum_{n=1}^{m-1}\frac{z^{b(m+n)}}{m+n}-\sum_{m=2}^{\infty}\frac{z^{am}}{m^{p+q-1}}\sum_{n=1}^{m-1}\frac{z^{bn}}{n}\right)\\
&= (-1)^p\binom{p+q-2}{p-1}\left(\sum_{m=1}^{\infty}\frac{z^{(a-b)m}}{m^{p+q-1}}\sum_{n=1}^{2m}\frac{z^{bn}}{n^l}-\frac{1}{2}\zeta_{p+q}(z^{a+b})\right.\\
&\quad\left.-\sum_{m=1}^{\infty}\frac{z^{(a-b)m}}{m^{p+q-1}}\sum_{n=1}^{m}\frac{z^{bn}}{n}-\left[\sum_{m=1}^{\infty}\frac{z^{am}}{m^{p+q-1}}\sum_{n=1}^{m}\frac{z^{bn}}{n}-\zeta_{p+q}(z^{a+b})\right]\right),
\end{aligned}
\tag{4.45}
$$

$$
L_1 = D_{11}+D_{21}+D_{31} \Rightarrow \tag{4.46}
$$

$$
\begin{aligned}
&\zeta_q(z^a)(\zeta_p(z^{b-a})-\zeta_p(z^b))-\sum_{m=1}^{\infty}\frac{z^{(b-a)m}}{m^p}\sum_{n=1}^{2m}\frac{z^{an}}{n^q}+\sum_{m=1}^{\infty}\frac{z^{bm}}{m^p}\sum_{n=1}^{m}\frac{z^{an}}{n^q}\\
&\quad= (-1)^p\sum_{l=1}^{p-1}(-1)^l\binom{p+q-1-l}{q-1}\sum_{m=1}^{\infty}\frac{z^{am}}{m^{p+q-l}}\sum_{n=1}^{m}\frac{z^{bn}}{n^l}\\
&\quad+(-1)^p\sum_{l=1}^{q}\binom{p+q-1-l}{p-1}\sum_{m=1}^{\infty}\frac{z^{(a-b)m}}{m^{p+q-l}}\sum_{n=1}^{2m}\frac{z^{bn}}{n^l}\\
&\quad+(-1)^{p+1}\sum_{l=1}^{q}\binom{p+q-1-l}{p-1}\sum_{m=1}^{\infty}\frac{z^{(a-b)m}}{m^{p+q-l}}\sum_{n=1}^{m}\frac{z^{bn}}{n^l}+\left(1-\frac{1}{2^q}\right)\zeta_{p+q}(z^{a+b}).
\end{aligned}
\tag{4.47}
$$

It is necessary to prove that

$$\sum_{l=1}^{p}(-1)^l\binom{p+q-1-l}{q-1}+\sum_{l=1}^{q}\frac{1}{2^l}\binom{p+q-1-l}{p-1}=(-1)^p2^{-q}. \tag{4.48}$$

Indeed, let

$$A=\sum_{l=1}^{q}\frac{1}{2^l}\binom{p+q-1-l}{p-1}=2^{-q}\sum_{r=0}^{q-1}2^r\binom{p-1+r}{p-1}=2^{-q}B, \tag{4.49}$$

$$\begin{aligned}
B&=\sum_{r=0}^{q-1}2^r\binom{p-1+r}{p-1}=\sum_{r=0}^{q-1}2^r\left[\binom{p+r}{r+1}-\binom{p-1+r}{r+1}\right]\\
&=\sum_{r=1}^{q}2^{r-1}\binom{p-1+r}{r}-\sum_{r=1}^{q}2^{r-1}\binom{p-2+r}{r}\\
&=\frac{1}{2}\left[-1+2^q\binom{p-1+q}{q}+B\right]-\frac{1}{2}\left[-1+2^q\binom{p-2+q}{q}+\sum_{r=0}^{q-1}2^r\binom{p-2+r}{r}\right]\\
&=\frac{1}{2}B+\frac{1}{2}2^q\binom{p-2+q}{q-1}-\frac{1}{2}\sum_{r=0}^{q-1}2^r\binom{p-2+r}{r}.
\end{aligned} \tag{4.50}$$

It follows that

$$\sum_{r=0}^{q-1}2^r\binom{p-1+r}{r}=2^q\binom{p-2+q}{q-1}-\sum_{r=0}^{q-1}2^r\binom{p-2+r}{r}, \tag{4.51}$$

and taking $p-1$ instead of p, we get

$$\sum_{r=0}^{q-1}2^r\binom{p-2+r}{r}=2^q\binom{p-3+q}{q-1}-\sum_{r=0}^{q-1}2^r\binom{p-3+r}{r}. \tag{4.52}$$

From (4.51) and (4.52) we can conclude that

$$\begin{aligned}
\sum_{r=0}^{q-1}2^r\binom{p-1+r}{r}&=2^q\left(\binom{p-1-1+q}{q-1}-\binom{p-1-2+q}{q-1}\right.\\
&\qquad\left.+\cdots+(-1)^{s+1}\binom{p-1-s+q}{q-1}\right)\\
&\qquad+(-1)^s\sum_{r=0}^{q-1}2^r\binom{p-1-s+r}{r}.
\end{aligned} \tag{4.53}$$

For $s=p-1$, we have

$$\sum_{r=0}^{q-1}2^r\binom{p-1+r}{r}=-2^q\sum_{l=1}^{p-1}(-1)^l\binom{p+q-1-l}{q-1}+(-1)^{p+1}\sum_{r=0}^{q-1}2^r$$

$$= -2^q \sum_{l=1}^{p} (-1)^l \binom{p+q-1-l}{q-1} + (-1)^p. \tag{4.54}$$

Multiplying this relation by 2^{-q}, we get (4.48). Now we also multiply relation (4.40) by z^{am+bn} and sum by $\sum_{n=2}^{\infty}\sum_{m=1}^{n-1}$:

$$\begin{aligned} L_2 &= \sum_{n=2}^{\infty}\sum_{m=1}^{n-1} \frac{z^{am+bn}}{n^p}\left(\frac{1}{(m+n)^q} - \frac{1}{m^q}\right) \\ &= \sum_{n=2}^{\infty} \frac{z^{(b-a)n}}{n^p} \sum_{m=1}^{n-1} \frac{z^{a(m+n)}}{(m+n)^q} - \sum_{n=2}^{\infty} \frac{z^{bn}}{n^p} \sum_{m=1}^{n-1} \frac{z^{am}}{m^q} \\ &= \sum_{m=1}^{\infty} \frac{z^{(b-a)m}}{m^p} \sum_{n=1}^{2m} \frac{z^{an}}{n^q} + \left(1 - \frac{1}{2^q}\right)\zeta_{p+q}(z^{a+b}) - \sum_{m=1}^{\infty} \frac{z^{(b-a)m}}{m^p} \sum_{n=1}^{m} \frac{z^{an}}{n^q} - \sum_{m=1}^{\infty} \frac{z^{bm}}{m^p} \sum_{n=1}^{m} \frac{z^{an}}{n^q}, \end{aligned} \tag{4.55}$$

$$\begin{aligned} D_{12} &= (-1)^p \sum_{l=2}^{p-1} (-1)^l \binom{p+q-1-l}{q-1} \sum_{m=1}^{\infty} \frac{z^{am}}{m^{p+q-l}} \sum_{n=m+1}^{\infty} \frac{z^{bn}}{n^l} \\ &= (-1)^p \sum_{l=2}^{p-1} (-1)^l \binom{p+q-1-l}{q-1} \cdot \left(\zeta_l(z^b)\zeta_{p+q-l}(z^a) - \sum_{m=1}^{\infty} \frac{z^{am}}{m^{p+q-l}} \sum_{n=1}^{m} \frac{z^{bn}}{n^l}\right), \end{aligned} \tag{4.56}$$

$$\begin{aligned} D_{22} &= (-1)^p \sum_{l=2}^{q} \binom{p+q-1-l}{p-1} \sum_{m=1}^{\infty} \frac{z^{(a-b)m}}{m^{p+q-l}} \sum_{n=m+1}^{\infty} \frac{z^{b(m+n)}}{(m+n)^l} \\ &= (-1)^p \sum_{l=2}^{q} \binom{p+q-1-l}{p-1} \cdot \left(\zeta_l(z^b)\zeta_{p+q-l}(z^{a-b}) - \sum_{m=1}^{\infty} \frac{z^{(ab)m}}{m^{p+q-l}} \sum_{n=1}^{2m} \frac{z^{bn}}{n^l}\right), \end{aligned} \tag{4.57}$$

$$\begin{aligned} D_{32} &= (-1)^p \binom{p+q-2}{p-1} \cdot \left(\sum_{m=1}^{\infty} \frac{z^{(a-b)m}}{m^{p+q-1}} \sum_{n=m+1}^{\infty} \frac{z^{b(m+n)}}{m+n} - \sum_{m=1}^{\infty} \frac{z^{am}}{m^{p+q-1}} \sum_{n=m+1}^{\infty} \frac{z^{bn}}{n}\right) \\ &= (-1)^p \binom{p+q-2}{p-1} \Bigg(\zeta_1(z^b)(\zeta_{p+q-1}(z^{a-b}) - \zeta_{p+q-1}(z^a)) \\ &\quad - \sum_{m=1}^{\infty} \frac{z^{(a-b)m}}{m^{p+q-1}} \sum_{n=1}^{2m} \frac{z^{bn}}{n} + \sum_{m=1}^{\infty} \frac{z^{am}}{m^{p+q-1}} \sum_{n=1}^{m} \frac{z^{bn}}{n}\Bigg) \quad (b \neq 0), \end{aligned} \tag{4.58}$$

$$\begin{aligned} D_{32} &= (-1)^p \binom{p+q-2}{p-1} \sum_{m=1}^{\infty} \frac{z^{am}}{m^{p+q-1}} \sum_{n=m+1}^{\infty} \left(\frac{1}{m+n} - \frac{1}{n}\right) = \text{(according to (4.A27))} \\ &= (-1)^p \binom{p+q-2}{p-1} \sum_{m=1}^{\infty} \frac{z^{am}}{m^{p+q-1}} \left(-\sum_{n=1}^{2m} \frac{1}{n} + \sum_{n=1}^{m} \frac{1}{n}\right) \\ &= (-1)^p \binom{p+q-2}{p-1} \left(-\sum_{m=1}^{\infty} \frac{z^{am}}{m^{p+q-1}} \sum_{n=1}^{2m} \frac{1}{n} + \sum_{m=1}^{\infty} \frac{z^{am}}{m^{p+q-1}} \sum_{n=1}^{m} \frac{1}{n}\right) \quad (b = 0). \end{aligned} \tag{4.59}$$

It follows that the expression D_{32} for $b = 0$ can be obtained from the expression D_{32} for $b \neq 0$ if $\zeta_1(z^b)(\zeta_{p+q-1}(z^{a-b}) - \zeta_{p+q-1}(z^a))$ is omitted from D_{32} for $b \neq 0$ and if in the obtained expression, we include $b = 0$. Since $L_1 = D_{12} + D_{22} + D_{32}$, it follows that

$$\begin{aligned}
&\sum_{m=1}^{\infty} \frac{z^{(b-a)m}}{m^p} \sum_{n=1}^{2m} \frac{z^{an}}{n^q} + \left(1 - \frac{1}{2^q}\right)\zeta_{p+q}(z^{a+b}) - \sum_{m=1}^{\infty} \frac{z^{(b-a)m}}{m^p} \sum_{n=1}^{m} \frac{z^{an}}{n^q} - \sum_{m=1}^{\infty} \frac{z^{bm}}{m^p} \sum_{n=1}^{m} \frac{z^{an}}{n^q} \\
&\quad = (-1)^{p+1} \sum_{l=1}^{p-1} (-1)^l \binom{p+q-1-l}{q-1} \sum_{m=1}^{\infty} \frac{z^{am}}{m^{p+q-1}} \sum_{n=1}^{m} \frac{z^{bn}}{n^l} \\
&\qquad + (-1)^p \sum_{l=2}^{p-1} (-1)^l \binom{p+q-1-l}{q-1} \zeta_l(z^b)\zeta_{p+q-l}(z^a) \\
&\qquad + (-1)^{p+1} \sum_{l=1}^{q} \binom{p+q-1-l}{p-1} \sum_{m=1}^{\infty} \frac{z^{(a-b)m}}{m^{p+q-1}} \sum_{n=1}^{2m} \frac{z^{bn}}{n^l} \\
&\qquad + (-1)^p \sum_{l=2}^{q} \binom{p+q-1-l}{p-1} \zeta_l(z^b)\zeta_{p+q-l}(z^{a-b}) \\
&\qquad + (-1)^p \binom{p+q-2}{p-1} \zeta_1(z^b)(\zeta_{p+q-1}(z^{a-b}) - \zeta_{p+q-1}(z^a)) \quad (b \neq 0).
\end{aligned} \tag{4.60}$$

In the case $b = 0$, we have a modified relation (4.60), in which the last term on the right-hand side of the equality is omitted, whereas $b = 0$ should be included in the expression thus obtained. Adding (4.47) and (4.60), we have

$$\begin{aligned}
&\sum_{l=1}^{q} \binom{p+q-1-l}{q-1} \sum_{m=1}^{\infty} \frac{z^{(a-b)m}}{m^{p+q-l}} \sum_{n=1}^{m} \frac{z^{bn}}{n^l} + (-1)^{p+1} \sum_{m=1}^{\infty} \frac{z^{(b-a)m}}{m^p} \sum_{n=1}^{m} \frac{z^{an}}{nq} \\
&\quad = (-1)^p \zeta_q(z^a)(\zeta_p(z^b) - \zeta_p(z^{b-a})) \\
&\qquad + \sum_{l=2}^{p-1} (-1)^l \binom{p+q-1-l}{q-1} \zeta_l(z^b)\zeta_{p+q-l}(z^a) \\
&\qquad + \sum_{l=2}^{q} \binom{p+q-1-l}{p-1} \zeta_l(z^b)\zeta_{p+q-l}(z^{a-b}) \\
&\qquad + \binom{p+q-2}{p-1} \zeta_1(z^b)(\zeta_{p+q-l}(z^{a-b}) - \zeta_{p+q-1}(z^a)) \quad (b \neq 0).
\end{aligned} \tag{4.61}$$

In the case $b = 0$, we obtain a slightly modified relation (4.61), which is derived directly from (4.61) by omitting the last term on the right-hand side of the equality and including $b = 0$. Now multiply relation (4.41) by z^{am+bn} and sum by $\sum_{n=1}^{\infty} \sum_{m=n+1}^{\infty}$:

$$\begin{aligned}
L_3 &= \sum_{n=1}^{\infty} \sum_{m=n+1}^{\infty} \frac{z^{am+bn}}{n^p} \left(\frac{1}{(m-n)^q} - \frac{1}{m^q} \right) \\
&= \sum_{n=1}^{\infty} \frac{z^{(a+b)n}}{n^p} \sum_{m=n+1}^{\infty} \frac{z^{a(m-n)}}{(m-n)^q} - \sum_{n=1}^{\infty} \frac{z^{bn}}{n^p} \sum_{m=n+1}^{\infty} \frac{z^{am}}{m^q}
\end{aligned}$$

$$= \zeta_q(z^a)\cdot(\zeta_p(z^{a+b}) - \zeta_p(z^b)) + \sum_{m=1}^{\infty}\frac{z^{bm}}{m^p}\sum_{n=1}^{m}\frac{z^{an}}{n^q}, \tag{4.62}$$

$$D_{13} = \sum_{l=2}^{p-1}\binom{p+q-1-l}{q-1}\sum_{m=2}^{\infty}\frac{z^{am}}{m^{p+q-l}}\sum_{n=1}^{m-1}\frac{z^{bn}}{n^l}$$

$$= \sum_{l=2}^{p-1}\binom{p+q-1-l}{q-1}\cdot\left(\sum_{m=1}^{\infty}\frac{z^{am}}{m^{p+q-l}}\sum_{n=1}^{m}\frac{z^{bn}}{n^l} - \zeta_{p+q}(z^{a+b})\right), \tag{4.63}$$

$$D_{23} = \sum_{l=2}^{q}\binom{p+q-1-l}{p-1}\sum_{m=2}^{\infty}\frac{z^{(a+b)m}}{m^{p+q-l}}\sum_{n=1}^{m-1}\frac{(\overline{z})^{b(m-n)}}{(m-n)^l}$$

$$= \sum_{l=2}^{q}\binom{p+q-1-l}{p-1}\left(\sum_{m=1}^{\infty}\frac{z^{(a+b)m}}{m^{p+q-l}}\sum_{n=1}^{m}\frac{(\overline{z})^{bn}}{n^l} - \zeta_{p+q}(z^a)\right), \tag{4.64}$$

$$D_{33} = \binom{p+q-2}{p-1}\left(\sum_{m=2}^{\infty}\frac{z^{(a+b)m}}{m^{p+q-1}}\sum_{n=1}^{m-1}\frac{(\overline{z})^{b(m-n)}}{m-n} + \sum_{m=2}^{\infty}\frac{z^{am}}{m^{p+q-1}}\sum_{n=1}^{m-1}\frac{z^{bn}}{n}\right)$$

$$= \binom{p+q-2}{p-1}\cdot\left(-\zeta_{p+q}(z^a) + \sum_{m=1}^{\infty}\frac{z^{(a+b)m}}{m^{p+q-1}}\sum_{n=1}^{m}\frac{(\overline{z})^{bn}}{n}\right.$$

$$\left. + \sum_{m=1}^{\infty}\frac{z^{am}}{m^{p+q-1}}\sum_{n=1}^{m}\frac{z^{bn}}{n} - \zeta_{p+q}(z^{a+b})\right). \tag{4.65}$$

Using the formulas

$$\sum_{l=1}^{p-1}\binom{p+q-1-l}{q-1} = \binom{p+q-1}{q} - 1,$$
$$\sum_{l=1}^{q}\binom{p+q-1-l}{p-1} = \binom{p+q-1}{q}, \tag{4.66}$$

from $L_3 = D_{13} + D_{23} + D_{33}$ it follows that

$$\zeta_q(z^a)\cdot[\zeta_p(z^{a+b}) - \zeta_p(z^b)] + \sum_{m=1}^{\infty}\frac{z^{bm}}{m^p}\sum_{n=1}^{m}\frac{z^{an}}{n^q}$$
$$= \sum_{l=1}^{p-1}\binom{p+q-1-l}{q-1}\sum_{m=1}^{\infty}\frac{z^{am}}{m^{p+q-l}}\sum_{n=1}^{m}\frac{z^{bn}}{n^l}$$
$$+ \sum_{l=1}^{q}\binom{p+q-1-l}{p-1}\sum_{m=1}^{\infty}\frac{z^{(a+b)m}}{m^{p+q-l}}\sum_{n=1}^{m}\frac{(\overline{z})^{bn}}{n^l}$$
$$- \left(\binom{p+q-1}{q} - 1\right)\zeta_{p+q}(z^{a+b}) - \binom{p+q-1}{q}\zeta_{p+q}(z^a). \tag{4.67}$$

Multiplying (4.41) by z^{am+bn} and summing the resulting product by $\sum_{n=2}^{\infty}\sum_{m=1}^{n-1}$, we get

$$L_4 = \sum_{n=2}^{\infty}\sum_{m=1}^{n-1} \frac{z^{am+bn}}{n^p}\left(\frac{1}{(m-n)^q} - \frac{1}{m^q}\right)$$
$$= (-1)^q \sum_{n=2}^{\infty} \frac{z^{(a+b)n}}{n^p} \sum_{m=1}^{n-1} \frac{(\overline{z})^{a(n-m)}}{(n-m)^q} - \sum_{n=2}^{\infty} \frac{z^{bn}}{n^p} \sum_{m=1}^{n-1} \frac{z^{am}}{m^q}$$
$$= (-1)^q\left(\sum_{m=1}^{\infty} \frac{z^{(a+b)m}}{m^p} \sum_{n=1}^{m} \frac{(\overline{z})^{an}}{n^q} - \zeta_{p+q}(z^b)\right) - \sum_{m=1}^{\infty} \frac{z^{bm}}{m^p} \sum_{n=1}^{m} \frac{z^{an}}{n^q} + \zeta_{p+q}(z^{a+b}), \tag{4.68}$$

$$D_{14} = \sum_{l=2}^{p-1} \binom{p+q-1-l}{q-1} \sum_{m=1}^{\infty} \frac{z^{am}}{m^{p+q-l}} \sum_{n=m+1}^{\infty} \frac{z^{bn}}{n^l}$$
$$= \sum_{l=2}^{p-1} \binom{p+q-1-l}{q-1}\left(\zeta_l(z^b)\zeta_{p+q-l}(z^a) - \sum_{m=1}^{\infty} \frac{z^{am}}{m^{p+q-l}} \sum_{n=1}^{m} \frac{z^{bn}}{n^l}\right), \tag{4.69}$$

$$D_{24} = \sum_{l=2}^{q} \binom{p+q-1-l}{p-1} \sum_{m=1}^{\infty} \frac{z^{(a+b)m}}{m^{p+q-l}} \sum_{n=m+1}^{\infty} \frac{z^{b(n-m)}}{(m-n)^l}$$
$$= \sum_{l=2}^{q} (-1)^l \binom{p+q-1-l}{p-1} \zeta_l(z^b)\zeta_{p+q-l}(z^{a+b}), \tag{4.70}$$

$$D_{34} = \binom{p+q-2}{p-1}\left(-\sum_{m=1}^{\infty} \frac{z^{(a+b)m}}{m^{p+q-1}} \sum_{n=m+1}^{\infty} \frac{z^{b(n-m)}}{n-m} + \sum_{m=1}^{\infty} \frac{z^{am}}{m^{p+q-1}} \sum_{n=m+1}^{\infty} \frac{z^{bn}}{n}\right)$$
$$= \binom{p+q-2}{p-1}\left(\zeta_1(z^b)(\zeta_{p+q-1}(z^a) - \zeta_{p+q-1}(z^{a+b})) - \sum_{m=1}^{\infty} \frac{z^{am}}{m^{p+q-1}} \sum_{n=1}^{m} \frac{z^{bn}}{n}\right) \quad (b \neq 0), \tag{4.71}$$

$$D_{34} = \binom{p+q-2}{p-1} \sum_{m=1}^{\infty} \frac{z^{am}}{m^{p+q-1}} \sum_{n=m+1}^{\infty} \left(\frac{1}{m-n} + \frac{1}{n}\right)$$
$$= \text{(with respect to (4.A27))} = -\binom{p+q-2}{p-1} \sum_{m=1}^{\infty} \frac{z^{am}}{m^{p+q-1}} \sum_{n=1}^{m} \frac{1}{n} \quad (b = 0). \tag{4.72}$$

The expression that specifies a value for D_{34} when $b = 0$ can be obtained from the expression D_{34} for $b \neq 0$ by omitting $\zeta_1(z^b)(\zeta_{p+q-1}(z^a) - \zeta_{p+q-1}(z^{a+b}))$ from D_{34} for $b \neq 0$ and including $b = 0$ in the obtained expression. Since $L_4 = D_{14} + D_{24} + D_{34}$, it follows that

$$(-1)^q\left(\sum_{m=1}^{\infty} \frac{z^{(a+b)m}}{m^p} \sum_{n=1}^{m} \frac{(\overline{z})^{am}}{n^q} - \zeta_{p+q}(z^b)\right) - \sum_{m=1}^{\infty} \frac{z^{bm}}{m^p} \sum_{n=1}^{m} \frac{z^{an}}{n^q} + \zeta_{p+q}(z^{a+b})$$
$$= -\sum_{l=1}^{p-1} \binom{p+q-1-l}{q-1} \sum_{m=1}^{\infty} \frac{z^{am}}{m^{p+q-l}} \sum_{n=1}^{m} \frac{z^{bn}}{n^l} + \sum_{l=2}^{p-1} \binom{p+q-1-l}{q-1} \zeta_l(z^b)\zeta_{p+q-l}(z^a)$$
$$+ \sum_{l=2}^{q} (-1)^l \binom{p+q-1-l}{p-1} \zeta_l(z^b)\zeta_{p+q-l}(z^{a+b})$$
$$+ \binom{p+q-2}{p-1} \zeta_1(z^b)(\zeta_{p+q-1}(z^a) - \zeta_{p+q-1}(z^{a+b})) \quad (b \neq 0). \tag{4.73}$$

In the case $b = 0$, relation (4.73) is transformed into a relation of a very similar form, which follows directly from the aforementioned relation by omitting the last term on the right-hand side of the equation and including $b = 0$ in the expression obtained. Analogously to the transformation conducted with respect to b, it is possible to do the same regarding a, provided that $q = 1$. The application of formula (4.A27) (given in the Appendix to this chapter) to the expression defining the sum L_1 (relation (4.42)) allows us to conclude the following: the expression L_1 for $a = 0$ (provided that $q = 1$) can be obtained from the expression L_1 for $a \neq 0$ by omitting $\zeta_q(z^a)(\zeta_p(z^{b-a}) - \zeta_p(z^b))$ from the expression L_1 for $a \neq 0$ and including $a = 0$ and $q = 1$ in the resulting expression. In the case of $a = 0$ and $q = 1$, there is a modified relation (4.47), which is obtained by omitting the first term on the left-hand side of the equation and including $a = 0$ and $q = 1$. In the case of $a = 0$ and $q = 1$, relation (4.61) is transformed into a similar one, where the first term on the right-hand side of the relation is omitted, and $a = 0$ and $q = 1$ are included in the obtained expression. By applying the already mentioned formula (Appendix to this chapter, formula (4.A27)) to the expression for L_3 (relation (4.62)) we conclude that the expression that defines the value of the sum L_3 for $a = 0$ (provided that $q = 1$) can also be derived from the expression L_3 for $a \neq 0$ if we exclude $\zeta_q(z^a)(\zeta_p(z^{a+b}) - \zeta_p(z^b))$ from it and include $a = 0$ and $q = 1$. In the case of $a = 0$ and $q = 1$, we have a relation that can be obtained directly from (4.67) if the first term on the left-hand side of the equation is excluded and $a = 0$ and $q = 1$ are included in the obtained expression. Adding (4.67) and (4.73), we obtain

$$\begin{aligned}\sum_{l=1}^{q}\binom{p+q-1-l}{p-1}\sum_{m=1}^{\infty}\frac{z^{(a+b)m}}{m^{p+q-l}}\sum_{n=1}^{m}\frac{(\overline{z})^{bn}}{n^l} &+ (-1)^{q+1}\sum_{m=1}^{\infty}\frac{z^{(a+b)m}}{m^p}\sum_{n=1}^{m}\frac{(\overline{z})^{an}}{n^q}\\ = \zeta_q(z^a)(\zeta_p(z^{a+b}) - \zeta_p(z^b)) &+ (-1)^{q+1}\zeta_{p+q}(z^b) + \binom{p+q-1}{p}\zeta_{p+q}(z^a)\\ &+ \binom{p+q-1}{q}\zeta_{p+q}(z^{a+b}) - \sum_{l=2}^{p-1}\binom{p+q-1-l}{q-1}\zeta_l(z^b)\zeta_{p+q-l}(z^a)\\ &- \sum_{l=2}^{q}(-1)^l\binom{p+q-1-l}{q-1}\zeta_l(z^b)\zeta_{p+q-l}(z^{a+b})\\ &+ \binom{p+q-2}{p-1}\zeta_1(z^b)(\zeta_{p+q-1}(z^{a+b}) - \zeta_{p+q-1}(z^a)).\end{aligned} \tag{4.74}$$

If $b = 0$, then relation (4.74) is transformed into a similar one when the last term on the right-hand side of the equation is omitted out and $b = 0$ is included. If $a = 0$ and $q = 1$, then we can obtain another modification of relation (4.74) by eliminating the first term on the right-hand side of equation (4.73) and including $a = 0$ and $q = 1$ in the obtained expression. If $b \neq 0$ and we include $-b$ instead of b in (4.74) and subtract (4.61) from the relation obtained, then we obtain

$$(-1)^p\sum_{m=1}^{\infty}\frac{(\overline{z})^{(a-b)m}}{m^p}\sum_{n=1}^{m}\frac{z^{an}}{n^q} + (-1)^{q+1}\sum_{m=1}^{\infty}\frac{z^{(a-b)m}}{m^p}\sum_{n=1}^{m}\frac{(\overline{z})^{an}}{n^q}$$

$$
\begin{aligned}
&= \zeta_q(z^a)(\zeta_p(z^{a-b}) - \zeta_p(\overline{z}^b)) + (-1)^p\zeta_q(z^a)(\zeta_p(z^{b-a}) - \zeta_p(z^b)) \\
&\quad + (-1)^{q+1}\zeta_{p+q}(\overline{z}^b) + \binom{p+q-1}{q}\zeta_{p+q}(z^a) + \binom{p+q-1}{q}\zeta_{p+q}(z^{a-b}) \\
&\quad - \sum_{l=2}^{p-1}\binom{p+q-1-l}{q-1}\zeta_l(\overline{z}^b)\zeta_{p+q-l}(z^a) - \sum_{l=2}^{p-1}(-1)^l\binom{p+q-1-l}{q-1}\zeta_l(z^b)\zeta_{p+q-l}(z^a) \\
&\quad - \sum_{l=2}^{q}\binom{p+q-1-l}{q-1}\zeta_l(z^b)\zeta_{p+q-l}(z^{a-b}) - \sum_{l=2}^{q}(-1)^l\binom{p+q-1-l}{q-1}\zeta_l(\overline{z}^b)\zeta_{p+q-l}(z^{a-b}) \\
&\quad + \binom{p+q-2}{p-1}(\zeta_{p+q-l}(z^a) - \zeta_{p+q-l}(z^{a-b}))(\zeta_1(z^b) - \zeta_1(\overline{z}^b)) \quad (a, b \neq 0).
\end{aligned} \tag{4.75}
$$

If $b = 0$, then a relation similar to relation (4.75) is obtained by eliminating the last term on the right-hand side of equation (4.75) and including $b = 0$. This relation can also be obtained by subtracting the modified relations (4.61) and (4.74), as described above. In the case of $a = 0$ ($q = 1$), we have another modification of relation (4.75), which follows directly from it if we omit the first two terms on the right-hand side of the equation and if we include $a = 0$ ($q = 1$) in the expression obtained. It can also be obtained by subtracting the modified relations (4.61) and (4.74). If $a = 1$ and $b = 2$, then from (4.75) we get

$$
\begin{aligned}
&(-1)^p F_{(p,q)}(z) + (-1)^{q+1}F_{(p,q)}(\overline{z}) \\
&\quad = \zeta_q(z)(\zeta_p(\overline{z}) - \zeta_p(\overline{z}^2)) + (-1)^p\zeta_q(z)(\zeta_p(z) - \zeta_p(z^2)) \\
&\qquad + (-1)^{q+1}\zeta_{p+q}(\overline{z}^2) + \binom{p+q-1}{p}\zeta_{p+q}(z) + \binom{p+q-1}{q}\zeta_{p+q}(\overline{z}) \\
&\qquad - \sum_{l=2}^{p-1}\binom{p+q-1-l}{q-1}\zeta_l(\overline{z}^2)\zeta_{p+q-l}(z) - \sum_{l=2}^{p-1}(-1)^l\binom{p+q-1-l}{q-1}\zeta_l(z^2)\zeta_{p+q-l}(z) \\
&\qquad - \sum_{l=2}^{q}\binom{p+q-1-l}{p-1}\zeta_l(z^2)\zeta_{p+q-l}(\overline{z}) - \sum_{l=2}^{q}(-1)^l\binom{p+q-1-l}{p-1}\zeta_l(\overline{z}^2)\zeta_{p+q-l}(\overline{z}) \\
&\qquad + \binom{p+q-2}{p-1}(\zeta_1(z^2) - \zeta_1(\overline{z}^2))(\zeta_{p+q-1}(z) - \zeta_{p+q-1}(\overline{z})) \quad (|z| = 1).
\end{aligned} \tag{4.76}
$$

We will here determine $\lim_{\substack{|z|=1\\ z\to 1}}$ for the above expression. First, we will prove that

$$
\lim_{\substack{|z|=1\\ z\to 1}} \zeta_q(z)(\zeta_p(\overline{z}) - \zeta_p(\overline{z}^2)) = 0. \tag{4.77}
$$

First of all,

$$
\lim_{\substack{|z|=1\\ z\to 1}} \frac{\zeta_1(z)}{\zeta_1(\overline{z})} = \lim_{\varphi\to 0^+} \frac{-\log(1-\mathrm{e}^{\varphi \mathrm{i}})}{-\log(1-\mathrm{e}^{-\varphi \mathrm{i}})} = \lim_{\varphi\to 0^+} \frac{\log(2\sin\frac{\varphi}{2}) + \frac{\varphi-\pi}{2}\mathrm{i}}{\log(2\sin\frac{\varphi}{2}) + \frac{\pi-\varphi}{2}\mathrm{i}} = 1, \tag{4.78}
$$

where the limits are taken at the points of the upper semicircle. For the points on the lower semicircle, we have

$$\lim_{\substack{|z|=1\\ z\to 1}} \frac{\zeta_1(z)}{-\zeta_1(\overline{z})} = \lim_{\varphi\to 0^+} \frac{-\log(1-e^{-\varphi i})}{-\log(1-e^{\varphi i})} = 1. \tag{4.79}$$

It follows that

$$\lim_{\substack{|z|=1\\ z\to 1}} \frac{\zeta_1(z)}{-\zeta_1(\overline{z})} = 1. \tag{4.80}$$

Further, we can write

$$\lim_{\substack{|z|=1\\ z\to 1}} \zeta_1(z)(\zeta_p(z)-\zeta_p(z)^2) = (\text{according to (4.A7)}) = \lim_{\substack{|z|=1\\ z\to 1}} \frac{\frac{\zeta_{p-1}(z)}{z} - \frac{2}{z}\zeta_{p-1}(z^2)}{\frac{-1}{(1-z)\log^2(1-z)}} = 0. \tag{4.81}$$

The value of the requested limits in the case of $q = 1$ and $p \geqslant 2$ follows from (4.80). The proof in the case of $q \geqslant 2$ is trivial, and there is no need to cite it. Further, since

$$\zeta_1(z^2) - \zeta_1(\overline{z}^2) + (\pi - 2\varphi)\mathrm{i}, \quad z = e^{\varphi \mathrm{i}}, \quad \varphi > 0, \tag{4.82}$$

$$\zeta_1(z^2) - \zeta_1(\overline{z}^2) + (2\varphi - \pi)\mathrm{i}, \quad z = e^{-\varphi \mathrm{i}}, \quad \varphi > 0, \tag{4.83}$$

the limit as z tends to 1 on the circle $|z| = 1$ of the last term on the right-hand side of (4.76) equals zero. Then from (4.76), provided that $p + q = 2k + 1$, it follows that

$$\begin{aligned} F_{(p,2k+1-p)}(1) &= \sum_{m=1}^{\infty} \frac{1}{m^p} \sum_{n=1}^{\infty} \frac{1}{n^{2k+1-p}} = \frac{1}{2}\left(1 + (-1)^p \binom{2k+1}{p}\right)\zeta_{2k+1}(1) \\ &+ (-1)^{p+1} \sum_{2\leqslant 2l\leqslant p-1} \binom{2k-2l}{2k-p} \zeta_{2l}(1)\zeta_{2k+1-2l}(1) \\ &+ (-1)^{p+1} \sum_{2\leqslant 2l\leqslant 2k+1-p} \binom{2k-2l}{p-1} \zeta_{2l}(1)\zeta_{2k+1-2l}(1) \quad (p = 2, 3, \ldots, 2k). \end{aligned} \tag{4.84}$$

By analogy we conclude that the limit as z tends to -1 on the circle $|z| = 1$ of the last term on the right-hand side of (4.75) equals zero. From (4.75), passing to the limit as z tends to -1 on the circle $|z| = 1$, with $p + q = 2k + 1$, we get

$$\begin{aligned} F_{(p,2k+1-p)}(-1) &= \sum_{m=1}^{\infty} \frac{(-1)^m}{m^p} \sum_{n=1}^{\infty} \frac{(-1)^n}{n^{2k+1-p}} = \frac{1}{2}\zeta_{2k+1}(1) \\ &+ \frac{(-1)^p}{2}\binom{2k+1}{p}\zeta_{2k+1}(-1) + \frac{1}{2}(1 + (-1)^p)\zeta_p(-1)\zeta_{2k+1-p}(-1) \\ &+ (-1)^{p+1} \sum_{2\leqslant 2l\leqslant p} \binom{2k-2l}{2k-p} \zeta_{2l}(1)\zeta_{2k+1-2l}(-1) \\ &+ (-1)^{p+1} \sum_{2\leqslant 2l\leqslant 2k+1-p} \binom{2k-2l}{p-1} \zeta_{2l}(1)\zeta_{2k+1-2l}(-1) \quad (p = 1, 2, \ldots, 2k). \end{aligned} \tag{4.85}$$

In fact, formula (4.85) is first determined for $p = 2, 3, \dots, 2k$, and then by using formula (4.85) for $p = 2k$ and formula (4.16), we get formula (4.85) for $p = 1$. Further, by the modified formula (4.75) for $b = 1$, we get

$$\begin{aligned}
&(-1)^p G_{(p,q)}(z) + (-1)^{q+1} G_{(p,q)}(z) \\
&\quad = \binom{p+q-1}{p}\zeta_{p+q}(1) \\
&\quad + \left((-1)^{q+1} + \binom{p+q-1}{q}\right)\zeta_{p+q}(\overline{z}) - \sum_{l=2}^{p-1}\binom{p+q-1-l}{q-1}\zeta_l(\overline{z})\zeta_{p+q-l}(1) \\
&\quad - \sum_{l=2}^{p-1}(-1)^l\binom{p+q-1-l}{q-1}\zeta_l(z)\zeta_{p+q-l}(1) - \sum_{l=2}^{q}\binom{p+q-1-l}{p-1}\zeta_l(z)\zeta_{p+q-l}(\overline{z}) \\
&\quad - \sum_{l=2}^{q}(-1)^l\binom{p+q-1-l}{p-1}\zeta_l(\overline{z})\zeta_{p+q-l}(\overline{z}) \\
&\quad + \binom{p+q-2}{p-1}(\zeta_{p+q-1}(1) - \zeta_{p+q-1}(\overline{z}))(\zeta_1(z) - \zeta_1(\overline{z})) \quad (|z| = 1).
\end{aligned} \tag{4.86}$$

For $p + q = 2k + 1$, from (4.86) for $z = -1$, it follows that

$$\begin{aligned}
G_{(p,2k+1-p)}(-1) &= \sum_{m=1}^{\infty}\frac{(-1)^m}{m^p}\sum_{n=1}^{m}\frac{1}{n^{2k+1-p}} = \frac{1}{2}\left(1 + (-1)^p\binom{2k}{p-1}\right)\zeta_{2k+1}(-1) \\
&+ \frac{(-1)^p}{2}\binom{2k}{p}\zeta_{2k+1}(-1) + (-1)^{p+1}\sum_{2\leqslant 2l\leqslant p-1}\binom{2k-2l}{2k-p}\zeta_{2l}(-1)\zeta_{2k+1-2l}(1) + r \\
&+ (-1)^{p+1}\sum_{2\leqslant 2l\leqslant 2k+1-p}\binom{2k-2l}{p-1}\zeta_{2l}(-1)\zeta_{2k+1-2l}(-1) \quad (p = 1, 2, \dots, 2k).
\end{aligned} \tag{4.87}$$

For $a = 1$ and $b = 1$, from (4.75) we have

$$\begin{aligned}
&(-1)^p H_{(p,q)}(z) + (-1)^{q+1} H_{(p,q)}(\overline{z}) \\
&\quad = \zeta_q(z)(\zeta_p(1) - \zeta_p(\overline{z})) + (-1)^p\zeta_q(z)(\zeta_p(1) - \zeta_p(z)) \\
&\quad + \binom{p+q-1}{q}\zeta_{p+q}(1) + \binom{p+q-1}{p}\zeta_{p+q}(z) + (-1)^{q+1}\zeta_{p+q}(\overline{z}) \\
&\quad - \sum_{l=2}^{p-1}\binom{p+q-1-l}{q-1}\zeta_l(\overline{z})\zeta_{p+q-l}(z) - \sum_{l=2}^{p-1}(-1)^l\binom{p+q-1-l}{q-1}\zeta_l(z)\zeta_{p+q-l}(z) \\
&\quad - \sum_{l=2}^{q}\binom{p+q-1-l}{p-1}\zeta_l(z)\zeta_{p+q-l}(1) - \sum_{l=2}^{q}(-1)^l\binom{p+q-1-l}{p-1}\zeta_l(\overline{z})\zeta_{p+q-l}(1) \\
&\quad + \binom{p+q-2}{p-1}(\zeta_{p+q-l}(z) - \zeta_{p+q-l}(1))(\zeta_1(z) - \zeta_1(\overline{z})) \quad (|z| = 1).
\end{aligned} \tag{4.88}$$

For $p+q=2k+1$, from (4.88) when $z=-1$, it follows that

$$\begin{aligned} H_{(p,2k+1-p)}(-1) &= \sum_{m=1}^{\infty}\frac{1}{m^p}\sum_{n=1}^{m}\frac{(-1)^n}{n^{2k+1-p}} \\ &= \zeta_p(1)\zeta_{2k+1-p}(-1)+\frac{1}{2}\left(1+(-1)^p\binom{2k}{p}\right)\zeta_{2k+1}(-1) \\ &\quad + \frac{(-1)^p}{2}\binom{2k}{p-1}\zeta_{2k+1}(1)+(-1)^{p+1}\sum_{2\leqslant 2l\leqslant 2k-p}\binom{2k-2l}{2k-p}\zeta_{2l}(-1)\zeta_{2k+1-2l}(1) \\ &\quad + (-1)^{p+1}\sum_{2\leqslant 2l\leqslant 2k-p}\binom{2k-2l}{p-1}\zeta_{2l}(-1)\zeta_{2k+1-2l}(1) \quad (p=2,3,\dots,2k). \end{aligned} \tag{4.89}$$

Further, from the modified relation (4.75) for $a=-1$, we obtain

$$\begin{aligned} &(-1)^p I_{(p,q)}(z)+(-1)^{q+1}I_{(p,q)}(\overline{z}) \\ &\quad = \zeta_q(\overline{z})(\zeta_p(\overline{z})-\zeta_p(1))+(-1)^p\zeta_q(\overline{z})(\zeta_p(z)-\zeta_p(1))+(-1)^{q+1}\zeta_{p+q}(1)+\binom{p+q}{p}\zeta_{p+q}(\overline{z}) \\ &\quad\quad - 2\sum_{2\leqslant 2l\leqslant p-1}\binom{p+q-1-2l}{q-1}\zeta_{2l}(1)\zeta_{p+q-2l}(\overline{z}) \\ &\quad\quad - 2\sum_{2\leqslant 2l\leqslant q}\binom{p+q-1-2l}{p-1}\zeta_{2l}(1)\zeta_{p+q-2l}(\overline{z}) \quad (|z|=1). \end{aligned} \tag{4.90}$$

For $a=2$ and $b=3$, from (4.75) it follows that

$$\begin{aligned} &(-1)^p J_{(p,q)}(z)+(-1)^{q+1}J_{(p,q)}(\overline{z}) \\ &\quad = \zeta_q(z^2)(\zeta_p(\overline{z})-\zeta_p(\overline{z}^3))+(-1)^p\zeta_q(z^2)(\zeta_p(z)-\zeta_p(z^3)) \\ &\quad\quad + (-1)^{q+1}\zeta_{p+q}(\overline{z}^3)+\binom{p+q-1}{p}\zeta_{p+q}(z^2)+\binom{p+q-1}{q}\zeta_{p+q}(\overline{z}) \\ &\quad\quad - \sum_{l=2}^{p-1}\binom{p+q-1-l}{q-1}\zeta_l(\overline{z}^3)\zeta_{p+q-l}(z^2) \\ &\quad\quad - \sum_{l=2}^{p-1}(-1)^l\binom{p+q-1-l}{q-1}\zeta_l(z^3)\zeta_{p+q-l}(z^2) \\ &\quad\quad - \sum_{l=2}^{q}\binom{p+q-1-l}{p-1}\zeta_l(z^3)\zeta_{p+q-l}(\overline{z}) \\ &\quad\quad - \sum_{l=2}^{q}(-1)^l\binom{p+q-1-l}{p-1}\zeta_l(\overline{z}^3)\zeta_{p+q-l}(\overline{z}) \\ &\quad\quad + \binom{p+q-2}{p-1}(\zeta_l(z^2)-\zeta_{p+q-l}(\overline{z}))(\zeta_1(z^3)-\zeta_1(\overline{z}^3)) \quad (|z|=1). \end{aligned} \tag{4.91}$$

In addition to the formulas obtained for $\omega_1(p,q)$, $\omega_2(p,q)$, $\omega_5(p,q)$, and $\omega_6(p,q)$ (formulas (4.11)–(4.13)), it is necessary to determine other functional relations, coefficients of the form $\omega_j(p,q)$, under certain conditions for the number $p+q$, as it was stated at the beginning of this chapter. Let

$$A(1)=\sum_{m=1}^{\infty}\frac{1}{(2m-1)^p m^q},\quad A(-1)=\sum_{m=1}^{\infty}\frac{(-1)^m}{(2m-1)^p m^q}. \tag{4.92}$$

From the formula

$$\begin{aligned}\frac{1}{y^p(x+y)^q}&=(-1)^p\sum_{l=2}^{p-1}(-1)^l\binom{p+q-1-2l}{q-1}\frac{1}{y^l x^{p+q-l}}\\&\quad+(-1)^p\sum_{l=1}^{q}\binom{p+q-1-l}{p-1}\frac{1}{(x+y)^l x^{p+q-l}}\end{aligned} \tag{4.93}$$

for $x=1$ and $y=2m-1$, using the relations

$$\begin{aligned}&\sum_{m=1}^{\infty}\frac{1}{(2m-1)^l}=\left(1-\frac{1}{2^l}\right)\zeta_l(1)\quad(l\geqslant 2),\\&\sum_{m=1}^{\infty}\frac{(-1)^m}{(2m-1)^l}=-\beta_l(-1)\quad(l\geqslant 1),\end{aligned} \tag{4.94}$$

we get

$$\begin{aligned}\sum_{m=1}^{\infty}\frac{1}{(2m-1)^p m^q}&=(-1)^p2^q\Bigg(\sum_{l=2}^{p}(-1)^l\binom{p+q-1-l}{q-1}\left(1-\frac{1}{2^l}\right)\zeta_l(1)\\&\quad+\sum_{l=2}^{q}\frac{1}{2^l}\binom{p+q-1-l}{p-1}\zeta_l(1)-\binom{p+q-2}{p-1}\log 2\Bigg),\end{aligned} \tag{4.95}$$

$$\begin{aligned}\sum_{m=1}^{\infty}\frac{1}{(2m-1)^p m^q}&=(-1)^p2^q\Bigg(-\sum_{l=1}^{p}(-1)^l\binom{p+q-1-l}{q-1}\beta_l(-1)\\&\quad+\sum_{l=1}^{q}\frac{1}{2^l}\binom{p+q-1-l}{p-1}\zeta_l(-1)\Bigg).\end{aligned} \tag{4.96}$$

For $\omega_j(p,q)$, we can form the relations

$$\omega_5(p,q)=-\omega_2(q,p)+\zeta_q(1)\zeta_p(-1)+\zeta_{p+q}(-1), \tag{4.97}$$

$$\omega_9(p,q)=-\omega_3(q,p)+\left(1-\frac{1}{2^p}\right)\zeta_q(1)\zeta_q(1)+A(1), \tag{4.98}$$

$$\omega_{13}(p,q)=-\omega_4(q,p)-\beta_p(-1)\zeta_q(1)+A(-1), \tag{4.99}$$

$$\omega_{10}(p,q)=-\omega_7(q,p)+\left(1-\frac{1}{2^p}\right)\zeta_p(1)\zeta_q(-1)+A(-1), \tag{4.100}$$

$$\omega_{14}(p,q) = -\omega_8(q,p) - \beta_p(-1)\zeta_q(-1) + A(1), \tag{4.101}$$

$$\omega_{15}(p,q) = -\omega_{12}(q,p) - \beta_p(-1)\left(1 - \frac{1}{2^p}\right)\zeta_q(-1) - \beta_{p+q}(-1), \tag{4.102}$$

as well as four relations for "symmetric coefficients":

$$\omega_1(p,q) + \omega_1(q,p) = \zeta_p(1)\zeta_q(1) + \zeta_{p+q}(1), \tag{4.103}$$

$$\omega_6(p,q) + \omega_6(q,p) = \zeta_p(-1)\zeta_q(-1) + \zeta_{p+q}(1), \tag{4.104}$$

$$\omega_{11}(p,q) + \omega_{11}(q,p) = \left(1 - \frac{1}{2^p}\right)\left(1 - \frac{1}{2^q}\right)\zeta_p(1)\zeta_q(1) + \left(1 - \frac{1}{2^{p+q}}\right)\zeta_{p+q}(1), \tag{4.105}$$

$$\omega_{16}(p,q) + \omega_{16}(q,p) = \beta_p(-1)\beta_q(-1) + \left(1 - \frac{1}{2^{p+q}}\right)\zeta_{p+q}(1). \tag{4.106}$$

Next,

$$\omega_1(p,q) = G_{(p,q)}(1) = \frac{1}{2^p}\sum_{m=1}^{\infty}\frac{1}{m^p}\sum_{n=1}^{2m}\frac{1}{n^q} + \sum_{m=1}^{\infty}\frac{1}{(2m-1)^p}\sum_{n=1}^{2m-1}\frac{1}{n^q}, \tag{4.107}$$

$$\omega_5(p,q) = G_{(p,q)}(-1) = \frac{1}{2^p}\sum_{m=1}^{\infty}\frac{1}{m^p}\sum_{n=1}^{2m}\frac{1}{n^q} - \sum_{m=1}^{\infty}\frac{1}{(2m-1)^p}\sum_{n=1}^{2m-1}\frac{1}{n^q}. \tag{4.108}$$

From the last relation it follows that

$$\omega_1(p,q) + \omega_5(p,q) = \frac{1}{2^{p-1}}\left(\frac{1}{2^q}\omega_1(p,q) + \omega_3(p,q)\right). \tag{4.109}$$

From (4.109) for $p + q = 2k + 1$ by (4.84) and (4.87) it follows that

$$\begin{aligned} &\omega_3(p, 2k+1-p) \\ &\quad = \sum_{m=1}^{\infty}\frac{1}{m^p}\sum_{n=1}^{m}\frac{1}{(2n-1)^{2k+1-p}} \\ &\quad = (-1)^p 2^{-2k-1+p}\left\{\frac{1}{2}(2^{2k+1}-1)\binom{2k}{p}\zeta_{2k+1}(1) - \sum_{2\leqslant 2l\leqslant p-1}(2^{2k+1-2l}-1)\binom{2k-2l}{2k-p}\zeta_{2l}(1)\zeta_{2k+1-2l}(1)\right. \\ &\qquad \left. - \sum_{2\leqslant 2l\leqslant 2k+1-p}(2^{2l}-1)(2^{2k+1-2l}-1)\cdots\binom{2k-2l}{p-1}\zeta_{2l}(1)\zeta_{2k+1-2l}(1)\right\} \quad (p = 2,3,\ldots,2k). \end{aligned} \tag{4.110}$$

Further, we have

$$\omega_1(p,q) = \frac{1}{2^p}\left(\frac{1}{2^q}\omega_1(p,q) + \omega_3(p,q)\right) + \frac{1}{2^q}(\omega_9(p,q) - A(1)) + \omega_{11}(p,q), \tag{4.111}$$

$$\omega_6(p,q) = \frac{1}{2^p}\left(\frac{1}{2^q}\omega_1(p,q) - \omega_3(p,q)\right) - \frac{1}{2^q}(\omega_9(p,q) - A(1)) + \omega_{11}(p,q), \tag{4.112}$$

from which it follows that

$$2\omega_{11}(p,q) = \left(1 - \frac{1}{2^{p+q-1}}\right)\omega_1(p,q) + \omega_6(p,q). \tag{4.113}$$

The formula for $\omega_{11}(p,q)$ follows from (4.113) for $p + q = 2k + 1$ and from (4.84) and (4.85), since

$$\omega_1(p,q) - \omega_5(p,q) = 2^{1-q}(\omega_9(p,q) - A(1)) + 2\omega_{11}(p,q). \tag{4.114}$$

From (4.114) and (4.113) for $p+q = 2k+1$ and from (4.84), (4.85), and (4.87) we obtain the formula for $\omega_9(p,q)$:

$$\begin{aligned}
\omega_{11}(p,q) &= \sum_{m=1}^{\infty} \frac{1}{(2m-1)^p} \sum_{n=1}^{m} \frac{1}{(2n-1)^{2k+1-p}} \\
&= (-1)^p 2^{-2k-1} \left\{ \frac{(-1)^p}{2} (2^{2k+1} - 1)\zeta_{2k+1}(1) \right. \\
&\quad - \frac{1+(-1)^p}{2} 2^{2k+1-p} (2^p - 1)\zeta_p(1)\zeta_{2k+1}(-1) \\
&\quad - \sum_{2 \leqslant 2l \leqslant p-1} (2^{2l} - 1)\binom{2k-2l}{2k-p} \zeta_{2l}(1)\zeta_{2k+1-2l}(1) \\
&\quad \left. - \sum_{2 \leqslant 2l \leqslant 2k+1-p} (2^{2l} - 1)\binom{2k-2l}{p-1} \zeta_{2l}(1)\zeta_{2k+1-2l}(1) \right\} \quad (p = 2, 3, \dots, 2k),
\end{aligned} \tag{4.115}$$

$$\begin{aligned}
\omega_9(p,q) &= \sum_{m=1}^{\infty} \frac{1}{(2m-1)^p} \sum_{n=1}^{m} \frac{1}{(n^{2k+1-p}} \\
&= (-1)^p 2^{-p} \left\{ \frac{1}{2} (2^{2k+1} - 1)\binom{2k}{p-1} \zeta_{2k+1}(1) \right. \\
&\quad + \frac{1+(-1)^p}{2} 2^{2k+1-p} (2^p - 1)\zeta_p(1)\zeta_{2k+1-p}(-1) \\
&\quad - \sum_{2 \leqslant 2l \leqslant p-1} (2^{2l} - 1)(2^{2k+1-2l} - 1)\binom{2k-2l}{2k-p} \zeta_{2l}(1)\zeta_{2k+1-2l}(1) \\
&\quad \left. - \sum_{2 \leqslant 2l \leqslant 2k+1-p} (2^{2k+1-2l} - 1)\binom{2k-2l}{p-1} \zeta_{2l}(1)\zeta_{2k+1-2l}(1) \right\} + A(1) \quad (p = 2, 3, \dots, 2k),
\end{aligned} \tag{4.116}$$

i. e.,

$$\mathfrak{Re}\, G_{(p,q)}(\mathrm{i}) = \frac{1}{2^{p+q}} \omega_5(p,q) + \frac{1}{2^p} \omega_7(p,q), \tag{4.117}$$

$$\mathfrak{Re}\, J_{(p,q)}(\mathrm{i}) = \frac{1}{2^{p+q}} \omega_2(p,q) + \frac{1}{2^q} (\omega_{10}(p,q) - A(-1)). \tag{4.118}$$

From (4.117), (4.87), and (4.86) for $p + q = 2k + 1$ we obtain (4.119) for $\omega_7(p,q)$, $p = 2, 3, \dots, 2k$. From (4.118), (4.89), and (4.91) for $p+q = 2k+1$ we obtain (4.120) for $\omega_{10}(p,q)$,

$p = 2, 3, \dots, 2k$. Formula (4.119) for $p = 1$ follows from (4.120) for $p = 2k$ and the relation between $\omega_7(p,q)$ and $\omega_{10}(p,q)$:

$$\begin{aligned}
&\omega_7(p, 2k+1-p)\\
&\quad= \sum_{m=1}^{\infty} \frac{(-1)^m}{m^p} \sum_{n=1}^{m} \frac{1}{(2n-1)^{2k+1-p}}\\
&\quad= (-1)^p 2^{-2k-1+p} \Bigg\{ \frac{1}{2}(2^{2k+1}-1)\binom{2k}{p}\zeta_{2k+1}(1)\\
&\quad- \sum_{2\leqslant 2l\leqslant p-1} (2^{2k+1-2l}-1)\binom{2k-2l}{2k-p}\zeta_{2l}(-1)\zeta_{2k+1-2l}(1)\\
&\quad- 2^{2k+1} \sum_{2\leqslant 2l\leqslant 2k+2-p} \binom{2k+1-2l}{p-1}\beta_{2l-1}(-1)\beta_{2k+2-2l}(-1)\Bigg\} \quad (p = 1, 2, \dots, 2k), \qquad (4.119)
\end{aligned}$$

$$\begin{aligned}
&\omega_{10}(p, 2k+1-p)\\
&\quad= \sum_{m=1}^{\infty} \frac{1}{(2m-1)^p} \sum_{n=1}^{m} \frac{(-1)^n}{n^{2k+1-p}}\\
&\quad= (-1)^p 2^{-p}\Bigg(\frac{1}{2}(2^{2k+1}-1)\binom{2k}{p-1}\zeta_{2k+1}(1) + \frac{1+(-1)^p}{2}(2^p-1)\zeta_p(1)\zeta_{2k+1-p}(-1)\\
&\quad- \sum_{2\leqslant 2l\leqslant 2k+1-p} (2^{2k+1-2l}-1)\binom{2k-2l}{p-1}\zeta_{2l}(-1)\zeta_{2k+1-2l}(1)\\
&\quad- 2^{2k+1} \sum_{2\leqslant 2l\leqslant p+1} \binom{2k+1-2l}{2k-p}\beta_{2l-1}(-1)\beta_{2k+2-2l}(-1)\Bigg) + A(-1) \quad (p = 2, \dots, 2k).
\end{aligned} \qquad (4.120)$$

From this it follows that

$$\mathfrak{Re}\, I_{(p,q)}(\mathrm{i}) = \frac{1}{2^{p+q}}\omega_6(p,q) + \omega_{16}(p,q). \qquad (4.121)$$

Based on (4.90), for $p + q = 2k+1$ and $z = \mathrm{i}$, using the expressions for $\zeta_l(\mathrm{i})$ and $\zeta_l(-\mathrm{i})$, we obtain $\mathfrak{Re}\, I_{(p,q)}(\mathrm{i})$, which, together with (4.85) and (4.121), leads to

$$\begin{aligned}
\omega_{16}(p, 2k+1-p) &= \sum_{m=1}^{\infty} \frac{(-1)^m}{(2m-1)^p} \sum_{n=1}^{m} \frac{(-1)^n}{(2n-1)^{2k+1-p}}\\
&= (-1)^p 2^{-2k-1}\Bigg(\frac{(-1)^p}{2}(2^{2k+1}-1)\zeta_{2k+1}(1)\\
&\quad+ \frac{(-1)^p-1}{2} 2^{2k+1}\beta_p(-1)\beta_{2k+1-p}(-1)\\
&\quad- \sum_{2\leqslant 2l\leqslant p} (2^{2l}-1)\binom{2k-2l}{2k-p}\zeta_{2l}(1)\zeta_{2k+1-2l}(-1)
\end{aligned}$$

$$-\sum_{2\leqslant 2l\leqslant 2k+1-p}(2^{2l}-1)\binom{2k-2l}{p-1}\zeta_{2l}(1)\zeta_{2k+1-2l}(-1)\Big)\quad (p=1,2,\dots,2k). \tag{4.122}$$

Note that formula (4.122) is first determined for $p = 2, 3, \dots, 2k$, and then, based on the expression thus obtained for $p = 2k$ and the relation between $\omega_{16}(p, q)$ and $\omega_{16}(q, p)$, we obtain (4.122) for the case $p = 1$. Next, for the series $N_{(p,q)}(z) = \sum_{m=1}^{\infty}\frac{z^{2m}}{m^p}\sum_{n=1}^{m}\frac{z^n}{n^q}$, from (4.75) for $a = 1$ and $b = 3$, we get the corresponding functional equation. In addition, we have the following formulas:

$$I_m G_{(p,q)}(\mathrm{i}) = -\frac{1}{2^q}(\omega_{13}(p,q) - A(-1)) - \omega_{15}(p,q), \tag{4.123}$$

$$I_m J_{(p,q)}(\mathrm{i}) = -\frac{1}{2^q}(\omega_{13}(p,q) - A(-1)) + \omega_{15}(p,q), \tag{4.124}$$

$$I_m H_{(p,q)}(\mathrm{i}) = -\frac{1}{2}\omega_4(p,q) - \omega_{12}(p,q), \tag{4.125}$$

$$I_m N_{(p,q)}(\mathrm{i}) = -\frac{1}{2^p}\omega_4(p,q) + \omega_{12}(p,q). \tag{4.126}$$

Formulas (4.127) and (4.128) for $p = 2, 3, \dots, 2k-1$ follow from (4.125) and (4.126), with $p + q = 2k$. Formulas (4.129) and (4.130) for $p = 2, 3, \dots, 2k-1$ follow from (4.123) and (4.124) with $p+q = 2k$. Furthermore, from (4.127) and (4.128) for $p = 2k-1$, together with the relations between ω_{12} and ω_{15}, ω_4 and ω_{13}, we obtain (4.129) and (4.130) for $p = 1$:

$$\begin{aligned}
&\omega_{12}(p, 2k-p)\\
&\quad=\sum_{m=1}^{\infty}\frac{1}{(2m-1)^p}\sum_{n=1}^{m}\frac{(-1)^n}{(2n-1)^{2k-p}}\\
&\quad=(-1)^p 2^{-2k-1}\Big((-1)^{p+1}2^{2k}\beta_{2k}(-1) + 2^{2k}((-1)^{p+1}\zeta_p(1) + \zeta_p(-1))\beta_{2k-p}(-1)\\
&\qquad+\sum_{2\leqslant 2l\leqslant 2k-p}2^{2l}\binom{2k-2l}{p-1}\zeta_{2k+1-2l}(1)\cdot\beta_{2l-1}(-1)\\
&\qquad-\sum_{2\leqslant 2l\leqslant p+1}2^{2l}\binom{2k-2l}{2k-1-p}\zeta_{2k+1-2l}(-1)\beta_{2l-1}(-1)\Big)\quad (p=2,\dots,2k-1),
\end{aligned}\tag{4.127}$$

$$\begin{aligned}
&\omega_4(p, 2k-p)\\
&\quad=\sum_{m=1}^{\infty}\frac{1}{m^p}\sum_{n=1}^{m}\frac{(-1)^n}{(2n-1)^{2k-p}}\\
&\quad=(-1)^p 2^{p-1}\Big(-\binom{2k-1}{p}\beta_{2k}(-1) - ((-1)^p\zeta_p(1) + \zeta_p(-1))\beta_{2k-p}(-1)\\
&\qquad+2\sum_{2\leqslant 2l\leqslant p}2^{-2l}\binom{2k-1-2l}{2k-1-p}\zeta_{2l}(-1)\beta_{2k-2l}(-1)
\end{aligned}$$

$$+ 2^{-2k} \sum_{2\leqslant 2l\leqslant 2k-p} (2^{2k+1} - 2^{2l})\binom{2k-2l}{p-1}\zeta_{2k+1-2l}(1)\beta_{2l-1}(-1)\Big) \quad (p = 2,\dots,2k-1), \tag{4.128}$$

$$\begin{aligned}
&\omega_{15}(p, 2k-p)\\
&= \sum_{m=1}^{\infty} \frac{(-1)^m}{(2m-1)^p} \sum_{n=1}^{m} \frac{1}{(2n-1)^{2k-p}}\\
&= (-1)^p 2^{-2k-1}\Big((-1)^{p+1}2^{2k}\binom{2k-1}{p-1}\beta_{2k}(-1) + (-1)^p 2^{2k}(2^{p+1-2k}-1)\zeta_{2k-p}(1)\beta_p(-1)\\
&\quad - 2^{2k}\zeta_{2k-p}(-1)\beta_p(-1) - \sum_{2\leqslant 2l\leqslant p} 2^{2l}\binom{2k-2l}{2k-1-p}\zeta_{2k+1-2l}(1)\beta_{2l-1}(-1)\\
&\quad + \sum_{2\leqslant 2l\leqslant 2k+1-p} 2^{2l}\binom{2k-2l}{p-1}\zeta_{2k+1-2l}(-1)\beta_{2l-1}(-1)\Big) \quad (p = 1,2,\dots,2k-1),
\end{aligned} \tag{4.129}$$

$$\begin{aligned}
&\omega_{13}(p, 2k-p)\\
&= \sum_{m=1}^{\infty} \frac{(-1)^m}{(2m-1)^p} \sum_{n=1}^{m} \frac{1}{n^{2k-p}}\\
&= (-1)^p 2^{-p-1}\Big(2^{2k}\binom{2k-1}{p-1}\beta_{2k}(-1)\\
&\quad + 2^{2k}((-1)^p(1-2^{-2k+1+p})\zeta_{2k-p}(1) + \zeta_{2k-p}(-1))\beta_p(-1)\\
&\quad - \sum_{2\leqslant 2l\leqslant p} (2^{2k+1} - 2^{2l})\binom{2k-2l}{2k-1-p}\zeta_{2k+1-2l}(1)\beta_{2l-1}(-1)\\
&\quad - \sum_{2\leqslant 2l\leqslant 2k-p} 2^{2k+1-2l}\binom{2k-1-2l}{p-1}\zeta_{2l}(-1)\beta_{2k-2l}(-1)\Big) + A(-1) \quad (p = 1,2,\dots,2k-1).
\end{aligned} \tag{4.130}$$

We can conclude that

$$I_m F_{(p,q)}(\mathrm{i}) = -\frac{1}{2^p}\omega_8(p,q) - \frac{1}{2^q}(\omega_{14}(p,q) - A(1)), \tag{4.131}$$

$$I_m I_{(p,q)}(\mathrm{i}) = \frac{1}{2^p}\omega_8(p,q) - \frac{1}{2^q}(\omega_{14}(p,q) - A(1)). \tag{4.132}$$

However, by the above it follows that

$$\omega_8(p,q) = 2^{p-1} I_m(I_{(p,q)}(\mathrm{i}) - F_{(p,q)}(\mathrm{i})), \tag{4.133}$$

$$\omega_{14}(p,q) = -2^{q-1} I_m(I_{(p,q)}(\mathrm{i}) - F_{(p,q)}(\mathrm{i})) + A(1). \tag{4.134}$$

From (4.77) and (4.90) for $p + q = 2k$ and for $z = \mathrm{i}$, from (4.133) and (4.134) it follows that

$$\omega_8(p, 2k-p) = \sum_{m=1}^{\infty} \frac{(-1)^m}{m^p} \sum_{n=1}^{m} \frac{(-1)^n}{(2n-1)^{2k-p}}$$

$$
\begin{aligned}
&= (-1)^p 2^{p-1}\Big(-\binom{2k-1}{p}\beta_{2k}(-1) - 2^{-p}(1+(-1)^p)\zeta_p(-1)\beta_{2k-p} \\
&\quad + 2\sum_{2\leqslant 2l\leqslant p} 2^{-2l}\binom{2k-1-2l}{2k-1-p}\zeta_{2l}(1)\beta_{2k-2l}(-1) \\
&\quad + 2\sum_{2\leqslant 2l\leqslant 2k-p} (1-2^{-2l})\binom{2k-1-2l}{p-1}\zeta_{2l}(1)\beta_{2k-2l}(-1)\Big) \quad (p=1,2,\dots,2k-1),
\end{aligned}
\tag{4.135}
$$

$$
\begin{aligned}
\omega_{14}(p,2k-p) &= \sum_{m=1}^{\infty}\frac{(-1)^m}{(2m-1)^p}\sum_{n=1}^{m}\frac{(-1)^n}{n^{2k-p}} \\
&= (-1)^p 2^{2k-1-p}\Big(\binom{2k-1}{p-1}\beta_{2k}(-1) + 2^{p-2k}(1-(-1)^p)\zeta_{2k-p}(-1)\beta_p(-1) \\
&\quad - 2\sum_{2\leqslant 2l\leqslant p} (1-2^{-2l})\binom{2k-1-2l}{2k-1-p}\zeta_{2l}(1)\beta_{2k-2l}(-1) \\
&\quad - 2\sum_{2\leqslant 2l\leqslant 2k-p} 2^{-2l}\binom{2k-1-2l}{p-1}\zeta_{2l}(1)\beta_{2k-2l}(-1)\Big) + A(1) \quad (p=1,2,\dots,2k-1).
\end{aligned}
\tag{4.136}
$$

Based on the determined values of $\omega_8(p,q)$ and $\omega_{14}(p,q)$ for $p = 2,\dots,2k-1$ and the relation between $\omega_8(p,q)$ and $\omega_{14}(q,p)$ for $p = 2k-1$, (4.136) for $p = 1$ follows from (4.135). In addition, based on $\omega_8(p,q)$ and $\omega_{14}(p,q)$ for $p = 2k-1$, (4.135) for $p = 1$ follows from (4.136). So the coefficients $\omega_j(p,q)$, with indices $j = 1,2,3,5,6,7,9,10,11,16$ are determined if $p+q = 2k+1$ and with indices $j = 4,8,12,13,14,15$ if $p+q = 2k$. Further, we present the following pairs of mutually inverse relations:

$$
\begin{aligned}
&\sum_{l=1}^{q}\binom{p+q-1-l}{p-1}\sum_{m=1}^{\infty}\frac{z^{(a-b)m}}{m^{p+q-l}}\sum_{n=1}^{m}\frac{z^{bn}}{n^l} + (-1)^{p+1}\sum_{m=1}^{\infty}\frac{z^{(b-a)m}}{m^p}\sum_{n=1}^{m}\frac{z^{an}}{n^q} \\
&\quad = (-1)^p\zeta_q(z^a)(\zeta_p(z^b)-\zeta_p(z^{b-a})) \\
&\qquad + \binom{p+q-2}{p-1}\zeta_1(z^b)(\zeta_{p+q-1}(z^{a-b})-\zeta_{p+q-1}(z^a)) \\
&\qquad + \sum_{l=2}^{p-1}(-1)^l\binom{p+q-1-l}{q-1}\zeta_l(z^b)\zeta_{p+q-l}(z^a) \\
&\qquad + \sum_{l=2}^{q}\binom{p+q-1-l}{p-1}\zeta_l(z^b)\zeta_{p+q-l}(z^{a-b}),
\end{aligned}
\tag{4.137}
$$

$$
\begin{aligned}
&\sum_{l=1}^{q}\binom{p+q-1-l}{p-1}\sum_{m=1}^{\infty}\frac{z^{(b-a)m}}{m^{p+q-l}}\sum_{n=1}^{m}\frac{z^{an}}{n^l} + (-1)^{p+1}\sum_{m=1}^{\infty}\frac{z^{(a-b)m}}{m^p}\sum_{n=1}^{m}\frac{z^{bn}}{n^q} \\
&\quad = (-1)^p\zeta_q(z^b)(\zeta_p(z^a)-\zeta_p(z^{a-b})) \\
&\qquad + \binom{p+q-2}{p-1}\zeta_1(z^a)\cdot(\zeta_{p+q-1}(z^{b-a})-\zeta_{p+q-1}(z^b))
\end{aligned}
$$

$$+\sum_{l=2}^{p-1}(-1)^l\binom{p+q-1-l}{q-1}\zeta_l(z^a)\zeta_{p+q-l}(z^b)+\sum_{l=2}^{q}\binom{p+q-1-l}{p-1}\zeta_l(z^a)\zeta_{p+q-l}(z^{b-a}), \tag{4.138}$$

$$\begin{aligned}\sum_{l=1}^{q}\binom{p+q-1-l}{p-1}&G_{(p+q-l,l)}(z)+(-1)^{p+1}I_{(p,q)}(\overline{z})\\ &=(-1)^p\zeta_q(z)(\zeta_p(1)-\zeta_p(\overline{z}))\\ &+\sum_{l=2}^{p-1}(-1)^l\binom{p+q-1-l}{q-1}\zeta_l(1)\zeta_{p+q-l}(z)\\ &+\sum_{l=2}^{q}\binom{p+q-1-l}{p-1}\zeta_l(1)\zeta_{p+q-l}(z),\end{aligned} \tag{4.139}$$

$$\begin{aligned}\sum_{l=1}^{q}\binom{p+q-1-l}{p-1}&I_{(p+q-l,l)}(\overline{z})+(-1)^{p+1}G_{(p,q)}(z)\\ &=\binom{p+q-2}{p-1}\zeta_1(z)(\zeta_{p+q-1}(\overline{z})-\zeta_{p+q-1}(1))\\ &+\sum_{l=2}^{p-1}(-1)^l\binom{p+q-1-l}{q-1}\zeta_{p+q-l}(1)\zeta_l(z)\\ &+\sum_{l=2}^{q}\binom{p+q-1-l}{p-1}\zeta_l(z)\zeta_{p+q-l}(\overline{z}).\end{aligned} \tag{4.140}$$

If we understand equality (4.137) as a relation by $\sum_{m=1}^{q}\frac{z^{(b-a)m}}{m^p}\sum_{n=1}^{m}\frac{z^{an}}{n^q}$ and if we solve it by that value, after the replacements $p\to p+q-l$ and $q\to l$, relation (4.138) becomes an identity. Relation (4.138) is, formally speaking, obtained from (4.137) after the replacements $a\to b$ and $b\to a$. Relation (4.137) is in fact relation (4.61); (4.139) and (4.140) are the already mentioned modifications of relation (4.61). Let us introduce the functions

$$\begin{gathered}K_{(p,q)}(z)=\sum_{m=1}^{\infty}\frac{z^m}{m^p}\sum_{n=1}^{m}\frac{(\overline{z})^{2n}}{n^q},\quad L_{(p,q)}(z)=\sum_{m=1}^{\infty}\frac{z^m}{m^p}\sum_{n=1}^{m}\frac{(\overline{z})^{3n}}{n^q},\\ M_{(p,q)}(z)=\sum_{m=1}^{\infty}\frac{z^{2m}}{m^p}\sum_{n=1}^{m}\frac{(\overline{z})^{n}}{n^q}\quad(|z|=1).\end{gathered} \tag{4.141}$$

From the following relations

$$F_{(p,q)}(z)+F_{(q,p)}(z)=\zeta_p(z)\zeta_q(z)+\zeta_{p+q}(z^2), \tag{4.142}$$

$$G_{(p,q)}(z)+H_{(q,p)}(z)=\zeta_p(z)\zeta_q(1)+\zeta_{p+q}(z), \tag{4.143}$$

$$M_{(p,q)}(z)+K_{(q,p)}(\overline{z})=\zeta_p(z^2)\zeta_q(\overline{z})+\zeta_{p+q}(z), \tag{4.144}$$

$$I_{(p,q)}(z)+I_{(q,p)}(\overline{z})=\zeta_p(z)\zeta_q(\overline{z})+\zeta_{p+q}(1) \tag{4.145}$$

we also have the following relations: for $a = 2$ and $b = 1$, from (4.137) and (4.138) it follows that

$$\sum_{l=1}^{q}\binom{p+q-1-l}{p-1}F_{(p+q-l,l)}(z)+(-1)^{p+1}K_{(p,q)}(\overline{z})$$
$$=(-1)^{p}\zeta_q(z^2)(\zeta_p(z)-\zeta_p(\overline{z}))+\binom{p+q-2}{p-1}\zeta_1(z)(\zeta_{p+q-1}(z)-\zeta_{p+q-1}(z^2))$$
$$+\sum_{l=2}^{p-1}(-1)^{l}\binom{p+q-1-l}{q-1}\zeta_l(z)\zeta_{p+q-1}(z^2)+\sum_{l=2}^{q}\binom{p+q-1-l}{p-1}\zeta_l(z)\zeta_{p+q-1}(z), \quad (4.146)$$

$$\sum_{l=1}^{q}\binom{p+q-1-l}{p-1}K_{(p+q-l,l)}(\overline{z})+(-1)^{p+1}F_{(p,q)}(z)$$
$$=(-1)^{p}\zeta_q(z)(\zeta_p(z^2)-\zeta_p(z))+\binom{p+q-2}{p-1}\zeta_1(z^2)(\zeta_{p+q-1}(\overline{z})-\zeta_{p+q-1}(z))$$
$$+\sum_{l=2}^{p-1}(-1)^{l}\binom{p+q-1-l}{q-1}\zeta_l(z^2)\zeta_{p+q-1}(z)+\sum_{l=2}^{q}\binom{p+q-1-l}{p-1}\zeta_l(z^2)\zeta_{p+q-l}(\overline{z}). \quad (4.147)$$

For $a = 1$ and $b = 1$, from (4.137) we have

$$\sum_{l=1}^{q}\binom{p+q-1-l}{p-1}H_{(p+q-l,l)}(z)+(-1)^{p+1}H_{(p,q)}(z)$$
$$=(-1)^{p}\zeta_q(z)\cdot(\zeta_p(z)-\zeta_p(1))+\binom{p+q-2}{p-1}\zeta_1(z)(\zeta_{p+q-1}(1)-\zeta_{p+q-1}(z))$$
$$+\sum_{l=2}^{p-1}(-1)^{l}\binom{p+q-1-l}{q-1}\zeta_l(z)\zeta_{p+q-1}(z)+\sum_{l=2}^{q}\binom{p+q-1-l}{p-1}\zeta_l(z)\zeta_{p+q-l}(1). \quad (4.148)$$

For $a = 3$ and $b = 2$, from (4.137) and (4.138) we have

$$\sum_{l=1}^{q}\binom{p+q-1-l}{p-1}J_{(p+q-l,l)}(z)+(-1)^{p+1}L_{(p,q)}(\overline{z})$$
$$=(-1)^{p}\zeta_q(z^3)\cdot(\zeta_p(z^2)-\zeta_p(\overline{z}))+\binom{p+q-2}{p-1}\zeta_1(z^2)(\zeta_{p+q-1}(z)-\zeta_{p+q-1}(z^3))$$
$$+\sum_{l=2}^{p-1}(-1)^{l}\binom{p+q-1-l}{q-1}\zeta_l(z^2)\zeta_{p+q-1}(z^3)+\sum_{l=2}^{q}\binom{p+q-1-l}{p-1}\zeta_l(z^2)\zeta_{p+q-l}(z), \quad (4.149)$$

$$\sum_{l=1}^{q}\binom{p+q-1-l}{p-1}L_{(p+q-l,l)}(\overline{z})+(-1)^{p+1}J_{(p,q)}(z)$$
$$=(-1)^{p}\zeta_q(z^2)\cdot(\zeta_p(z^3)-\zeta_p(z))+\binom{p+q-2}{p-1}\zeta_1(z^3)(\zeta_{p+q-1}(\overline{z})-\zeta_{p+q-1}(z^2))$$
$$+\sum_{l=2}^{p-1}(-1)^{l}\binom{p+q-1-l}{q-1}\zeta_l(z^3)\zeta_{p+q-l}(z^2)+\sum_{l=2}^{q}\binom{p+q-1-l}{p-1}\zeta_l(z^3)\zeta_{p+q-l}(\overline{z}). \quad (4.150)$$

For $a = 1$ and $b = -1$, from (4.137) we have

$$\sum_{l=1}^{q}\binom{p+q-1-l}{p-1}M_{(p+q-l,l)}(z)+(-1)^{p+1}M_{(p,q)}(\overline{z})$$
$$=(-1)^p\zeta_q(z)\cdot(\zeta_p(\overline{z})-\zeta_p(\overline{z}^2))+\binom{p+q-2}{p-1}\zeta_1(\overline{z})(\zeta_{p+q-1}(z^2)-\zeta_{p+q-1}(z))$$
$$+\sum_{l=2}^{p-1}(-1)^l\binom{p+q-1-l}{q-1}\zeta_l(\overline{z})\zeta_{p+q-1}(z)+\sum_{l=2}^{q}\binom{p+q-1-l}{p-1}\zeta_l(\overline{z})\zeta_{p+q-l}(z^2). \tag{4.151}$$

From (4.148) and (4.143) we have

$$\binom{p+q-2}{p-1}\cdot(G_{(1,p+q-1)}(z)-\zeta_{p+q-1}(1)\zeta_1(z))$$
$$+\sum_{l=2}^{q}\binom{p+q-1-l}{p-1}G_{(l,p+q-l)}(z)+(-1)^{p+1}G_{(q,p)}(z)$$
$$=\left((-1)^{p+1}+\binom{p+q-1}{p}\right)\zeta_{p+q}(z)$$
$$+\binom{p+q-2}{p-1}\zeta_1(z)(\zeta_{p+q-1}(z)-\zeta_{p+q-1}(1))$$
$$-\sum_{l=2}^{p}(-1)^l\binom{p+q-1-l}{q-1}\zeta_l(z)\zeta_{p+q-l}(z). \tag{4.152}$$

Since by (4.143)

$$\lim_{\substack{|z|\leqslant 1\\ z\to 1}}(G_{(1,p+q-1)}(z)-\zeta_{p+q-1}(1)\zeta_1(z))=\zeta_{p+q}(1)-G_{(p+q-1,1)}(1), \tag{4.153}$$

in equality (4.152), we can pass to the limit as $z \to 1$ along the circle $|z| = 1$. For $a = 1$ and $b = -1$, from (4.75) and (4.151) it follows that

$$\sum_{l=1}^{q}\binom{p+q-1-l}{p-1}M_{(p+q-l,l)}(z)+(-1)^{q+1}M_{(p,q)}(z)$$
$$=\left((-1)^{q+1}+\binom{p+q-1}{p}\right)\zeta_{q+q}(z)$$
$$+\binom{p+q-1}{q}\zeta_{p+q}(z^2)+\zeta_q(z)(\zeta_p(z^2)-\zeta_p(z))$$
$$+\binom{p+q-2}{p-1}\zeta_1(z)(\zeta_{p+q-1}(z^2)-\zeta_{p+q-1}(z))$$
$$-\sum_{l=2}^{p-1}\binom{p+q-1-l}{q-1}\zeta_l(z)\zeta_{p+q-l}(z)-\sum_{l=2}^{q}(-1)^l\binom{p+q-1-l}{p-1}\zeta_l(z)\zeta_{p+q-l}(z^2). \tag{4.154}$$

Further, we will invert some of the previously derived relations. As the first step in this direction, we will prove the following formulas:

$$\sum_{l=s}^{q}(-1)^l\binom{p+q-1-l}{p-1}\binom{p+q-1-s}{p+q-1-l}=(-1)^q\delta_{0,q-s},\tag{4.155}$$

$$\sum_{l=1}^{q}(-1)^l\binom{p+q-1-l}{p-1}\sum_{s=1}^{l}\binom{p+q-1-s}{p+q-1-l}f_s=(-1)^q f_q,\tag{4.156}$$

$$\sum_{l=1}^{q}(-1)^l\binom{p+q-1-l}{p-1}\sum_{s=1}^{p+q-l}\binom{p+q-1-s}{l-1}f_s=-\sum_{l=1}^{p}\binom{p+q-1-l}{q-1}f_{p+q-l}.\tag{4.157}$$

Proof. First of all, notice that (4.156) follows from (4.155) by changing the order of summation. Therefore it remains to prove (4.155) and (4.157):

$$\begin{aligned}
&\sum_{l=s}^{q}(-1)^l\binom{p+q-1-l}{p-1}\binom{p+q-1-s}{p+q-1-l}\\
&\quad=\sum_{l=s}^{q}(-1)^l\frac{(p+q-1-l)!}{(p-1)!(q-l)!}\frac{(p+q-1-s)!}{(p+q-1-l)!(l-s)!}\\
&\quad=\binom{p+q-1-s}{p-1}\sum_{l=0}^{q-s}(-1)^l\binom{q-s}{l-s}\\
&\quad=(-1)^s\binom{p+q-1-s}{p-1}\sum_{l=0}^{q-s}(-1)^l\binom{q-s}{l}\\
&\quad=(-1)^s\binom{p+q-1-s}{p-1}\delta_{0,q-s}=(-1)^q\delta_{0,q-s}.
\end{aligned}\tag{4.158}$$

Next, let

$$A=\sum_{l=1}^{q}(-1)^l\binom{p+q-1-l}{p-1}\sum_{s=1}^{p+q-l}\binom{p+q-1-s}{l-1}f_s.\tag{4.159}$$

Then

$$A=A+A_2,\tag{4.160}$$

$$\begin{aligned}
A_1&=\sum_{s=1}^{p-1}f_s\sum_{l=1}^{q}(-1)^l\binom{p+q-1-l}{p-1}\binom{p+q-1-s}{l-1},\\
A_2&=\sum_{s=p}^{p+q-1}f_s\sum_{l=1}^{p+q-s}(-1)^l\binom{p+q-1-l}{p-1}\binom{p+q-1-s}{l-1},
\end{aligned}\tag{4.161}$$

$$B_s=\sum_{l=1}^{q}(-1)^l\binom{p+q-1-l}{p-1}\binom{p+q-1-s}{l-1}\quad(1\leqslant s\leqslant p-1),\quad A_1=\sum_{s=1}^{p-1}f_sB_s,\tag{4.162}$$

$$B_s = -C_s, \quad C_s = \sum_{l=0}^{q-1} (-1)^l \binom{p+q-2-l}{p-1} \binom{p+q-1-s}{l}, \tag{4.163}$$

$$\begin{aligned} C_s &= \sum_{l=0}^{q-1} (-1)^l \binom{p+q-3-l}{p-2} \binom{p+q-2-s}{l} \\ &= \sum_{l=0}^{q-1} (-1)^l \binom{p+q-2-r-l}{p-1-r} \binom{p+q-1-r-s}{l} = |r = p-s| \\ &= \sum_{l=0}^{q-1} (-1)^l \binom{q+s-2-l}{s-1} \binom{q-1}{l} \\ &= (-1)^{q-1} \sum_{l=0}^{q-1} (-1)^l \binom{s-1+l}{s-1} \binom{q-1}{l} \\ &= (-1)^{q-1} (-1)^r \sum_{l=0}^{q-1-r} (-1)^l \binom{s-1+l}{s-1-r} \binom{q-1-r}{l}. \end{aligned} \tag{4.164}$$

If $s \leqslant q$, then $r = s-1$, and so

$$C_s = (-1)^{q-1}(-1)^{s-1} \sum_{l=0}^{q-s} (-1)^l \binom{q-s}{l} = (-1)^{q+s} \delta_{0,q-s} = \delta_{0,q-s}. \tag{4.165}$$

If $s \geqslant q$, then $r = q-1$, so it follows that

$$C_s = \binom{s-1}{s-q}, \tag{4.166}$$

that is,

$$A_1 = -\sum_{s=q}^{p-1} \binom{s-1}{s-q} f_s \tag{4.167}$$

(this formula also contains the result $A_1 = 0$ for $p \leqslant q$),

$$D_s = \sum_{l=1}^{p+q-s} (-1)^l \binom{p+q-1-l}{p-1} \binom{p+q-1-s}{l-1} \quad (p \leqslant s \leqslant p+q-1), \quad A_2 = \sum_{s=p}^{p+q-1} f_s D_s, \tag{4.168}$$

$$D_s = -E_s, \quad E_s = \sum_{l=0}^{p+q-s-1} (-1)^l \binom{p+q-2-l}{p-1} \binom{p+q-1-s}{l}, \tag{4.169}$$

$$E_s = \sum_{l=0}^{p+q-1-s-r} (-1)^l \binom{p+q-2-r-l}{p-1-r} \binom{p+q-1-s-r}{l}. \tag{4.170}$$

If $s \leqslant q$, then $r = p-1$, so $E_s = \delta_{0,q-s}$. For $s \geqslant q$, $r = p+q-s-1$, so it follows that $E_s = \binom{s-1}{s-q}$. Accordingly, we can conclude that

$$A_2 = -\sum_{s=\max\{p,q\}}^{p+q-1} \binom{s-1}{s-q} f_s, \tag{4.171}$$

from which it follows that

$$A = -\sum_{s=q}^{p+q-1} \binom{s-1}{s-q} f_s = -\sum_{l=1}^{p} \binom{p+q-1-l}{q-1} f_{p+q-l}, \tag{4.172}$$

thus proving formula (4.157). □

Relation (4.137) (i. e., (4.61)) can be presented in a slightly different form, which is more suitable for inversion:

$$\begin{aligned}
&\sum_{l=1}^{q} \binom{p+q-1-l}{p-1} \sum_{m=1}^{\infty} \frac{z^{(a-b)m}}{m^{p+q-l}} \sum_{n=1}^{m} \frac{z^{bn}}{n^l} + (-1)^{p+1} \sum_{m=1}^{\infty} \frac{z^{(b-a)m}}{m^p} \sum_{n=1}^{m} \frac{z^{an}}{n^q} \\
&\quad = (-1)^{p+1} \zeta_p(z^{b-a}) \zeta_q(z^a) + \sum_{l=1}^{q} \binom{p+q-1-l}{p-1} \zeta_{p+q-l}(z^{a-b}) \zeta_l(z^b) \\
&\quad + \sum_{l=1}^{p} (-1)^l \binom{p+q-1-l}{q-1} \zeta_{p+q-l}(z^a) \zeta_l(z^b).
\end{aligned} \tag{4.173}$$

In this equation, making the replacements $l \to s$, $p \to p+q-l$, $q \to l$, we get

$$\begin{aligned}
&\sum_{s=1}^{l} \binom{p+q-1-s}{p+q-1-l} \sum_{m=1}^{\infty} \frac{z^{(a-b)m}}{m^{p+q-s}} \sum_{n=1}^{m} \frac{z^{bn}}{n^s} + (-1)^{p+q+1+l} \sum_{m=1}^{\infty} \frac{z^{(b-a)m}}{m^{p+q-l}} \sum_{n=1}^{m} \frac{z^{an}}{n^l} \\
&\quad = (-1)^{p+q+1+l} \zeta_{p+q-l}(z^{b-a}) \zeta_l(z^a) \\
&\quad + \sum_{s=1}^{l} \binom{p+q-1-s}{p+q-1-l} \zeta_{p+q-s}(z^{a-b}) \zeta_s(z^b) \\
&\quad + \sum_{s=1}^{p+q-l} (-1)^s \binom{p+q-1-s}{l-1} \zeta_{p+q-s}(z^a) \zeta_s(z^b).
\end{aligned} \tag{4.174}$$

Applying the summation $\sum_{l=1}^{q} (-1)^l \binom{p+q-1-l}{p-1}$ to equation (4.174), we obtain

$$L_1 = \left| f_s = \sum_{m=1}^{\infty} \frac{z^{(b-a)m}}{m^{p+q-s}} \sum_{n=1}^{m} \frac{z^{bn}}{n^s},\ (4.156) \right| = (-1)^q \sum_{m=1}^{\infty} \frac{z^{(a-b)m}}{m^p} \sum_{n=1}^{m} \frac{z^{bn}}{n^q}, \tag{4.175}$$

$$L_2 = (-1)^{p+q+1} \sum_{l=1}^{q} \binom{p+q-1-l}{p-1} \sum_{m=1}^{\infty} \frac{z^{(b-a)m}}{m^{p+q-l}} \sum_{n=1}^{m} \frac{z^{an}}{n^l}, \tag{4.176}$$

$$D_1 = (-1)^{p+q+1} \sum_{l=1}^{q} \binom{p+q-1-l}{p-1} \zeta_{p+q-l}(z^{b-a}) \zeta_l(z^a), \tag{4.177}$$

$$D_2 = \left| f_s = \zeta_{p+q-s}(z^{a-b}) \zeta_s(z^b),\ (4.156) \right| = (-1)^q \zeta_p(z^{a-b}) \zeta_q(z^b), \tag{4.178}$$

$$D_3 = \left|f_s = (-1)^s \zeta_{p+q-s}(z^a)\zeta_s(z^b) \text{ (by (4.157))}\right|$$
$$= (-1)^{p+q+1} \sum_{l=1}^{p} (-1)^l \binom{p+q-1-l}{q-1} \zeta_{p+q-l}(z^b)\zeta_l(z^a). \tag{4.179}$$

Equating $L_1 + L_2$ with $D_1 + D_2 + D_3$, we get (4.138). Therefore (4.137) has been transformed into (4.138) using the inversion formulas (4.156) and (4.157). So we proved that formulas (4.156) and (4.157) are inversion formulas, which translate (4.137) and (4.138) into each other. We will further present another example of the application of formulas (4.156) and (4.157) to equation (4.48). For this reason, we have to prove the following formula:

$$\sum_{l=1}^{p} 2^l \binom{p+q-1-l}{q-1} + \sum_{l=1}^{q} 2^l \binom{p+q-1-l}{p-1} = 2^{p+q}. \tag{4.180}$$

Proof. By (4.48) we have

$$\sum_{s=1}^{l} (-1)^s \binom{p+q-1-s}{p+q-1-l} + \sum_{s=1}^{p+q-l} \frac{1}{2^s} \binom{p+q-1-s}{l-1} = (-1)^l 2^{-(p+q-l)}. \tag{4.181}$$

Applying the summation of the form $\sum_{l=1}^{q}(-1)^l\binom{p+q-1-l}{p-l}$ to the above relation,

$$L_1 = \left|f_s = (-1)^s,\ (4.156)\right| = 1, \tag{4.182}$$

$$L_2 = \left|f_s = \frac{1}{2^s},\ (4.157)\right| = -2^{-(p+q)} \sum_{l=1}^{p} 2^l \binom{p+q-1-l}{q-1}, \tag{4.183}$$

$$D = 2^{-(p+q)} \sum_{l=1}^{q} 2^l \binom{p+q-1-l}{p-1}, \tag{4.184}$$

we directly obtain (4.180). □

From the formulas given above it is now possible to derive additional formulas. From (4.148), for $p = 2k-1$ and $q = 1$, we get

$$H_{(2k-1,1)}(z) = \frac{(-1)^{k+1}}{2}\zeta_k^2(z) + \zeta_1(z)(\zeta_{2k-1}(1) - \zeta_{2k-1}(z)) + \sum_{l=2}^{k}(-1)^l \zeta_l(z)\zeta_{2k-l}(z) \quad (k \geq 2). \tag{4.185}$$

By this relation and (4.143) it follows that

$$G_{(1,2k-1)}(z) = \frac{(-1)^k}{2}\zeta_k^2(z) + \zeta_{2k}(z) + \zeta_1(z)\zeta_{2k-1}(z) - \sum_{l=2}^{k}(-1)^l \zeta_l(z)\zeta_{2k-l}(z) \quad (k \geq 2). \tag{4.186}$$

From (4.154), for $q = 1$, we have

$$M_{(p,1)}(z) = \zeta_{p+1}(z) + \frac{p}{2}\zeta_{p+1}(z^2) + \zeta_1(z)(\zeta_p(z^2) - \zeta_p(z)) - \frac{1}{2}\sum_{l=2}^{p-1} \zeta_l(z)\zeta_{p+1-l}(z). \tag{4.187}$$

From (4.187) with $z = 1$ and from the equality $M_{(p,1)}(1) = G_{(p,1)}(1)$ we get the well-known formula

$$G_{(p,1)}(1) = \frac{1}{2}(p+q)\zeta_{p+1}(1) - \frac{1}{2}\sum_{l=2}^{p-1} \zeta_l(1)\zeta_{p+1-l}(1) \quad (p \geqslant 2). \tag{4.188}$$

From (4.187) with $z = -1$ and from the equality $M_{(p,1)}(-1) = H_{(p,1)}(-1)$ we obtain

$$H_{(p,1)}(-1) = \zeta_{p+1}(1) + \frac{p}{2}\zeta_{p+1}(1) + \zeta_1(-1)(\zeta_p(1) - \zeta_p(-1)) - \frac{1}{2}\sum_{l=2}^{p-1} \zeta_l(-1)\zeta_{p+1-l}(-1) \quad (p \geqslant 2). \tag{4.189}$$

By this relation and (4.143) it follows that

$$G_{(1,p)}(-1) = -\frac{p}{2}\zeta_{p+1}(1) + \zeta_1(-1)\zeta_p(-1) + \frac{1}{2}\sum_{l=2}^{p-1} \zeta_l(-1)\zeta_{p+1-l}(-1) \quad (p \geqslant 2). \tag{4.190}$$

Taking $z = -1$ in (4.185) and $p = 2k - 1$ in (4.189), we get

$$\sum_{l=1}^{k-1} \zeta_{2l}(-1)\zeta_{2k-2l}(-1) = \zeta_{2k}(-1) + \frac{2k-1}{2}\zeta_{2k}(1) \quad (k \geqslant 1). \tag{4.191}$$

Further, let

$$\begin{aligned}
\Phi_{p,q}^{(a,b)}(z) &= \zeta_q(z^a)(\zeta_p(z^{a-b}) - \zeta_p(\overline{z}^b)) + (-1)^p\zeta_q(z^a)(\zeta_p(z^{b-a}) - \zeta_p(z^b)) \\
&\quad + (-1)^{q+1}\zeta_{p+q}(\overline{z}^b) + \binom{p+q-1}{p}\zeta_{p+q}(z^a) + \binom{p+q-1}{q}\zeta_{p+q}(z^{a-b}) \\
&\quad - \sum_{l=2}^{p-1}\binom{p+q-1-l}{q-1}\zeta_l(\overline{z}^b)\zeta_{p+q-l}(z^a) - \sum_{l=2}^{p-1}(-1)^l\binom{p+q-1-l}{q-1}\zeta_l(z^b)\zeta_{p+q-l}(z^a) \\
&\quad - \sum_{l=2}^{q}\binom{p+q-1-l}{p-1}\zeta_l(z^b)\zeta_{p+q-l}(z^{a-b}) - \sum_{l=2}^{q}(-1)^l\binom{p+q-1-l}{p-1}\zeta_l(\overline{z}^b)\zeta_{p+q-l}(z^{a-b}) \\
&\quad + \binom{p+q-2}{p-1}(\zeta_{p+q-1}(^a) - \zeta_{p+q-1}(z^{a-b}))(\zeta_1(z^b) - \zeta_1(\overline{z}^b)), \\
&\quad |z| = 1, \ \ a, b \in \mathbb{Z}, \ \ a \neq 0, \ \ b \neq 0,
\end{aligned} \tag{4.192}$$

where the expression $\Phi_{(p,q)}^{(a,b)}(z)$ for $a \neq 0$ and $b = 0$ is equal to the right-hand side of equation (4.192), from which the last term is eliminated (let us mark this relation as the first modification of relation (4.192)), whereas the expression $\Phi_{(p,q)}^{(a,b)}(z)$ for $a = 0$ and

$b \neq 0$ is equal to the right-hand side of equation (4.192) from which the first two terms are eliminated (let us mark this relation as the second modification of relation (4.192)). From (4.75) we have

$$(-1)^p \sum_{m=1}^{\infty} \frac{(\overline{z})^{(a-b)m}}{m^p} \sum_{m=1}^{m} \frac{(z)^{an}}{n^q} + (-1)^{q+1} \sum_{m=1}^{\infty} \frac{z^{(a-b)m}}{m^p} \sum_{m=1}^{m} \frac{(\overline{z})^{an}}{n^q} = \Phi^{(a,b)}_{(p,q)}(z). \tag{4.193}$$

After conjugating expression (4.193), we get

$$(-1)^{p+q+1}\Phi^{(a,b)}_{(p,q)}(z) = \Phi^{(a,b)}_{(p,q)}(\overline{z}). \tag{4.194}$$

From the last equation, for $z = 1$ and $z = -1$, we have

$$\Phi^{(a,b)}_{(p,2k-p)}(1) = 0, \tag{4.195}$$

$$\Phi^{(a,b)}_{(p,2k-p)}(-1) = 0. \tag{4.196}$$

Therefore by (4.195) we obtain

$$\begin{aligned} &\sum_{2\leqslant 2l\leqslant p-1} \binom{2k-1-2l}{2k-1-p} \zeta_{2l}(1)\zeta_{2k-2l}(1) + \sum_{2\leqslant 2l\leqslant 2k-p} \binom{2k-1-2l}{p-1} \zeta_{2l}(1)\zeta_{2k-2l}(1) \\ &= \frac{1}{2}\left[(-1)^{p+1} + \binom{2k}{p}\right]\zeta_{2k}(1). \end{aligned} \tag{4.197}$$

From (4.196), for $a = 2$ and $b = 1$, it follows that

$$\begin{aligned} &\sum_{2\leqslant 2l\leqslant p-1} \binom{2k-1-2l}{2k-1-p} \zeta_{2l}(-1)\zeta_{2k-2l}(1) + \sum_{2\leqslant 2l\leqslant 2k-p} \binom{2k-1-2l}{p-1} \zeta_{2l}(-1)\zeta_{2k-2l}(-1) \\ &= \frac{1}{2}\left[(-1)^{p+1} + \binom{2k-1}{p-1}\right]\zeta_{2k}(-1) + \frac{1}{2}\binom{2k-1}{p}\zeta_{2k}(1). \end{aligned} \tag{4.198}$$

By the first modification of relation (4.192) and (4.194), for $z = -1$, we have

$$\begin{aligned} &\sum_{2\leqslant 2l\leqslant p-1} \binom{2k-1-2l}{2k-1-p} \zeta_{2l}(1)\zeta_{2k-2l}(-1) + \sum_{2\leqslant 2l\leqslant 2k-p} \binom{2k-1-2l}{p-1} \zeta_{2l}(1)\zeta_{2k-2l}(-1) \\ &= \frac{(-1)^{p+1}}{2}\zeta_{2k}(1) + \frac{1}{2}\binom{2k}{p}\zeta_{2k}(-1) + \frac{1}{2}[1 + (-1)^p]\zeta_{2k-p}(-1)[\zeta_p(-1) - \zeta_p(1)]. \end{aligned} \tag{4.199}$$

From (4.197), for $p = 2k - 1$, we get

$$\sum_{l=1}^{k-1} \zeta_{2l}(1)\zeta_{2k-2l}(1) = \frac{2k+1}{2}\zeta_{2k}(1) \quad (k \geqslant 2), \tag{4.200}$$

and by (4.198) for $p = 2k - 1$, it follows that

$$\sum_{l=1}^{k-1} \zeta_{2l}(-1)\zeta_{2k-2l}(1) = k\zeta_{2k}(-1) + \frac{1}{2}\zeta_{2k}(1) \quad (k \geqslant 1). \tag{4.201}$$

Furthermore, for $p = k$, from (4.197) we obtain

$$\begin{aligned}\sum_{2\leqslant 2l\leqslant k} &\binom{2k-1-2l}{k-1}\zeta_{2l}(1)\zeta_{2k-2l}(1)\\ &= \frac{1}{4}[1+(-1)^k]\zeta_k^2(1) + \frac{1}{4}\left[(-1)^{k+1} + \binom{2k}{k}\right]\zeta_{2k}(1) \quad (k \geqslant 2),\end{aligned} \tag{4.202}$$

whereas for $p = k$, (4.199) gives

$$\begin{aligned}\sum_{2\leqslant 2l\leqslant k} &\binom{2k-1-2l}{k-1}\zeta_{2l}(1)\zeta_{2k-2l}(-1)\\ &= \frac{1}{4}[1+(-1)^k]\zeta_k^2(-1) + \frac{(-1)^{k+1}}{4}\zeta_{2k}(1) + \frac{1}{4}\binom{2k}{k}\zeta_{2k}(-1) \quad (k \geqslant 1).\end{aligned} \tag{4.203}$$

Further, we introduce the values $\Omega_j(p,q)$ $(j = 1, 2, \dots, 16)$ by the following formulas:

$$\Omega_j(p,q) = \begin{cases} \omega_j(p,q), & p+q = 2k+1,\\ 0, & p+q = 2k \end{cases} \quad (j = 1,2,3,5,6,7,9,10,11,16), \tag{4.204}$$

$$\Omega_j(p,q) = \begin{cases} 0, & p+q = 2k+1,\\ \omega_j(p,q), & p+q = 2k \end{cases} \quad (j = 4,8,12,13,14,15). \tag{4.205}$$

We will prove that

$$\begin{aligned}\Omega_j(p,q) = (-1)^p\Bigg\{&\frac{1}{2}\left[(-1)^{q+1} + \binom{p+q}{p}\right]\zeta_{p+q}(1)\\ &- \sum_{2\leqslant 2l\leqslant p-1}\binom{p+q-1-2l}{q-1}\zeta_{2l}(1)\zeta_{p+q-2l}(1)\\ &- \sum_{2\leqslant 2l\leqslant q}\binom{p+q-1-2l}{p-1}\zeta_{2l}(1)\zeta_{p+q-2l}(1)\Bigg\},\end{aligned} \tag{4.206}$$

$$\begin{aligned}\Omega_2(p,q) = (-1)^p\Bigg\{&\frac{1}{2}\binom{p+q-1}{q}\zeta_{p+q}(1)\\ &+ \frac{1}{2}\left[(-1)^{q+1} + \binom{p+q-1}{p}\right]\zeta_{p+q}(-1)\\ &+ \frac{1}{2}[(-1)^p - (-1)^q]\zeta_p(1)\zeta_q(-1)\\ &- \sum_{2\leqslant 2l\leqslant p}\binom{p+q-1-2l}{q-1}\zeta_{2l}(-1)\zeta_{p+q-2l}(-1)\\ &- \sum_{2\leqslant 2l\leqslant q-1}\binom{p+q-1-2l}{p-1}\zeta_{2l}(-1)\zeta_{p+q-2l}(1)\Bigg\},\end{aligned} \tag{4.207}$$

$$\begin{aligned}\Omega_3(p,q) = (-1)^p 2^{-q}\Bigg\{&\frac{1}{2}(2^{p+q}-1)\binom{p+q-1}{p}\zeta_{p+q}(1)\\&-\sum_{2\leqslant 2l\leqslant p-1}(2^{p+q-2l}-1)\binom{p+q-1-2l}{q-1}\zeta_{2l}(1)\zeta_{p+q-2l}(1)\\&-\sum_{2\leqslant 2l\leqslant q}(2^{2l}-1)(2^{p+q-2l}-1)\binom{p+q-1-2l}{p-1}\zeta_{2l}(1)\zeta_{p+q-2l}(1)\Bigg\},\end{aligned}\tag{4.208}$$

$$\begin{aligned}\Omega_4(p,q) = (-1)^p 2^{-q-1}\Bigg\{&-2^{p+q}\binom{p+q-1}{p}\beta_{p+q}(-1)\\&-2^{p+q}[(-1)^p\zeta_p(1)+\zeta_p(-1)]\beta_q(-1)\\&+2^{p+q}\frac{(-1)^p-(-1)^q}{2}[\zeta_p(1)-\zeta_p(-1)]\beta_q(-1)\\&+\sum_{2\leqslant 2l\leqslant p}2^{p+q+1-2l}\binom{p+q-1-2l}{q-1}\zeta_{2l}(-1)\beta_{p+q-2l}(-1)\\&+\sum_{2\leqslant 2l\leqslant q}(2^{p+q+1}-2^{2l})\binom{p+q-2l}{p-1}\beta_{2l-1}(-1)\zeta_{p+q+1-2l}(1)\Bigg\},\end{aligned}\tag{4.209}$$

$$\begin{aligned}\Omega_5(p,q) = (-1)^p\Bigg\{&\frac{1}{2}\left[(-1)^{q+1}+\binom{p+q-1}{q}\right]\zeta_{p+q}(-1)\\&+\frac{1}{2}\binom{p+q-1}{p}\zeta_{p+q}(1)-\sum_{2\leqslant 2l\leqslant p-1}\binom{p+q-1-2l}{q-1}\zeta_{2l}(-1)\zeta_{p+q-2l}(-1)\\&-\sum_{2\leqslant 2l\leqslant q}\binom{p+q-1-2l}{p-1}\zeta_{2l}(-1)\zeta_{p+q-2l}(-1)\Bigg\},\end{aligned}\tag{4.210}$$

$$\begin{aligned}\Omega_6(p,q) = (-1)^p\Bigg\{&\frac{(-1)^{q+1}}{2}\zeta_{p+q}(1)+\frac{1}{2}\binom{p+q}{p}\zeta_{p+q}(-1)\\&+\frac{1}{2}[1+(-1)^p]\zeta_p(-1)\zeta_q(-1)\\&-\sum_{2\leqslant 2l\leqslant p}\binom{p+q-1-2l}{q-1}\zeta_{2l}(1)\zeta_{p+q-2l}(-1)\\&-\sum_{2\leqslant 2l\leqslant q}\binom{p+q-1-2l}{p-1}\zeta_{2l}(1)\zeta_{p+q-2l}(-1)\Bigg\},\end{aligned}\tag{4.211}$$

$$\begin{aligned}\Omega_7(p,q) = (-1)^p 2^{-q}\Bigg\{&\frac{1}{2}(2^{p+q}-1)\binom{p+q-1}{p}\zeta_{p+q}(1)\\&-\sum_{2\leqslant 2l\leqslant p-1}(2^{p+q-2l}-1)\binom{p+q-1-2l}{q-1}\zeta_{2l}(-1)\zeta_{p+q-2l}(1)\\&-2^{p+q}\sum_{2\leqslant 2l\leqslant q+1}\binom{p+q-2l}{p-1}\beta_{2l-1}(-1)\beta_{p+q+1-2l}(-1)\Bigg\},\end{aligned}\tag{4.212}$$

$$\Omega_8(p,q) = (-1)^p 2^p\Bigg\{-\frac{1}{2}\binom{p+q-1}{p}\beta_{p+q}(-1)$$

$$
\begin{aligned}
&- 2^{-p}\frac{1+(-1)^p}{2}\zeta_p(-1)\beta_q(-1) \\
&+ \sum_{2\leqslant 2l\leqslant p} 2^{-2l}\binom{p+q-1-2l}{q-1}\zeta_{2l}(1)\beta_{p+q-2l}(-1) \\
&+ \sum_{2\leqslant 2l\leqslant q} (1-2^{-2l})\binom{p+q-1-2l}{p-1}\zeta_{2l}(1)\beta_{p+q-2l}(-1)\Big\},
\end{aligned} \tag{4.213}
$$

$$
\begin{aligned}
\Omega_9(p,q) = {} & (-1)^p 2^{-p}\Big\{\frac{1}{2}(2^{p+q}-1)\binom{p+q-1}{q}\zeta_{p+q}(1) \\
&+ 2^p\frac{(-1)^p-(-1)^q}{2}A(1) + \frac{1+(-1)^p}{2}2^q(2^p-1)\zeta_p(1)\zeta_q(-1) \\
&- \sum_{2\leqslant 2l\leqslant p-1} (2^{2l}-1)(2^{p+q-2l}-1)\binom{p+q-1-2l}{q-1}\zeta_{2l}(1)\zeta_{p+q-2l}(1) \\
&- \sum_{2\leqslant 2l\leqslant q} (2^{p+q-2l}-1)\binom{p+q-1-2l}{p-1}\zeta_{2l}(1)\zeta_{p+q-2l}(1)\Big\},
\end{aligned} \tag{4.214}
$$

$$
\begin{aligned}
\Omega_{10}(p,q) = {} & (-1)^p 2^{-p}\Big\{\frac{1}{2}(2^{p+q}-1)\binom{p+q-1}{q}\zeta_{p+q}(1) \\
&+ \frac{1+(-1)^p}{2}(2^p-1)\zeta_p(1)\zeta_q(-1) + 2^p\frac{(-1)^p-(-1)^q}{2}A(-1) \\
&- 2^{p+q}\sum_{2\leqslant 2l\leqslant p+1} \binom{p+q-2l}{q-1}\beta_{2l-1}(-1)\beta_{p+q+1-2l}(-1) \\
&- \sum_{2\leqslant 2l\leqslant q} (2^{p+q-2l}-1)\binom{p+q-1-2l}{p-1}\zeta_{2l}(-1)\zeta_{p+q-2l}(1)\Big\},
\end{aligned} \tag{4.215}
$$

$$
\begin{aligned}
\Omega_{11}(p,q) = {} & (-1)^p 2^{-(p+q)}\Big\{\frac{(-1)^{q+1}}{2}(2^{p+q}-1)\zeta_{p+q}(1) \\
&- \frac{1+(-1)^p}{2}2^q(2^p-1)\zeta_p(1)\zeta_q(-1) \\
&- \sum_{2\leqslant 2l\leqslant p-1} (2^{2l}-1)\binom{p+q-1-2l}{q-1}\zeta_{2l}(1)\zeta_{p+q-2l}(1) \\
&- \sum_{2\leqslant 2l\leqslant q} (2^{2l}-1)\binom{p+q-1-2l}{p-1}\zeta_{2l}(1)\zeta_{p+q-2l}(1)\Big\},
\end{aligned} \tag{4.216}
$$

$$
\begin{aligned}
\Omega_{12}(p,q) = {} & (-1)^p 2^{-p-q-1}\Big\{(-1)^{q+1}2^{p+q}\beta_{p+q}(-1) \\
&+ 2^{p+q}\Big[\Big(1+\frac{(-1)^p-(-1)^q}{2}\Big)\zeta_p(-1) - \frac{(-1)^p+(-1)^q}{2}\zeta_p(1)\Big]\beta_q(-1) \\
&- \sum_{2\leqslant 2l\leqslant p+1} 2^{2l}\binom{p+q-2l}{q-1}\beta_{2l-1}(-1)\zeta_{p+q+1-2l}(-1) \\
&+ \sum_{2\leqslant 2l\leqslant q} 2^{2l}\binom{p+q-2l}{p-1}\beta_{2l-1}(-1)\zeta_{p+q+1-2l}(1)\Big\},
\end{aligned} \tag{4.217}
$$

$$\begin{aligned}\Omega_{13}(p,q) &= (-1)^p 2^{-p-1}\Bigg\{2^{p+q}\binom{p+q-1}{q}\beta_{p+q}(-1)\\ &\quad + [-1+(-1)^p]2^{p+1}(2^{q-1}-1)\zeta_q(1)\beta_p(-1)\\ &\quad - \sum_{2\leqslant 2l\leqslant p}(2^{p+q+1}-2^{2l})\binom{p+q-2l}{q-1}\beta_{2l-1}(-1)\zeta_{p+q+1-2l}(1)\\ &\quad - \sum_{2\leqslant 2l\leqslant q}2^{p+q+1-2l}\binom{p+q-1-2l}{p-1}\zeta_{2l}(-1)\beta_{p+q-2l}(-1)\Bigg\},\end{aligned} \tag{4.218}$$

$$\begin{aligned}\Omega_{14}(p,q) &= (-1)^p 2^{q}\Bigg\{\frac{1}{2}\binom{p+q-1}{q}\beta_{p+q}(-1)\\ &\quad + 2^{-q}\frac{1-(-1)^p}{2}\zeta_q(-1)\beta_p(-1) + 2^{-q}\frac{(-1)^p+(-1)^q}{2}A(1)\\ &\quad - \sum_{2\leqslant 2l\leqslant p}(1-2^{2l})\binom{p+q-1-2l}{q-1}\zeta_{2l}(1)\beta_{p+q-2l}(-1)\\ &\quad - \sum_{2\leqslant 2l\leqslant q}2^{-2l}\binom{p+q-1-2l}{p-1}\zeta_{2l}(1)\beta_{p+q-2l}(-1)\Bigg\},\end{aligned} \tag{4.219}$$

$$\begin{aligned}\Omega_{15}(p,q) &= (-1)^p 2^{-p-q-1}\Bigg\{(-1)^{q+1}2^{p+q}\beta_{p+q}(-1)\\ &\quad - 2^{p+q}[1-(-1)^p]\zeta_q(-1)\beta_p(-1)\\ &\quad - \sum_{2\leqslant 2l\leqslant p}2^{2l}\binom{p+q-2l}{q-1}\beta_{2l-1}(-1)\zeta_{p+q+1-2l}(1)\\ &\quad + \sum_{2\leqslant 2l\leqslant q+1}2^{2l}\binom{p+q-2l}{p-1}\beta_{2l-1}(-1)\zeta_{p+q+1-2l}(-1)\Bigg\},\end{aligned} \tag{4.220}$$

$$\begin{aligned}\Omega_{16}(p,q) &= (-1)^p 2^{-(p+q)}\Bigg\{\frac{(-1)^{q+1}}{2}(2^{p+q}-1)\zeta_{p+q}(1)\\ &\quad + 2^{p+q}\frac{(-1)^p-1}{2}\beta_p(-1)\beta_q(-1)\\ &\quad - \sum_{2\leqslant 2l\leqslant p}(2^{2l}-1)\binom{p+q-1-2l}{q-1}\zeta_{2l}(1)\zeta_{p+q-2l}(-1)\\ &\quad - \sum_{2\leqslant 2l\leqslant q}(2^{2l}-1)\binom{p+q-1-2l}{p-1}\zeta_{2l}(1)\zeta_{p+q-2l}(-1)\Bigg\}.\end{aligned} \tag{4.221}$$

In addition to these 16 relations for $\omega_j(p,q)$, relations (4.206)–(4.221) define 16 terms that by (4.204)–(4.205) are all equal to zero.

Proof. To prove formulas (4.206)–(4.221), we proceed as follows:
- (4.206) is obtained by comparing (4.84) and (4.197).
- (4.210) is obtained by comparing (4.87) and (4.198).
- (4.211) is obtained by comparing (4.85) and (4.199).

- (4.207) can be obtained from the formula

$$\Omega_2(p,q) = (-1)^{p+q}\Omega_5(q,p) + \frac{1-(-1)^{p+q}}{2}\zeta_p(1)\zeta_q(-1) + \frac{1-(-1)^{p+q}}{2}\zeta_{p+q}(-1). \quad (4.222)$$

In the case $p+q = 2k+1$, we have

$$\begin{aligned}(-1)^{p+q}\Omega_5(q,p) &+ \frac{1-(-1)^{p+q}}{2}\zeta_p(1)\zeta_q(-1) + \frac{1-(-1)^{p+q}}{2}\zeta_{p+q}(-1)\\ &= -\Omega_5(q,p) + \zeta_p(1)\zeta_q(-1) + \zeta_{p+q}(-1)\\ &\overset{(4.210)}{=} -\omega_5(q,p) + \zeta_p(1)\zeta_q(-1) + \zeta_{p+q}(-1) = \omega_2(p,q). \end{aligned} \quad (4.223)$$

For $p+q = 2k$,

$$(-1)^{p+q}\Omega_5(q,p) + \frac{1-(-1)^{p+q}}{2}\zeta_p(1)\zeta_q(-1) + \frac{1-(-1)^{p+q}}{2}\zeta_{p+q}(-1) = \Omega_5(q,p) \overset{(4.204)}{=} 0. \quad (4.224)$$

Formula (4.207) follows from (4.222) and (4.210). Since the proving technique is the same in all cases, we will only list the auxiliary relations regarding the proof of the formulas for $\Omega_{j(p,q)}$. Formula (4.208) can be obtained from

$$\Omega_3(p,q) = (2^{p-1} - 2^{-q})\Omega_1(p,q) + 2^{p-1}\Omega_5(p,q). \quad (4.225)$$

To prove formula (4.225), we use (4.204) and (4.109). Formula (4.208) follows from (4.225), (4.206), and (4.210). Formula (4.214) can be obtained from

$$\Omega_9(p,q) = (-1)^{p+q}\Omega_3(q,p) + \frac{1-(-1)^{p+q}}{2}(1-2^{-p})\zeta_p(1)\zeta_q(1) + \frac{1-(-1)^{p+q}}{2}A(1). \quad (4.226)$$

To prove the last formula, we use (4.204). Formula (4.214) follows from (4.226) and (4.208). Formula (4.216) can be obtained from

$$\Omega_{11}(p,q) = \frac{1}{2}\{(1-2^{-p-q+1})\Omega_1(p,q) + \Omega_6(p,q)\}. \quad (4.227)$$

To prove (4.227), we use (4.204) and (4.113). Formula (4.216) follows from (4.227), (4.206), and (4.211). Formula (4.212) can be obtained from

$$\Omega_7(p,q) = 2^{p-1}[1-(-1)^{p+q}][\Re\mathfrak{e}\, G_{(p,q)}(i) - 2^{-p-q}\Omega_5(p,q)]. \quad (4.228)$$

To prove (4.228), we use the formulas (4.204) and (4.117). Formula (4.212) follows from (4.228), (4.210), and (4.86). Formula (4.215) can be obtained from

$$\Omega_{10}(p,q) = (-1)^{p+q}\Omega_7(q,p) + \frac{1-(-1)^{p+q}}{2}(1-2^{-p})\zeta_p(1)\zeta_q(-1) + \frac{1-(-1)^{p+q}}{2}A(-1). \quad (4.229)$$

To prove formula (4.229), we use (4.204). Formula (4.215) follows from (4.229) and (4.212). Formula (4.221) can be obtained from

$$\Omega_{16}(p,q) = \frac{1}{2}[1-(-1)^{p+q}]\cdot[\Re\mathfrak{e}\, I_{(p,q)}(i) - 2^{-p-q}\Omega_6(p,q)]. \tag{4.230}$$

Formula (4.230) can be verified on the basis of (4.204) and (4.121). Formula (4.221) follows from (4.230), (4.211), and (4.90). So the formulas for the coefficients $\Omega_j(p,q)$ for j from the first set of indices are proved. Let us now prove the formulas for the coefficients $\Omega_j(p,q)$, where j is from the other set of indices. Formula (4.220) can be obtained from

$$\Omega_{15}(p,q) = \frac{1}{4}[1+(-1)^{p+q}]I_m(J_{(p,q)}(\mathrm{i}) - G_{(p,q)}(\mathrm{i})) \tag{4.231}$$

using (4.205), (4.124), and (4.123) for verification. Formula (4.220) follows from (4.231), (4.91), and (4.86). Formula (4.218) follows from the equality

$$\Omega_{13}(p,q) = [1+(-1)^{p+q}]\cdot\left\{-2^{q-2}\cdot I_m(J_{(p,q)}(i) + G_{(p,q)}(i)) + \frac{1}{2}A(-1)\right\}. \tag{4.232}$$

To determine the above relation, we use formulas (4.205), (4.124), and (4.123). Formula (4.218) follows from (4.232), (4.91), and (4.86). The formula (4.217) follows from

$$\Omega_{12}(p,q) = (-1)^{p+q+1}\Omega_{15}(q,p) + \frac{(-1)^{p+q+1}}{2}(1-2^{-p})\zeta_p(1)\beta_q(-1) + \frac{(-1)^{p+q+1}-1}{2}\beta_{p+q}(-1), \tag{4.233}$$

whereas formula (4.233) follows from (4.205). Formula (4.217) follows from (4.233) and (4.220). Formula (4.209) can be derived from

$$\Omega_4(p,q) = (-1)^{p+q+1}\Omega_{13}(q,p) + \frac{(-1)^{p+q+1}}{2}\zeta_p(1)\beta_q(-1) - \frac{(-1)^{p+q+1}-1}{2}A(-1), \tag{4.234}$$

and to verify this relation, we use (4.205). Formula (4.209) is derived from (4.234) and (4.218). Formula (4.213) is obtained from

$$\Omega_8(p,q) = 2^{p-2}[1+(-1)^{p+q}]\cdot[I_m I_{(p,q)}(\mathrm{i}) - I_m F_{(p,q)}(\mathrm{i})], \tag{4.235}$$

which can be easily verified with the help of (4.205) and (4.133). Formula (4.213) follows from (4.235), (4.90), and (4.76). Formula (4.219) is derived from

$$\Omega_{14}(p,q) = (-1)^{p+q+1}\Omega_8(q,p) - \frac{1-(-1)^{p+q+1}}{2}\zeta_q(-1)\beta_p(-1) + \frac{1-(-1)^{p+q+1}}{2}A(1), \tag{4.236}$$

which follows from (4.205). Formula (4.219) follows from (4.236) and (4.213). This proves the formulas for the coefficients $\Omega_j(p,q)$ for j from the other set of indices, formulas (4.206)–(4.221). □

At the end of this chapter, we also mention the following formulas:

$$
\begin{aligned}
&(-1)^{q+1}G_{(p,q)}(-1)+\sum_{l=1}^{p}\binom{p+q-1-l}{q-1}G_{(l,p+q-l)}(-1)\\
&\quad=\left[(-1)^{q+1}+\binom{p+q-1}{q}\right]\zeta_{(p+q)}(-1)-\sum_{l=1}^{q}(-1)^{l}\binom{p+q-1-l}{p-1}\zeta_{l}(-1)\zeta_{p+q-l}(-1),
\end{aligned}
\tag{4.237}
$$

$$
\begin{aligned}
&(-1)^{p+1}G_{(p,q)}(1)+\sum_{l=1}^{q}\binom{p+q-1-l}{p-1}G_{(p+q-l,l)}(1)\\
&\quad=\sum_{l=2}^{p-1}(-1)^{l}\binom{p+q-1-l}{q-1}\zeta_{l}(1)\zeta_{p+q-l}(1)+\sum_{l=2}^{q}\binom{p+q-1-l}{p-1}\zeta_{l}(1)\zeta_{p+q-l}(1),
\end{aligned}
\tag{4.238}
$$

$$
\begin{aligned}
&(-1)^{q+1}G_{(p,q)}(1)+\sum_{l=1}^{q}\binom{p+q-1-l}{p-1}G_{(p+q-l,l)}(1)\\
&\quad=\left[(-1)^{q+1}+\binom{p+q}{p}\right]\zeta_{p+q}(1)-\sum_{l=2}^{p-1}\binom{p+q-1-l}{q-1}\zeta_{l}(1)\zeta_{p+q-l}(1)\\
&\qquad-\sum_{l=2}^{q}(-1)^{l}\binom{p+q-1-l}{p-1}\zeta_{l}(1)\zeta_{p+q-l}(1),
\end{aligned}
\tag{4.239}
$$

$$
\begin{aligned}
&(-1)^{p+1}G_{(p,q)}(-1)+\sum_{l=1}^{q}\binom{p+q-1-l}{p-1}F_{(p+q-l,l)}(-1)\\
&\quad=\sum_{l=1}^{p-1}(-1)^{l}\binom{p+q-1-l}{q-1}\zeta_{l}(-1)\zeta_{p+q-l}(1)+\sum_{l=1}^{q}\binom{p+q-1-l}{p-1}\zeta_{l}(-1)\zeta_{p+q-l}(-1).
\end{aligned}
\tag{4.240}
$$

Formula (4.237) follows from (4.152) for $z=-1, p\to q, q\to p$. Formula (4.238) follows from (4.152) for $z=1$ and from (4.142) for $z=1$. Formula (4.239) follows from (4.154) for $z=1$, and formula (4.240) follows from (4.146) for $z=-1$.

4.2 Appendix

As it was given before,

$$
\begin{aligned}
\zeta_k(z)&=\sum_{n=1}^{\infty}\frac{z^n}{n^k}, \quad F_{(p,q)}(z)=\sum_{m=1}^{\infty}\frac{z^m}{m^p}\sum_{n=1}^{m}\frac{z^n}{n^q}, \quad |z|\leqslant 1,\\
G_{(p,q)}(z)&=\sum_{m=1}^{\infty}\frac{z^m}{m^p}\sum_{n=1}^{m}\frac{1}{n^q}, \quad H_{(p,q)}(z)=\sum_{m=1}^{\infty}\frac{1}{m^p}\sum_{n=1}^{m}\frac{z^n}{n^q}.
\end{aligned}
\tag{4.A1}
$$

The series that determines $\zeta_k(z)$ ($k\in\mathbb{N}$) for $k\geqslant 2$ converges for $|z|\leqslant 1$, whereas for $k=1$, it converges for $|z|\leqslant 1, z\neq 1$. For $k=0$, we introduce $\zeta_0(z)=\frac{z}{1-z}$, $\zeta_0(-1)=-\frac{1}{2}$. For $z=x$, the sign $\zeta_k(x)$ originates from *Legendre*. For $k\in\mathbb{N}$, $k\geqslant 2$, we define

$$
\zeta_k(1)=\zeta(k), \quad \zeta_k(-1)=\zeta(-k),
\tag{4.A2}
$$

where the last relation holds for both $k = 1$ and $k = 0$. Further,

$$\zeta_k(-1) = -\left(1 - \frac{1}{2^{k-1}}\right)\zeta_k(1) \quad (k \geqslant 2); \quad \zeta_1(-1) = -\log 2, \tag{4.A3}$$

$$\zeta_1(z) = -\log(1-z) = \sum_{n=1}^{\infty} \frac{z^n}{n} \quad (|z| \leqslant 1,\ z \neq 1), \tag{4.A4}$$

$$\zeta_k(z) + \zeta_k(-z) = \frac{1}{2^{k-1}}\zeta_k(z^2) \quad (|z| \leqslant 1,\ k \geqslant 2), \tag{4.A5}$$

$$\zeta_1(z) + \zeta_1(-z) = \zeta_1(z^2) \quad (|z| \leqslant 1,\ z \neq 1). \tag{4.A6}$$

We can easily verify that

$$\zeta_1'(z) = \frac{1}{z}\zeta_{k-1}(z), \quad \zeta_k'(-z) = \frac{1}{2}\zeta_{k-1}(-z). \tag{4.A7}$$

For the series $\sum_{m=1}^{\infty} \frac{z^{rm}}{m^p} \sum_{n=1}^{m} \frac{z^{sn}}{n^q}$ (where r, s are nonnegative integers, and p, q are natural numbers), we will prove the following:

(a) For $p \geqslant 2$, the series uniformly converges for all $|z| \leqslant 1$.

(b) For $p = 1$, the series $\sum_{m=1}^{\infty} \frac{(-1)^m}{m^p} \sum_{n=1}^{m} \frac{1}{n^q}$ converges.

Proof. (a) Let

$$a_m(z) = \frac{z^{rm}}{m^p} \sum_{n=1}^{m} \frac{z^{sn}}{n^q} \quad (m \geqslant 1). \tag{4.A8}$$

Then for $|z| \leqslant 1$, we have

$$|a_m(z)| \leqslant \frac{|z|^{rm}}{m^p} \sum_{n=1}^{m} \frac{|z|^{sn}}{n^q} \leqslant \frac{1}{m^p} \sum_{n=1}^{m} \frac{1}{n^q} \leqslant \frac{1}{m^p} \sum_{n=1}^{m} \frac{1}{n} = b_m, \tag{4.A9}$$

$$\begin{aligned}\lim_{m\to\infty} m\left(\frac{b_m}{b_{m+1}} - 1\right) &= \lim_{m\to\infty} m\left(\frac{\frac{1}{m^p}\sum_{n=1}^{m}\frac{1}{n}}{\frac{1}{(m+1)^p}\sum_{n=1}^{m+1}\frac{1}{n}} - 1\right)\\ &= \lim_{m\to\infty} \frac{m\{[(1+\frac{1}{m})^p - 1]\sum_{n=1}^{m}\frac{1}{n} - \frac{1}{m+1}\}}{\sum_{n=1}^{m+1}\frac{1}{n}}\\ &= \lim_{m\to\infty} \frac{[p + O(\frac{1}{m})](\sum_{n=1}^{m}\frac{1}{n} - \log m) + [p + O(\frac{1}{m})]\log m - \frac{m}{m+1}}{(\sum_{n=1}^{m}\frac{1}{n} - \log m) + \log m + \frac{1}{m+1}} = p.\end{aligned} \tag{4.A10}$$

Since $p \geqslant 2$, by the *Rabe* criterion a number series converges, and then by the *Weierstrass* criterion the starting series converges uniformly for $|z| \leqslant 1$, thereby defining a continuous function in this region $|z| \leqslant 1$. That is why it was possible in the previous text to determine the coefficients of the form $F_{(p,q)}(1)$, $p + q = 2k + 1$, by determining the limits as z tends to 1 by the values along the circle $|z| = 1$. It is the same in the case of the point $z = -1$. For $p = 1$, the corresponding coefficients were obtained over the coefficients for $p \geqslant 2$.

(b) Let

$$C_m = \frac{1}{m}\sum_{n=1}^{m}\frac{1}{n^q}. \tag{4.A11}$$

Then

$$C_m > C_{m+1} \Leftrightarrow \left(\frac{1}{m} - \frac{1}{m+1}\right)\sum_{n=1}^{m}\frac{1}{n^q} > \frac{1}{(m+1)^{q+1}} \Leftrightarrow \sum_{n=1}^{m}\frac{1}{n^q} > \frac{m}{(m+1)^q}, \tag{4.A12}$$

$$\frac{1}{n^q} > \frac{1}{(m+1)^q} \quad (n = 1, 2, \ldots, m) \implies \sum_{n=1}^{m}\frac{1}{n^q} > \frac{m}{(m+1)^q}, \tag{4.A13}$$

$$\lim_{m\to\infty}\frac{1}{m}\sum_{n=1}^{m}\frac{1}{n^q} \leqslant \lim_{m\to\infty}\frac{1}{m}\sum_{n=1}^{m}\frac{1}{n} = \lim_{m\to\infty}\frac{1}{m}\left\{\left(\sum_{n=1}^{m}\frac{1}{n} - \log m\right) + \log m\right\} = 0. \tag{4.A14}$$

It follows that by the *Leibniz* criterion the series converges. □

We have the following equality:

$$F_{(p,0)}(z) = \frac{z}{1-z}[\zeta_p(z) - \zeta_p(z^2)]. \tag{4.A15}$$

Obviously,

$$F_{(p,q)}(1) = G_{(p,q)}(1) = H_{(p,q)}(1). \tag{4.A16}$$

Further, we have

$$G_{(p,0)}(z) = \zeta_{p-1}(z), \tag{4.A17}$$

$$H_{(p,0)}(z) = \frac{z}{1-z}[\zeta_p(1) - \zeta_p(z)], \tag{4.A18}$$

$$F_{(1,0)}(z) = \frac{z}{1-z}\log(1+z) = -\zeta_0(z)\zeta_1(-z), \tag{4.A19}$$

$$G_{(1,0)}(z) = \zeta_0(z) = \frac{z}{1-z}, \tag{4.A20}$$

$$F_{(p,q)}(z) = \frac{1}{z}F_{(p-1,q)}(z) + \frac{1}{2}F_{(p,q-1)}(z), \tag{4.A21}$$

$$G_{(p,q)}(z) = \frac{1}{z}G_{(p-1,q)}(z), \tag{4.A22}$$

$$H_{(p,q)}(z) = \frac{1}{z}H_{(p,q-1)}(z), \tag{4.A23}$$

$$F_{(p,q)}(z) = \sum_{m=1}^{\infty}\frac{z^m}{m^p}\sum_{n=1}^{m}\frac{z^n}{n^q} = \sum_{k=2}^{\infty}z^k\sum_{\frac{k}{2}\leqslant m\leqslant k-1}\frac{1}{m^p(k-m)^q}$$

$$= \sum_{k=2}^{\infty} z^k \sum_{1\leqslant m\leqslant \frac{k}{2}} \frac{1}{m^q(k-m)^p}, \tag{4.A24}$$

$$F_{(p,q)}(z) + F_{(q,p)}(z) = \zeta_{(p+q)}(z^2) + \zeta_p(z)\zeta_q(z)A, \tag{4.A25}$$

$$H_{(p,q)}(z) + G_{(q,p)}(z) = \zeta_{p+q}(z) + \zeta_p(1)\zeta_q(z), \tag{4.A26}$$

$$\sum_{n=1}^{\infty} \frac{1}{n(m+n)} = \frac{1}{m}\sum_{n=1}^{m}\frac{1}{n}. \tag{4.A27}$$

The last equality follows from

$$\begin{aligned}
\sum_{n=1}^{\infty} \frac{1}{n(m+n)} &= \frac{1}{m}\lim_{k\to\infty}\sum_{n=1}^{k}\left(\frac{1}{n} - \frac{1}{m=n}\right) \\
&= \frac{1}{m}\lim_{k\to\infty}\left\{\left(\sum_{n=1}^{m}\frac{1}{n} - \log k\right) + \log\frac{k}{m+k} - \left(\sum_{n=1}^{m+k}\frac{1}{n} - \log(m+k)\right) + \sum_{n=1}^{m}\frac{1}{n}\right\} \\
&= \frac{1}{m}\sum_{n=1}^{m}\frac{1}{n},
\end{aligned} \tag{4.A28}$$

where we used the definition of *Euler*'s constant γ. In addition, we have

$$\begin{aligned}
\frac{1}{n^b(m+n)^a} &= (-1)^b \sum_{l=1}^{a}\binom{a+b-1-l}{b-1}\frac{1}{m^{a+b-l}(m+n)^l} \\
&\quad + (-1)^b\sum_{l=1}^{b}(-1)^l\binom{a+b-1-l}{a-1}\frac{1}{m^{a+b-l}n^l}.
\end{aligned} \tag{4.A29}$$

Proof. Let

$$\frac{1}{y^b(x+y)^a} = \sum_{k=1}^{b}\frac{B_k}{y^k} + \sum_{l=1}^{a}\frac{A_l}{(x+y)^l}. \tag{4.A30}$$

From here it follows that

$$(x+y)^{-a} = \sum_{k=1}^{b} B_k y^{b-k} + \left(\sum_{l=1}^{a}(x+y)^{-l}A_l\right)y^b. \tag{4.A31}$$

If we regard this expression as a function of y, for $s = 0, 1, \ldots, b-1$, we have

$$\begin{aligned}
((x+y)^{-a})_y^{(s)}\big|_{y=0} &= (-a)(-a-1)\cdots[-a-(s-1)](x+y)^{-a-s}\big|_{y=0} \\
&= (-1)^s\frac{(a+s-1)!}{(a-1)!}x^{-a-s},
\end{aligned} \tag{4.A32}$$

$$\left(\sum_{k=1}^{b} B_k y^{b-k}\right)_y^{(s)}\Bigg|_{y=0} = \sum_{k=1}^{b-s}(b-k)(b-k-1)\cdots[b-k-(s-1)]\,y^{b-k-s}B_k\big|_{y=0}$$

$$= s!B_{b-s}, \tag{4.A33}$$

$$\left(\sum_{l=1}^{a}(x+y)^{-l}A_l y^b\right)_y^{(s)}\Bigg|_{y=0} = 0, \tag{4.A34}$$

$$(-1)^s\frac{(a+s-1)!}{(a-1)!}x^{-a-s} = s!B_{b-s} \quad (s=0,1,2,\ldots,b-1), \tag{4.A35}$$

$$B_s = (-1)^{b-s}\binom{a+b-1-s}{a-1}x^{-a-b+s} \quad (s=1,2,\ldots,b). \tag{4.A36}$$

We can write

$$y^{-b} = \left(\sum_{k=1}^{b}B_k y^{-k}\right)(x+y)^a + \sum_{l=1}^{a}A_l(x+y)^{a-l}. \tag{4.A37}$$

It follows that for $s = 0,1,\ldots,a-1$,

$$\begin{aligned}(y^{-b})_y^{(s)}|_{y=-x} &= (-b)(-b-1)\cdots[-b-(s-1)]\,y^{-b-s}|_{y=-x}\\ &= (-1)^b\frac{(b+s-1)!}{(b-1)!}x^{-b-s},\end{aligned} \tag{4.A38}$$

$$\begin{aligned}\left(\sum_{l=1}^{a}A_l(x+y)^{a-l}\right)_y^{(s)}\Bigg|_{y=-x} &= \sum_{l=1}^{a-s}(a-l)(a-l-1)\cdots[a-l-(s-1)]A_l(x+y)^{a-l-s}|_{y=-x}\\ &= s!A_{a-s},\end{aligned} \tag{4.A39}$$

$$\left(\sum_{k=1}^{b}B_k y^{-k}(x+y)^a\right)_y^{(s)}\Bigg|_{y=-x} = 0, \tag{4.A40}$$

$$(-1)^b\frac{(b+s-1)!}{(b-1)!}x^{-b-s} = s!A_{a-s} \quad (s=0,1,\ldots,a-1), \tag{4.A41}$$

$$A_s = (-1)^b\binom{a+b-1-s}{b-1}x^{-a-b+s} \quad (s=1,2,\ldots,a). \tag{4.A42}$$

Finally, relation (4.A29) is proved by taking $x = m$ and $y = n$. □

Further, we have

$$F_{(0,q)}(z) = \frac{\zeta_q(z^2)}{1-z}, \tag{4.A43}$$

$$G_{(0,q)}(z) = \frac{\zeta_q(z^2)}{1-z}, \tag{4.A44}$$

$$G_{(1,1)}(z) = \zeta_2(z) + \frac{1}{2}\zeta_1^2(z) \quad (|z| \leqslant 1,\ z \neq 1). \tag{4.A45}$$

It has already been proven that there exists $G_{(1,1)}(-1)$, and then from (4.A45) it follows that

$$G_{(1,1)}(-1) = -\frac{1}{2}\zeta_2(1) + \frac{1}{2}\log^2 2. \tag{4.A46}$$

For $z = \pm\mathrm{i}$, from (4.A45) we get

$$\begin{aligned} G_{(1,1)}(\mathrm{i}) + G_{(1,1)}(-\mathrm{i}) &= \zeta_2(\mathrm{i}) + \zeta_2(-\mathrm{i}) + \frac{1}{2}[\zeta_1^2(\mathrm{i}) + \zeta_1^2(-\mathrm{i})] \\ &= \left(\frac{1}{2^2}\zeta_2(-1) + \mathrm{i}\beta_2(-1)\right) + \left(\frac{1}{2^2}\zeta_2(-1) - \mathrm{i}\beta_2(-1)\right) \\ &\quad + \frac{1}{2}\left\{\left(\frac{1}{2}\zeta_1(-1) + \mathrm{i}\beta_1(-1)\right)^2 + \left(\frac{1}{2}\zeta_1(-1) - \mathrm{i}\beta_1(-1)\right)^2\right\} \\ &= \frac{1}{2}\zeta_2(-1) + \frac{1}{2}\left\{\frac{1}{2}\zeta_1^2(-1) - 2\beta_1^2(-1)\right\} = -\frac{5}{8}\zeta_2(1) + \frac{1}{4}\log^2 2. \end{aligned} \tag{4.A47}$$

From the relation

$$G_{(1,1)}(z) + G_{(1,1)}(-z) = \frac{1}{2}G_{(1,1)}(z^2) + \sum_{n=1}^{\infty}\frac{z^{2m}}{m}\sum_{n=1}^{m}\frac{1}{2n-1} \quad (|z| < 1), \tag{4.A48}$$

by taking $z = i$ we get

$$\sum_{m=1}^{\infty}\frac{(-1)^m}{m}\sum_{n=1}^{m}\frac{1}{2n-1} = G_{(1,1)}(\mathrm{i}) + G_{(1,1)}(-\mathrm{i}) - \frac{1}{2}G_{(1,1)}(-1) = -\frac{3}{8}\zeta_2(1), \tag{4.A49}$$

$$\sum_{m=1}^{\infty}\frac{(-1)^m}{m}\sum_{n=1}^{m}\frac{1}{2n-1} = -\frac{3}{8}\zeta_2(1) = \omega_7(1,1). \tag{4.A50}$$

From the equality

$$\omega_7(1,2k-1) = \sum_{m=1}^{\infty}\frac{(-1)^m}{m}\sum_{n=1}^{m}\frac{1}{(2n-1)^{2k-1}} = G_{(1,2k-1)}(\mathrm{i}) + G_{(1,2k-1)}(-\mathrm{i}) - \frac{1}{2^{2k-1}}G_{(1,2k-1)}(-1), \tag{4.A51}$$

using (4.186) with $z = \mathrm{i}$, $z = -\mathrm{i}$, $z = -1$, after rearranging, we obtain

$$\omega_7(1,2k-1) = \sum_{m=1}^{\infty}\frac{(-1)^m}{m}\sum_{n=1}^{\infty}\frac{1}{(2n-1)^{2k-1}} = (-1)^k\beta_k^2(-1) + 2\sum_{l=1}^{k-1}(-1)^l\beta_l(-1)\beta_{2k-l}(-1), \tag{4.A52}$$

where

$$\beta_k(-1) = \sum_{n=0}^{\infty}\frac{(-1)^n}{(2n+1)^k}, \quad \beta_1(-1) = \frac{\pi}{4}, \quad \beta_3(-1) = \frac{\pi^3}{32}, \tag{4.A53}$$

$$\beta_2(-1) = \sum_{n=0}^{\infty}\frac{(-1)^n}{(2n+1)^2} = G, \tag{4.A54}$$

and G is *Catalan*'s constant ($G = 0,1915965594\ldots$), and for $k = 2$,

$$\omega_7(1,3) = \sum_{m=1}^{\infty} \frac{(-1)^m}{m} \sum_{n=1}^{m} \frac{1}{(2n-1)^3} = G^2 - \frac{45}{32}\zeta_4(1). \tag{4.A55}$$

Further, from (4.200) for $k = 2, 3, 4, 5$, we have

$$\left.\begin{aligned} \zeta_2^2(1) &= \frac{5}{2}\zeta_4(1), \\ \zeta_2(1)\zeta_4(1) &= \frac{7}{4}\zeta_6(1), \\ \zeta_2(1)\zeta_6(1) &= \frac{5}{3}\zeta_8(1), \\ \zeta_4^2(1) &= \frac{7}{6}\zeta_8(1), \\ \zeta_2(1)\zeta_8(1) &= \frac{33}{20}\zeta_{10}(1), \\ \zeta_4(1)\zeta_6(1) &= \frac{11}{10}\zeta_{10}(1). \end{aligned}\right\} \tag{4.A56}$$

By these formulas from (4.239) we obtain

$$\left.\begin{aligned} G_{(3,1)}(1) &= \frac{5}{4}\zeta_4(1), \\ G_{(2,2)}(1) &= \frac{7}{4}\zeta_4(1), \\ G_{(5,1)}(1) &= \frac{7}{4}\zeta_6(1) - \frac{1}{2}\zeta_3^2(1), \\ G_{(4,2)}(1) &= -\frac{1}{3}\zeta_6(1) + \zeta_3^2(1), \\ G_{(3,3)}(1) &= \frac{1}{2}\zeta_6(1) + \frac{1}{2}\zeta_3^2(1), \\ G_{(2,4)}(1) &= \frac{37}{12}\zeta_6(1) - \zeta_3^2(1), \\ G_{(7,1)}(1) &= \frac{9}{4}\zeta_8(1) - \zeta_3(1)\zeta_5(1), \\ 5G_{(6,2)}(1) + 2G_{(5,3)}(1) &= -\frac{21}{4}\zeta_8(1) + 10\zeta_3(1)\zeta_5(1), \\ G_{(4,4)}(1) &= \frac{13}{12}\zeta_8(1), \\ G_{(9,1)}(1) &= \frac{11}{4}\zeta_{10}(1) - \zeta_3(1)\zeta_7(1) - \frac{1}{2}\zeta_5^2(1), \\ 7G_{(8,2)}(1) + 2G_{(7,3)}(1) &= -\frac{33}{2}\zeta_{10}(1) + 14\zeta_3(1)\zeta_7(1) + 8\zeta_5^2(1), \\ G_{(7,3)}(1) + G_{(6,4)}(1) &= \frac{31}{10}\zeta_{10}(1) - \zeta_5^2(1), \\ G_{(5,5)}(1) &= \frac{1}{2}\zeta_{10}(1) + \frac{1}{2}\zeta_5^2(1). \end{aligned}\right\} \tag{4.A57}$$

From (4.237) it follows that

$$\begin{aligned} G_{(1,3t)}(-1) &= \frac{19}{14}\zeta_4(-1) + \zeta_1(-1)\zeta_3(-1), \\ 2G_{(3,1)}(-1) + G_{(2,2)}(-1) &= \frac{37}{14}\zeta_4(-1). \end{aligned} \tag{4.A58}$$

The following formula was discovered by *Hardy*:

$$\sum_{n=1}^{\infty}(-1)^{n-1}\left(1+\frac{1}{2}+\cdots+\frac{1}{n}-\gamma-\log n\right) = \frac{1}{2}(\log\pi-\gamma). \tag{4.A59}$$

Proof. For the series $\{a_n\}$ defined by

$$a_n = 1+\frac{1}{2}+\cdots+\frac{1}{n}-\log n, \tag{4.A60}$$

it is known that it is decreasing and bounded by γ from below, so the series $\{b_n\}$ defined by

$$b_n = 1+\frac{1}{2}+\cdots+\frac{1}{n}-\gamma-\log n \tag{4.A61}$$

is decreasing and bounded by zero from below. Thus by the *Leibniz* criterion the series (4.A59) converges, and

$$\begin{aligned} s_{2n} &= \sum_{k=1}^{2n}(-1)^{k-1}b_k = (1-\gamma-\log 1) - \left(1+\frac{1}{2}-\gamma-\log 2\right) + \cdots \\ &\quad + \left(1+\frac{1}{2}+\cdots+\frac{1}{2n-1}-\gamma-\log(2n-1)\right) \\ &\quad - \left(1+\frac{1}{2}+\cdots+\frac{1}{2n}-\gamma-\log(2n)\right) \\ &= \left(-\frac{1}{2}+\log\frac{2}{1}\right)+\cdots = \left(1+\frac{1}{3}-\gamma+\log\frac{2}{1}-\log 3\right)+\cdots \\ &= \left(-\frac{1}{2}-\frac{1}{4}+\log\frac{2}{1}+\log\frac{4}{3}\right)+\cdots = \cdots \\ &= -\frac{1}{2}-\frac{1}{4}-\cdots-\frac{1}{2n}+\log\frac{(2n)!!}{(2n-1)!!} \\ &= -\frac{1}{2}\left(1+\frac{1}{2}+\cdots+\frac{1}{n}-\log n\right)+\left(-\frac{1}{2}\log n+\log\frac{(2n)!!}{(2n-1)!!}\right). \end{aligned} \tag{4.A62}$$

From *Wallis*'s formula it follows that

$$\pi = \lim_{n\to\infty}\frac{1}{n}\left(\frac{(2n)!!}{(2n-1)!!}\right)^2, \tag{4.A63}$$

so we have

$$\lim_{n\to\infty}\left(-\frac{1}{2}\log n+\log\frac{(2n)!!}{(2n-1)!!}\right)=\frac{1}{2}\log\pi, \tag{4.A64}$$

and thus

$$S=\lim_{n\to\infty}s_{2n}=-\frac{\gamma}{2}+\frac{1}{2}\log\pi. \tag{4.A65}$$

□

Let us have a look at the series

$$\sum_{n=1}^{\infty}\frac{(-1)^{n-1}}{n^{\alpha}}\left(1+\frac{1}{2}+\cdots+\frac{1}{n}-\log n\right)\quad(\alpha\in R). \tag{4.A66}$$

The series $\sum_{n=1}^{\infty}\frac{(-1)^{n-1}}{n^{\alpha}}$ by the *Leibniz* criterion converges for $\alpha>0$.

The series $\{1+\frac{1}{2}+\cdots+\frac{1}{n}-\log n\}$ decreases and is bounded from below by γ. Then the series (4.A66) by *Abel*'s criterion converges for $\alpha>0$. Next, from

$$\lim_{n\to\infty}\frac{\frac{1}{n^{\alpha}}(1+\frac{1}{2}+\cdots+\frac{1}{n}-\log n)}{\frac{1}{n^{\alpha}}}=\gamma \tag{4.A67}$$

we conclude that the series

$$\sum_{n=1}^{\infty}\frac{1}{n^{\alpha}}\left(1+\frac{1}{2}+\cdots+\frac{1}{n}-\log n\right) \tag{4.A68}$$

converges for $\alpha>1$ and diverges for $\alpha\leqslant 1$. Therefore the series (4.A66) converges absolutely for $\alpha>1$ and conditionally for $0<\alpha<1$. For $\alpha=1$, we have the following formula:

$$\sum_{n=1}^{\infty}\frac{(-1)^{n-1}}{n}\left(1+\frac{1}{2}+\cdots+\frac{1}{n}-\log n\right)=\frac{1}{2}\zeta_2(1)+\gamma\log 2-\log^2 2. \tag{4.A69}$$

In relation to (4.A69), by the inequality

$$0<\sum_{k=1}^{\infty}\frac{1}{k}-\log n-\gamma<\frac{1}{n} \tag{4.A70}$$

we have the absolute convergence of the series

$$\sum_{n=1}^{\infty}\frac{(-1)^{n-1}}{n}\left(\sum_{k=1}^{\infty}\frac{1}{k}-\log n-\gamma\right). \tag{4.A71}$$

For the series

$$A^*(z) = \sum_{n=1}^{\infty} \frac{z^n}{n}\left(\sum_{k=1}^{\infty} \frac{1}{k} - \log n - \gamma\right), \tag{4.A72}$$

in the region $|z| < 1$, by (4.A45) we have the following decomposition:

$$A^*(z) = \zeta_2(z) + \frac{1}{2}\zeta_1^2(z) + (\zeta_s(z))'_{s|s=1} + \gamma\log(1-z). \tag{4.A73}$$

Passing to the limit as z tends to -1 (along the real axis from 0 to -1), we calculate $A^*(-1)$. The value of $A^*(-1)$ is expressed over the sum $\sum_{n=1}^{\infty} \frac{(-1)^n \log n}{n}$, which is

$$\sum_{n=1}^{\infty} \frac{(-1)^n \log n}{n} = \gamma \log 2 - \frac{1}{2}\log^2 2. \tag{4.A74}$$

In short, (4.A74) is obtained from

$$(1-2^{1-s})\zeta_s(1) = \sum_{n=1}^{\infty} \frac{(-1)^{n-1}}{n^s} \quad (\mathfrak{Re}\, s > 0) \tag{4.A75}$$

differentiating by s, passing to the limit as s tends to 1, and using

$$\zeta_s(1) = \zeta(s) = \frac{1}{s-1} + \sum_{n=0}^{\infty} \gamma_n (s-1)^n \tag{4.A76}$$

around the point $s = 1$, with $\gamma_0 = \gamma$, because

$$\lim_{s\to\infty}\left(\zeta(s) - \frac{1}{s-1}\right) = \gamma. \tag{4.A77}$$

Formula (4.A69) is obtained by adding $\gamma \log 2$ to the expression $-A^*(-1)$.

For the series (4.A68), where $\alpha = k \geqslant 2$, $k \in \mathbb{N}$, we have the following formula:

$$\sum_{n=1}^{\infty} \frac{1}{n^k}\left(1 + \frac{1}{2} + \cdots + \frac{1}{n} - \log n\right) = \frac{1}{2}(k+2)\zeta_{k+1}(1) - \frac{1}{2}\sum_{l=2}^{k-1} \zeta_l(1)\cdot\zeta_{k+1-l}(1) + \zeta_k'(1) \quad (k \geq 2). \tag{4.A78}$$

Proof. The relation follows from (4.188) and from

$$\sum_{n=1}^{\infty} \frac{1}{n^s} = \zeta_s(1) = \zeta(s) \quad (\mathfrak{Re}\, s > 1) \tag{4.A79}$$

by differentiating by s and then taking $s = k$, where

$$\zeta_k'(1) = (\zeta_s(1))'_s|_{s=k} = \zeta'(k). \tag{4.A80}$$

Further, for the series (4.A66), we can write

$$\sum_{n=1}^{\infty}\frac{(-1)^n}{n^{2k}}\left(1+\frac{1}{2}+\cdots+\frac{1}{n}-\log n\right)$$
$$=\frac{1}{2}(2k+1)\zeta_{2k+1}(-1)+\frac{1}{2}\zeta_{2k+1}(1)$$
$$-\sum_{l=1}^{k-1}\zeta_{2l}(-1)\zeta_{2k+1-2l}(1)-2^{1-2k}\zeta_{2k}(1)\log 2+(1-2^{1-2k})\zeta'_{2k}(1). \tag{4.A81}$$

This formula follows from (4.87) for $p = 2k$ and from formula (4.A75) by differentiating by s and then taking $s = 2k$. □

We will also provide the following formulas:

$$F_{(1,1)}(-1) = \frac{1}{2}\zeta_2(1) + \frac{1}{2}\log^2 2, \tag{4.A82}$$
$$G_{(2,1)}(1) = 2\zeta_3(1), \tag{4.A83}$$
$$G_{(2,1)}(-1) = -\frac{5}{8}\zeta_3(1), \tag{4.A84}$$
$$G_{(1,2)}(-1) = -\zeta_3(1) + \frac{1}{2}\zeta_2(1)\log 2, \tag{4.A85}$$
$$F_{(2,1)}(-1) = -\frac{5}{8}\zeta_3(1) + \frac{3}{2}\zeta_2(1)\log 2, \tag{4.A86}$$
$$F_{(1,2)}(-1) = \frac{13}{8}\zeta_3(1) - \zeta_2(1)\log 2, \tag{4.A87}$$
$$H_{(2,1)}(-1) = \frac{1}{4}\zeta_3(1) - \frac{3}{2}\zeta_2(1)\log 2, \tag{4.A88}$$
$$F_{(2,2)}(-1) = \frac{13}{16}\zeta_4(1), \tag{4.A89}$$
$$G_{(1,3)}(-1) = -\frac{19}{16}\zeta_4(1) + \frac{3}{4}\zeta_3(1)\log 2, \tag{4.A90}$$
$$H_{(3,1)}(-1) = -\frac{7}{4}\zeta_3(1)\log 2 + \frac{5}{16}\zeta_4(1), \tag{4.A91}$$
$$G_{(3,1)}(-1) = -\frac{37}{32}\zeta_4(1) - \frac{1}{2}G_{(2,2)}(-1), \tag{4.A92}$$
$$F_{(1,3)}(-1) = \frac{35}{32}\zeta_4(1) - \zeta_3(1)\log 2 - \frac{1}{2}G_{(2,2)}(-1), \tag{4.A93}$$
$$F_{(3,1)}(-1) = -\frac{3}{32}\zeta_4(1) + \frac{7}{4}\zeta_3(1)\log 2 + \frac{1}{2}G_{(2,2)}(-1), \tag{4.A94}$$
$$H_{(2,2)}(-1) = -\frac{17}{8}\zeta_4(1) - G_{(2,2)}(-1). \tag{4.A95}$$

Formulas (4.A82)–(4.A95) can be obtained as follows:
(4.A82) follows from (4.142) for $p = 1, q = 1, z = -1$.
(4.A89) follows from (4.142) for $p = 2, q = 2, z = -1$.

(4.A95) follows from (4.142) for $p = 2$, $q = 2$, $z = -1$.

From (4.188) for $p = 2$ we have (4.A83), whereas applying (4.189), for $p = 2$, we obtain (4.A88).

(4.A91) follows from (4.189) for $p = 3$, whereas (4.A85) is derived from (4.190) for $p = 2$.

From (4.190) for $p = 3$ we have (4.A90), and (4.87) for $p = 2$ and $q = 1$ gives (4.A84).

(4.A85) is derived from (4.87) for $p = 1$ and $q = 2$.

From (4.85) for $p = 2$ and $q = 1$ we derive (4.A86), and (4.85) for $p = 1$ and $q = 2$ leads to (4.A87).

From (4.152) for $p = 1$ and $q = 3$, together with (4.A90), we derive (4.A92).

(4.A94) follows from (4.146) for $p = 3$, $q = 1$, $z = -1$, together with (4.A92), and (4.A93) is derived from (4.142) for $z = -1$ using (4.A94).

5 The basic triple sums up to order 5

5.1 About basic triple sums up to order 5

In this chapter, by applying contour integration, we obtain various formulas for multiple sums, such as a formula for

$$\sum_{m=1}^{\infty} \frac{\cos m\alpha}{m^{2k}} \sum_{n=1}^{m} \frac{1}{n} \quad (0 \leqslant \alpha \leqslant \pi) \tag{5.1}$$

and a particular value of this sum for $\alpha = \frac{\pi}{3}$. We also obtain a formula for the following sum:

$$\sum_{m=1}^{\infty} \sum_{n=1}^{\infty} \sum_{p=1}^{\infty} \frac{\cos(m+n+p)\alpha}{mnp(m+n+p)} \quad (0 \leqslant \alpha \leqslant \pi). \tag{5.2}$$

In addition to these sums, we also calculate the sums of the form

$$\sum_{m=1}^{\infty} \sum_{n=1}^{\infty} \sum_{p=1}^{\infty} \frac{(-1)^{\lambda m+\mu n+\nu p}}{mnp(m+n+p)} \quad (\lambda, \mu, \nu \in \{0,1\}) \tag{5.3}$$

for all combinations of parameters. Furthermore, we derive expressions for all the sums of the form

$$\sum_{m=1}^{\infty} \frac{\lambda^m}{m^2} \sum_{n=1}^{\infty} \frac{\mu^n}{n} \sum_{p=1}^{\infty} \frac{\nu^p}{p} \quad (\lambda, \mu, \nu \in \{-1,1\}). \tag{5.4}$$

In addition to these formulas, we obtain formulas for the following sums:

$$\sum_{m=1}^{\infty} \sum_{n=1}^{\infty} \sum_{p=1}^{\infty} \frac{1}{mnp(m+n+p)^3}, \quad \sum_{m=1}^{\infty} \frac{1}{m^3} \sum_{n=1}^{\infty} \frac{1}{n} \sum_{p=1}^{\infty} \frac{1}{p}, \quad \sum_{m=1}^{\infty} \frac{1}{m^2} \sum_{n=1}^{\infty} \frac{1}{n^2} \sum_{p=1}^{\infty} \frac{1}{p}. \tag{5.5}$$

Noteworthy is the relation between the formulas

$$\sum_{m=1}^{\infty} \frac{1}{m^2} \sum_{n=1}^{\infty} \frac{1}{n} = 2\zeta_3(1) \tag{5.6}$$

and

$$\sum_{m=1}^{\infty} \frac{1}{m^2} \sum_{n=1}^{\infty} \frac{1}{n^2} \sum_{p=1}^{\infty} \frac{1}{p} = 2\zeta_5(1). \tag{5.7}$$

At this point, it is important to mention that a special value of the results given in this chapter is related to the fact that the values of the *Riemann* ζ function are present in all these results.

https://doi.org/10.1515/9783112233276-005

The function $f(z) = \frac{\log^2(1-z)}{z}$, where $\log(1-z)$ matches $\log(1-x)$ for real $x < 1$, will be integrated along the contour C shown in Figure 5.1, where $0 < \alpha < \pi$, and all arcs are circular.

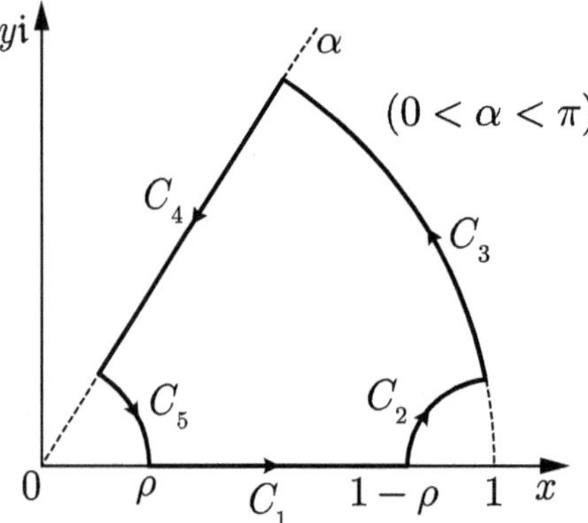

Figure 5.1: Contour of integration.

By the *Cauchy* theorem

$$\int_C f(z)\,\partial z = 0. \tag{5.8}$$

It follows that

$$\begin{aligned}\lim_{\rho\to 0} I_1 &= \lim_{\rho\to 0}\int_{C_1} f(z)\,\partial z = \lim_{\rho\to 0}\int_{\rho}^{1-\rho}\frac{\log^2(1-x)}{x}\,\partial x\\ &= \int_0^1 \frac{\log^2(1-x)}{x}\,\partial x = \int_0^1\frac{\log^2 x}{1-x}\,\partial x = \sum_{n=1}^{\infty}\int_0^1 x^{n-1}\log^2 x\,\partial x\\ &= -2\sum_{n=1}^{\infty}\frac{1}{n}\int_0^1 x^{n-1}\log x\,\partial x = 2\sum_{n=1}^{\infty}\frac{1}{n^3} = 2\zeta_3(1).\end{aligned} \tag{5.9}$$

On the curve C_2, we introduce the parameterization $z = 1 + \rho e^{\varphi i}$, $\varphi \downarrow_{\varphi_0}^{\pi}$, $\varphi_0 = \varphi_0(\rho)$:

$$\begin{aligned}&\lim_{\rho\to 0}\varphi_0(\rho) = \frac{\pi}{2}, \quad 1 - z = -\rho e^{\varphi i} = \rho e^{(\varphi-\pi)i},\\ &\lim_{\rho\to 0} I_2 = -\lim_{\rho\to 0}\int_{\varphi_0}^{\pi}\frac{[\log\rho + (\varphi-\pi)i]^2}{1+\rho e^{\varphi i}} i\rho e^{\varphi i}\,\partial\varphi = 0.\end{aligned} \tag{5.10}$$

On the curve C_5,

$$I_5 = -\mathrm{i}\int_0^{\alpha} \log^2(1-\rho\mathrm{e}^{\varphi\mathrm{i}})\,\partial\varphi, \quad \lim_{\rho\to 0} I_5 = 0. \tag{5.11}$$

On the curve C_3, $z = \mathrm{e}^{\varphi\mathrm{i}}$, $\varphi\downarrow_{\varphi_1}^{\alpha}$, $\varphi_1 = \varphi_1(\rho)$, $\lim_{\rho\to 0}\varphi_1(\rho) = 0$:

$$\lim_{\rho\to 0} I_3 = \mathrm{i}\int_0^{\alpha} \log^2(1-\mathrm{e}^{\varphi\mathrm{i}})\,\partial\varphi. \tag{5.12}$$

Since

$$\log(1-\mathrm{e}^{\varphi i}) = -\sum_{n=1}^{\infty}\frac{\cos n\varphi}{n} + \frac{\varphi-\pi}{2}\mathrm{i} \quad (0<\varphi<\pi), \tag{5.13}$$

we have

$$\begin{aligned}\mathfrak{Re}\Big(\lim_{\rho\to 0} I_3\Big) &= \sum_{n=1}^{\infty}\frac{1}{n}\int_0^{\alpha}(\varphi-\pi)\cos n\varphi\,\partial\varphi \\ &= -(\pi-\alpha)\sum_{k=1}^{\infty}\frac{\sin k\alpha}{k^2} + \sum_{k=1}^{\infty}\frac{\cos k\alpha}{k^3} - \zeta_3(1).\end{aligned} \tag{5.14}$$

On the curve C_4, $z = x\mathrm{e}^{\alpha\mathrm{i}}$, $x\downarrow_{\rho}^{1}$, and accordingly we have

$$I_4 = -\int_{\rho}^{1}\frac{\log^2(1-x\mathrm{e}^{\alpha\mathrm{i}})}{x}\,\partial x, \quad \lim_{\rho\to 0} I_4 = -\int_0^{1}\frac{\log^2(1-x\mathrm{e}^{\alpha\mathrm{i}})}{x}\,\partial x, \tag{5.15}$$

$$\begin{aligned}\log(1-x\mathrm{e}^{\alpha\mathrm{i}}) &= -\sum_{n=1}^{\infty}\frac{x^n\mathrm{e}^{n\alpha\mathrm{i}}}{n}, \quad (-1\leqslant x\leqslant 1) \quad \Longrightarrow \\ \lim_{\rho\to 0} I_4 &= -\sum_{m=1}^{\infty}\sum_{n=1}^{\infty}\frac{\mathrm{e}^{(m+n)\alpha\mathrm{i}}}{mn(m+n)} \quad \Longrightarrow \\ \mathfrak{Re}\Big(\lim_{\rho\to 0} I_4\Big) &= -\sum_{m=1}^{\infty}\sum_{n=1}^{\infty}\frac{\cos(m+n)\alpha}{mn(m+n)} = -\sum_{k=2}^{\infty}\frac{\cos k\alpha}{k}\sum_{m=1}^{k-1}\frac{1}{m(k-m)} \\ &= -\sum_{k=2}^{\infty}\frac{\cos k\alpha}{k^2}\cdot\sum_{m=1}^{k-1}\Big(\frac{1}{m}+\frac{1}{k-m}\Big) = -2\sum_{k=2}^{\infty}\frac{\cos k\alpha}{k^2}\sum_{m=1}^{k-1}\frac{1}{m} \\ &= -2\sum_{k=1}^{\infty}\frac{\cos k\alpha}{k^2}\sum_{m=1}^{k}\frac{1}{m} + 2\sum_{k=1}^{\infty}\frac{\cos k\alpha}{k^3}.\end{aligned} \tag{5.16}$$

By these formulas and relation (5.8) it follows that

$$2\sum_{m=1}^{\infty}\frac{\cos m\alpha}{m^2}\sum_{n=1}^{m}\frac{1}{n}=3\sum_{m=1}^{\infty}\frac{\cos m\alpha}{m^3}-(\pi-\alpha)\sum_{m=1}^{\infty}\frac{\sin m\alpha}{m^2}+\zeta_3(1)\quad(0\leqslant\alpha\leqslant\pi).\tag{5.17}$$

Relation (5.17) is also valid for $\alpha=0$ and $\alpha=\pi$.
We will prove the following two formulas:

$$\sum_{m=1}^{\infty}\frac{\cos m\alpha}{m^{2k}}=\sum_{j=0}^{k}(-1)^j\frac{(\alpha-\pi)^{2j}}{(2j)!}\zeta_{2k-2j}(-1)\quad(0\leqslant\alpha\leqslant\pi),\tag{5.18}$$

$$\sum_{m=1}^{\infty}\frac{\sin m\alpha}{m^{2k+1}}=\sum_{j=0}^{k}(-1)^j\frac{(\alpha-\pi)^{2j+1}}{(2j+1)!}\zeta_{2k-2j}(-1)\quad(0\leqslant\alpha\leqslant\pi),\tag{5.19}$$

Proof. We have

$$\begin{aligned}I_m\Big(\lim_{\rho\to0}I_3\Big)&=\sum_{m=1}^{\infty}\sum_{n=1}^{\infty}\frac{1}{mn}\int_0^{\alpha}\cos m\varphi\cos n\varphi\,\partial\varphi-\int_0^{\alpha}\Big(\frac{\alpha-\pi}{2}\Big)^2\partial\varphi\\&=\frac{1}{2}\sum_{m=1}^{\infty}\sum_{n=1}^{\infty}\frac{\sin(m+n)\alpha}{mn(m+n)}+\frac{1}{2}\sum_{m=1}^{\infty}\sum_{n=m+1}^{\infty}\frac{\sin(m-n)\alpha}{mn(m-n)}\\&\quad+\frac{1}{2}\sum_{m=2}^{\infty}\sum_{n=1}^{m-1}\frac{\sin(m-n)\alpha}{mn(m-n)}+\frac{\alpha}{2}\zeta_2(1)-\frac{1}{12}\big[(\alpha-\pi)^3+\pi^3\big].\end{aligned}\tag{5.20}$$

Let the half-sums on the right-hand side of the above equality be denoted by D_1, D_2, D_3. Then we can write

$$D_1=\frac{1}{2}\sum_{k=2}^{\infty}\frac{\sin k\alpha}{k}\sum_{m=1}^{k-1}\frac{1}{m(k-m)}=\sum_{m=2}^{\infty}\frac{\sin m\alpha}{m^2}\sum_{n=1}^{m-1}\frac{1}{n}=\sum_{m=1}^{\infty}\frac{\sin m\alpha}{m^2}\sum_{n=1}^{m}\frac{1}{n}-\sum_{m=2}^{\infty}\frac{\sin m\alpha}{m^3},\tag{5.21}$$

$$D_2=\frac{1}{2}\sum_{m=1}^{\infty}\sum_{n=1}^{\infty}\frac{\sin n\alpha}{mn(m+n)}=\frac{1}{2}\sum_{n=1}^{\infty}\frac{\sin n\alpha}{n}\sum_{m=1}^{\infty}\frac{1}{m(m+nn)}=\frac{1}{2}\sum_{m=1}^{\infty}\frac{\sin m\alpha}{m^2}\sum_{n=1}^{m}\frac{1}{n},\tag{5.22}$$

$$D_3=D_2,\quad I_m\Big(\lim_{\rho\to0}I_4\Big)=-2D_1.\tag{5.23}$$

It follows that

$$\sum_{m=1}^{\infty}\frac{\sin m\alpha}{m^3}=\frac{1}{2}\cdot\frac{(\alpha-\pi)^3}{3!}-\frac{1}{2}(\alpha-\pi)\zeta_2(1)\quad(0\leqslant\alpha\leqslant\pi),\tag{5.24}$$

which is a well-known relation. Integrating along α, from (5.24) we obtain

$$2\sum_{m=1}^{\infty}\frac{\cos m\alpha}{m^4}=-\frac{(\alpha-\pi)^4}{4!}+\frac{(\alpha-\pi)^2}{2!}\zeta_2(1)-\frac{7}{4}\zeta_4(1).\tag{5.25}$$

The sequential integration of this equation along α leads to

$$2\sum_{m=1}^{\infty}\frac{\cos m\alpha}{m^{2k}}=\sum_{j=0}^{k}\frac{(\alpha-\pi)^{2j}}{(2j)!}C_j^{(2k)}, \tag{5.26}$$

i. e.,

$$2\sum_{m=1}^{\infty}\frac{\cos m\alpha}{m^{2k+2}}=\sum_{j=0}^{k+1}\frac{(\alpha-\pi)^{2j}}{(2j)!}C_j^{(2k+2)}. \tag{5.27}$$

Differentiating (5.27) twice along α, we obtain

$$2\sum_{m=1}^{\infty}\frac{\cos m\alpha}{m^{2k}}=-\sum_{j=1}^{k+1}\frac{(\alpha-\pi)^{2j-2}}{(2j-2)!}C_j^{(2k+2)}=-\sum_{j=1}^{k}\frac{(\alpha-\pi)^{2j}}{(2j)!}C_{j+1}^{(2k+2)} \tag{5.28}$$

$$=\sum_{j=1}^{k}\frac{(\alpha-\pi)^{2j}}{(2j)!}C_{j+1}^{(2k)} \implies C_{j+1}^{(2k+2)}=-C_j^{(2k)}=\cdots=(-1)^{j+1}C_0^{(2k-2j)} \implies$$

$$C_j^{(2k)}=(-1)^jC_0^{(2k-2j)}. \tag{5.29}$$

From (5.26), for $\alpha=\pi$, we have

$$C_0^{(2l)}=2\zeta_{2l}(-1), \tag{5.30}$$

and by (5.29) and (5.30) it follows that

$$C_j^{(2k)}=2(-1)^j\zeta_{2k-2j}(-1). \tag{5.31}$$

Formula (5.18) follows from (5.26) and (5.31). Taking $k+1$ in (5.18) instead of k and then differentiating the obtained relation along α, we obtain (5.19). Note that we used $\zeta_0(-1)=-\frac{1}{2}$ in relations (5.18) and (5.19). □

In particular, for $\alpha=0$, from (5.18) we have

$$\zeta_{2k}(1)=\sum_{j=0}^{k}(-1)^j\frac{\pi^{2j}}{(2j)!}\zeta_{2k-2j}(-1), \tag{5.32}$$

and for $\alpha=\frac{\pi}{2}$,

$$\zeta_{2k}(-1)=\sum_{j=0}^{k}(-1)^j2^{2k-2j}\frac{\pi^{2j}}{(2j)!}\zeta_{2k-2j}(-1). \tag{5.33}$$

By (5.19) it follows that for $\alpha = \frac{\pi}{2}$,

$$\beta_{2k+1}(-1) = -\sum_{j=0}^{k}(-1)^j \frac{(\frac{\pi}{2})^{2j+1}}{(2j+1)!}\zeta_{2k-2j}(-1), \tag{5.34}$$

where

$$\beta_r(-1) = -\sum_{m=1}^{k}\frac{(-1)^m}{(2m-1)^r}. \tag{5.35}$$

Using the formulas:

$$\zeta_r(-1) = (2^{1-r}-1)\zeta_r(1), \quad \zeta_r(-1) = \frac{1}{(2r)!}(-1)^{r-1}2^{2r-1}\pi^{2r}B_{2r}, \tag{5.36}$$

for the numbers B_n and E_n defined by the relations

$$\begin{aligned} \frac{t}{e^t-1} &= \sum_{n=0}^{\infty} B_n \frac{t^n}{n!} \quad (|t| < 2\pi), \\ \frac{1}{\cosh t} &= \sum_{n=0}^{\infty} E_n \frac{t^n}{n!} \quad (|t| < \infty), \end{aligned} \tag{5.37}$$

from (5.32) we obtain

$$\sum_{j=0}^{k}(1-2^{2j-1}\binom{2k}{2j})B_{2j} = 2^{2k-1}B_{2k}. \tag{5.38}$$

This, together with the basic relation

$$\sum_{j=0}^{k}\binom{2k}{2j}B_{2j} = k + B_{2k}, \tag{5.39}$$

leads to

$$\sum_{j=0}^{k}2^{2j}\binom{2k}{2j}B_{2j} = 2k + (2-2^{2k})B_{2k}. \tag{5.40}$$

The last relation is in fact relation (24) on page 154 of the book by *Nielsen N. Traité élémentaire de nombres de Bernoulli*, Paris, 1923. Using (5.33), it follows that

$$(2^{2k-1}-1)B_{2k} = \sum_{j=0}^{k}2^{2j}(2^{2j-1}-1)\binom{2k}{2j}B_{2j}, \tag{5.41}$$

which, together with (5.40), gives

$$\sum_{j=0}^{k}2^{4j}\binom{2k}{2j}B_{2j} = 4k - (2^{2k}-2)B_{2k}. \tag{5.42}$$

This relation also appears as relation (38) on page 156 of the same book. From (5.34) we conclude that

$$\beta_{2k+1}(-1) = (-1)^{k-1}\left(\frac{\pi}{2}\right)^{2k+1} \frac{1}{(2k+1)!}\sum_{j=0}^{k} 2^{2j}(2^{2j-1}-1)\binom{2k+1}{2j}B_{2j}. \tag{5.43}$$

By (5.19), for $\alpha = 0$, it follows that

$$\sum_{j=0}^{k}(-1)^j \frac{\pi^{2j+1}}{(2j+1)!}\zeta_{2k-2j}(-1) = 0, \tag{5.44}$$

that is, after rearranging,

$$\sum_{j=0}^{k}(2^{2j-1}-1)\binom{2k+1}{2j}B_{2j} = 0. \tag{5.45}$$

This relation, together with

$$\sum_{j=0}^{k}\binom{2k+1}{2j}B_{2j} = \frac{2k+1}{2}, \tag{5.46}$$

leads to

$$\sum_{j=0}^{k}2^{2j}\binom{2k+1}{2j}B_{2j} = 2k+1. \tag{5.47}$$

Provided that

$$\sum_{j=0}^{k}2^{4j}\binom{2k+1}{2j}B_{2j} = (2k+1)(2-E_{2k}), \tag{5.48}$$

from (5.43), (5.47), and (5.48) we get

$$\beta_{2k+1}(-1) = \frac{(-1)^k}{2(2k)!}\left(\frac{\pi}{2}\right)^{2k+1}E_{2k}. \tag{5.49}$$

By sequential integration of relation (5.17) along α we get the following relation:

$$2\sum_{m=1}^{\infty}\frac{\cos m\alpha}{m^{2k}}\sum_{n=1}^{m}\frac{1}{n} = (2k+1)\sum_{m=1}^{\infty}\frac{\cos m\alpha}{m^{2k+1}} + (\alpha-\pi)\sum_{m=1}^{\infty}\frac{\sin m\alpha}{m^{2k}} + \sum_{j=0}^{k-1}d_{2j}^{(2k)}\frac{(\alpha-\pi)^{2j}}{(2j)!}, \tag{5.50}$$

that is,

$$2\sum_{m=1}^{\infty}\frac{\cos ma}{m^{2k+2}}\sum_{n=1}^{m}\frac{1}{n}=(2k+3)\sum_{m=1}^{\infty}\frac{\cos ma}{m^{2k+3}}+(a-\pi)\sum_{m=1}^{\infty}\frac{\sin ma}{m^{2k+2}}+\sum_{j=0}^{k}d_{2j}^{(2k+2)}\frac{(a-\pi)^{2j}}{(2j)!}. \quad (5.51)$$

Differentiating the last relation twice along a and comparing the result with (5.50), we obtain

$$d_{2j+2}^{(2k+2)}=-d_{2j}^{(2k)}=\cdots=(-1)^{j+1}d_0^{(2k-2j)} \implies \quad (5.52)$$

$$d_{2j}^{(2k)}=(-1)^j d_0^{(2k-2j)}. \quad (5.53)$$

By (5.50), for $a=\pi$, it follows that

$$d_0^{(2k)}=2G_{(2k,1)}(-1)-(2k+1)\zeta_{2k+1}(-1). \quad (5.54)$$

By the already proven formula (4.87), for $p=2k$, we have

$$2G_{(2k,1)}(-1)=(2k+1)\zeta_{2k+1}(-1)-2\sum_{r=0}^{k-1}\zeta_{2r}(-1)\zeta_{2k+1-2r}(1). \quad (5.55)$$

By (5.54), (5.55), and (5.53) we have

$$\begin{aligned}\sum_{j=0}^{k-1}d_{2j}^{(2k)}\frac{(a-\pi)^{2j}}{(2j)!}&=-2\sum_{j=0}^{k-1}(-1)^j\frac{(a-\pi)^{2j}}{(2j)!}\sum_{r=0}^{k-1-j}\zeta_{2r}(-1)\zeta_{2k+1-2r-2j}(1)\\&=-2\sum_{j=0}^{k-1}(-1)^j\frac{(a-\pi)^{2j}}{(2j)!}\sum_{s=j}^{k-1}\zeta_{2s-2j}(-1)\zeta_{2k+1-2s}(1)\\&=-2\sum_{s=0}^{k-1}\zeta_{2k+1-2s}(1)\sum_{j=0}^{s}(-1)^j\frac{(a-\pi)^{2j}}{(2j)!}\zeta_{2s-2j}(-1)\\&\overset{(5.18)}{=}-2\sum_{s=1}^{k-1}\zeta_{2k+1-2s}(1)\sum_{m=1}^{\infty}\frac{\cos ma}{m^{2s}}+\zeta_{2k+1}(1). \quad (5.56)\end{aligned}$$

By the above and (5.50) we get

$$\begin{aligned}2\sum_{m=1}^{\infty}\frac{\cos ma}{m^{2k}}\sum_{n=1}^{m}\frac{1}{n}&=(2k+1)\sum_{m=1}^{\infty}\frac{\cos ma}{m^{2k+1}}+(a-\pi)\sum_{m=1}^{\infty}\frac{\sin ma}{m^{2k}}\\&\quad-2\sum_{s=1}^{k-1}\zeta_{2k+1-2s}(1)\sum_{m=1}^{\infty}\frac{\cos ma}{m^{2s}}+\zeta_{2k+1}(1). \quad (5.57)\end{aligned}$$

Further, we can conclude that

$$\sum_{m=1}^{\infty}\frac{\cos m\frac{\pi}{3}}{m^r}=\frac{1}{3^r}\zeta_r(-1)+\frac{1}{2}\sum_{m=1}^{\infty}\frac{(-1)^m}{(3m-1)^r}-\frac{1}{2}\sum_{m=1}^{\infty}\frac{(-1)^m}{(3m-2)^r}, \quad (5.58)$$

$$\zeta_r(-1) = \frac{1}{3^r}\zeta_r(-1) - \sum_{m=1}^{\infty} \frac{(-1)^m}{(3m-1)^r} + \sum_{m=1}^{\infty} \frac{(-1)^m}{(3m-2)^r}, \tag{5.59}$$

from which it follows that

$$\sum_{m=1}^{\infty} \frac{\cos m\frac{\pi}{3}}{m^r} = \frac{1-3^{r-1}}{2\cdot 3^{r-1}} \sum_{m=1}^{\infty} \frac{\sin m\alpha}{m^{2k}} \zeta_r(-1). \tag{5.60}$$

From (5.57) and (5.60) we have

$$\begin{aligned} 2\sum_{m=1}^{\infty} \frac{\cos m\frac{\pi}{3}}{m^{2k}} \sum_{n=1}^{m} \frac{1}{n} = &-\frac{2\pi}{3} \sum_{m=1}^{\infty} \frac{\sin m\frac{\pi}{3}}{m^{2k}} + \frac{2k+1}{2}(3^{-2k}-1)\zeta_{2k+1}(-1) \\ &+ \zeta_{2k+1}(1) + \sum_{s=1}^{k-1} 1 - 3^{1-2s})\zeta_{2s}(-1)\zeta_{2k+1-2s}(1), \end{aligned} \tag{5.61}$$

i. e.,

$$\sum_{m=1}^{\infty} \frac{\sin m\frac{\pi}{3}}{m^{2k}} = -\frac{\sqrt{3}}{2} \sum_{m=1}^{\infty} \frac{(-1)^m}{(3m-1)^{2k}} - \frac{\sqrt{3}}{2} \sum_{m=1}^{\infty} \frac{(-1)^m}{(3m-2)^{2k}}, \tag{5.62}$$

$$\zeta_{2k}(-1) = \frac{1}{3^{2k}}\zeta_{2k}(-1) - \sum_{m=1}^{\infty} \frac{(-1)^m}{(3m-1)^{2k}} + \sum_{m=1}^{\infty} \frac{(-1)^m}{(3m-2)^{2k}}. \tag{5.63}$$

By these relations it follows that

$$\begin{aligned} 2\sum_{m=1}^{\infty} \frac{\cos m\frac{\pi}{3}}{m^{2k}} \sum_{n=1}^{m} \frac{1}{n} = &\frac{2\pi}{\sqrt{3}} \sum_{m=1}^{\infty} \frac{(-1)^m}{(3m-1)^{2k}} + \frac{\pi}{\sqrt{3}}(1-3^{-2k})\zeta_{2k}(-1) \\ &+ \frac{2k+1}{2}(3^{-2k}-1)\zeta_{2k+1}(-1) + \zeta_{2k+1}(1) \\ &+ \sum_{s=1}^{k-1}(1-3^{1-2s})\zeta_{2s}(-1)\zeta_{2k+1-2s}(1). \end{aligned} \tag{5.64}$$

Further, we will integrate the function $f(z) = \frac{\log^k(1-z)}{z}$ along the contour shown in Figure 5.1 for integer $k \geqslant 2$. By analogy we obtain

$$\lim_{\rho\to 0} I_1 = \int_0^1 \frac{\log^k(1-x)}{x}\,\partial x = \int_0^1 \frac{\log^k x}{1-x}\,\partial x = \sum_{n=0}^{\infty} \int_0^1 x^n \log^k x\,\partial x = (-1)^k k!\zeta_{k+1}(1), \tag{5.65}$$

$$\lim_{\rho\to 0} I_2 = 0, \quad \lim_{\rho\to 0} I_5 = 0, \quad \lim_{\rho\to 0} I_3 = \mathrm{i}\int_0^{\alpha} \left(\log 2\sin\frac{\varphi}{2} + \frac{\varphi-\pi}{2}\mathrm{i}\right)\partial\varphi, \tag{5.66}$$

$$I_4 = -\int_{\rho}^{1} \frac{\log^k(1-x\mathrm{e}^{\alpha\mathrm{i}})}{x}\,\partial x, \quad \lim_{\rho\to 0} I_4 = -\int_0^1 \frac{\log^k(1-x\mathrm{e}^{\alpha\mathrm{i}})}{x}\,\partial x, \tag{5.67}$$

$$\lim_{\rho\to 0} I_4 = (-1)^{k+1} \sum_{m_1=1}^{\infty}\sum_{m_2=1}^{\infty}\cdots\sum_{m_k=1}^{\infty} \frac{e^{(m_1+m_2+\cdots+m_k)\alpha i}}{m_1 m_2\cdots m_k(m_1+m_2+\cdots+m_k)}. \tag{5.68}$$

By the *Cauchy* theorem, i. e., (5.8), it follows that

$$\begin{aligned}0 = (-1)^k k!\zeta_{k+1}(1) + \mathrm{i}\int_0^{\alpha}\left(\log 2\sin\frac{\varphi}{2} + \frac{\varphi-\pi}{2}\mathrm{i}\right)^k \partial\varphi \\ + (-1)^{k+1} \sum_{m_1=1}^{\infty}\sum_{m_2=1}^{\infty}\cdots\sum_{m_k=1}^{\infty} \frac{e^{(m_1+m_2+\cdots+m_k)\alpha i}}{m_1 m_2\cdots m_k(m_1+m_2+\cdots+m_k)}.\end{aligned} \tag{5.69}$$

By the last relation, as $\alpha \to 0$, we get

$$\sum_{m_1=1}^{\infty}\sum_{m_2=1}^{\infty}\cdots\sum_{m_k=1}^{\infty} \frac{1}{m_1 m_2\cdots m_k(m_1+m_2+\cdots+m_k)} = k!\zeta_{k+1}(1). \tag{5.70}$$

For $k = 2$, from (5.70) it follows that

$$\sum_{m=1}^{\infty}\sum_{n=1}^{\infty}\frac{1}{mn(m+n)} = 2\zeta_3(1). \tag{5.71}$$

This result was obtained by *Euler* in 1775 and again by *Mordel* in 1958. For $k = 3$, from (5.70) it follows that

$$\sum_{m=1}^{\infty}\sum_{n=1}^{\infty}\sum_{p=1}^{\infty}\frac{1}{mnp(m+n+p)} = 6\zeta_4(1), \tag{5.72}$$

and, accordingly,

$$\begin{aligned}6\zeta_4(1) &= \sum_{m=1}^{\infty}\sum_{n=1}^{\infty}\frac{1}{mn}\sum_{p=1}^{\infty}\frac{1}{p(m+n+p)} = \sum_{m=1}^{\infty}\sum_{n=1}^{\infty}\frac{1}{mn(m+n)}\sum_{p=1}^{\infty}\frac{1}{p} \\ &= \sum_{r=2}^{\infty}\frac{1}{r}\sum_{p=1}^{r}\frac{1}{p}\sum_{q=2}^{r-1}\frac{1}{q(r-q)} \\ &= 2\sum_{r=2}^{\infty}\frac{1}{r^2}\sum_{p=1}^{r}\frac{1}{p}\sum_{q=2}^{r-1}\frac{1}{q} = 2\sum_{m=1}^{\infty}\frac{1}{m^2}\left(\sum_{n=1}^{m}\frac{1}{n}\right)^2 - 2G_{(3,1)}(1).\end{aligned} \tag{5.73}$$

By this relation and (4.A57) we obtain

$$\sum_{m=1}^{\infty}\frac{1}{m^2}\left(\sum_{n=1}^{m}\frac{1}{n}\right)^2 = \frac{17}{4}\zeta_4(1), \tag{5.74}$$

so

$$\left(\sum_{n=1}^{m}\frac{1}{n}\right)^2 = 2\sum_{n=1}^{m}\frac{1}{n}\sum_{p=1}^{n}\frac{1}{p} - \sum_{n=1}^{m}\frac{1}{n^2}, \tag{5.75}$$

and by (5.75) and (4.A86) we have

$$\sum_{m=1}^{\infty}\frac{1}{m^2}\sum_{n=1}^{m}\frac{1}{n}\sum_{p=1}^{n}\frac{1}{p} = 3\zeta_4(1). \tag{5.76}$$

We will prove the convergence of the following series (which is a condition for deriving the relations below):

$$\sum_{m=1}^{\infty}\frac{1}{m^{\lambda}}\sum_{n=1}^{m}\frac{1}{n^{\mu}}\sum_{p=1}^{n}\frac{1}{p^{\nu}} \quad (\lambda,\mu,\nu\in\mathbb{N},\ \lambda\geqslant 2,\ \mu\geqslant 1,\ \nu\geqslant 1). \tag{5.77}$$

Proof. We have

$$\begin{aligned}\sum_{m=1}^{\infty}\frac{1}{m^{\lambda}}\sum_{n=1}^{m}\frac{1}{n^{\mu}}\sum_{p=1}^{n}\frac{1}{p^{\nu}} &\leqslant \sum_{m=1}^{\infty}\frac{1}{m^{2}}\sum_{n=1}^{m}\frac{1}{n^{\mu}}\sum_{p=1}^{n}\frac{1}{p^{\nu}} \leqslant \sum_{m=1}^{\infty}\frac{1}{m^{2}}\sum_{n=1}^{m}\frac{1}{n}\sum_{p=1}^{n}\frac{1}{p^{\nu}}\\ &\leqslant \sum_{m=1}^{\infty}\frac{1}{m^{2}}\sum_{n=1}^{m}\frac{1}{n}\sum_{p=1}^{n}\frac{1}{p}\end{aligned} \tag{5.78}$$

$$a_m = \frac{1}{m^2}\sum_{n=1}^{m}\frac{1}{n}\sum_{p=1}^{n}\frac{1}{p}, \tag{5.79}$$

$$\begin{aligned}\lim_{m\to\infty} m\left(\frac{a_m}{a_{m+1}}-1\right) &= \lim_{m\to\infty} m\left(\frac{(1+\frac{1}{m})^2\sum_{n=1}^{m}\frac{1}{n}\sum_{p=1}^{n}\frac{1}{p}}{\sum_{n=1}^{m+1}\frac{1}{n}\sum_{p=1}^{n}\frac{1}{p}}-1\right)\\ &= \lim_{m\to\infty} m\left(\left(1+\frac{1}{m}\right)^2\frac{\sum_{n=1}^{m}\frac{1}{n}\sum_{p=1}^{n}\frac{1}{p}-\frac{1}{m+1}\sum_{p=1}^{m+1}\frac{1}{p}}{\sum_{n=1}^{m+1}\frac{1}{n}\sum_{p=1}^{n}\frac{1}{p}}-1\right)\\ &= \lim_{m\to\infty} m\left\{m\left[\left(1+\frac{1}{m}\right)^2-1\right]-\frac{m+1}{m}\frac{\sum_{p=1}^{m+1}\frac{1}{p}}{\sum_{n=1}^{m+1}\frac{1}{n}\sum_{p=1}^{n}\frac{1}{p}}\right\}\\ &\overset{*}{=} 2-\lim_{m\to\infty} m\frac{\sum_{n=1}^{m+1}\frac{1}{n}-\sum_{p=1}^{n}\frac{1}{p}}{\sum_{n=1}^{m+1}\frac{1}{n}\sum_{p=1}^{n}\frac{1}{p}-\sum_{n=1}^{m}\frac{1}{n}\sum_{p=1}^{n}\frac{1}{p}}\\ &= 2-\lim_{m\to\infty} m\frac{\frac{1}{m+1}}{\frac{1}{m+1}\sum_{p=1}^{m+1}\frac{1}{p}} = 2.\end{aligned}$$

By the *Raabe* criterion, the series given above converges, and by the comparison criteria we obtain the stated result. For the equality indicated by the asterisk $*$, *Stolz*'s theorem was used to calculate the limit of the quotient of two series. □

For $k = 3$, taking the real part of (5.69), we get

$$0 = -6\zeta_4(1) - \frac{3}{2}\int_0^\alpha (\varphi - \pi)\log^2 2\sin\frac{\varphi}{2}\,\partial\varphi + \int_0^\alpha \frac{(\varphi-\pi)^3}{8}\,\partial\varphi + \sum_{m=1}^\infty\sum_{n=1}^\infty\sum_{p=1}^\infty \frac{\cos(m+n+p)\alpha}{mnp(m+n+p)}, \tag{5.80}$$

$$A = \int_0^\alpha (\varphi-\pi)\log^2 2\sin\frac{\varphi}{2}\,\partial\varphi = \int_0^\alpha (\varphi-\pi)\sum_{m=1}^\infty\sum_{n=1}^\infty \frac{\cos m\varphi\cos n\varphi}{mn}\,\partial\varphi$$
$$= \frac{1}{2}\sum_{m=1}^\infty\sum_{n=1}^\infty \frac{1}{mn}\cdot\int_0^\alpha (\varphi-\pi)[\cos(m+n)\,\partial\varphi + \cos(m-n)]\,\partial\varphi = A_1 + A_2, \tag{5.81}$$

$$A_1 = \frac{1}{2}\sum_{m=1}^\infty\sum_{n=1}^\infty \frac{1}{mn}\int_0^\alpha (\varphi-\pi)\cos(m+n)\varphi\,\partial\varphi$$
$$= \frac{1}{2}\sum_{m=1}^\infty\sum_{n=1}^\infty \frac{1}{mn}\left\{(\alpha-\pi)\frac{\sin(m+n)\alpha}{m=n} + \frac{\cos(m+n)\alpha}{(m+n)^2} - \frac{1}{(m+n)^2}\right\}$$
$$= \frac{1}{2}(\alpha-\pi)\sum_{k=2}^\infty \frac{\sin k\alpha}{k}\sum_{n=1}^{k-1}\frac{1}{n(k-n)} + \frac{1}{2}\sum_{k=2}^\infty\frac{\cos k\alpha}{k^2}\sum_{n=1}^{k-1}\frac{1}{n(k-n)} - \frac{1}{2}\sum_{k=2}^\infty\frac{1}{k^2}\sum_{n=1}^{k-1}\frac{1}{n(k-n)}$$
$$= (\alpha-\pi)\sum_{m=1}^\infty\frac{\sin m\alpha}{m^2}\sum_{n=1}^m\frac{1}{n} - (\alpha-\pi)\sum_{m=1}^\infty\frac{\sin m\alpha}{m^3} + \sum_{m=1}^\infty\frac{\cos m\alpha}{m^3}\sum_{n=1}^m\frac{1}{n}$$
$$- \sum_{m=1}^\infty\frac{\cos m\alpha}{m^4} - G_{(3,1)}(1) + \zeta_4(1), \tag{5.82}$$

$$A_2 = \frac{1}{2}\sum_{m=1}^\infty\sum_{n=1}^\infty \frac{1}{mn}\int_0^\alpha (\varphi-\pi)\cos(m-n)\varphi\,\partial\varphi$$
$$= \frac{1}{2}\sum_{m=1}^\infty\frac{1}{m^2}\int_0^\alpha(\varphi-\pi)\,\partial\varphi + \frac{1}{2}\sum_{m=1}^\infty\sum_{\substack{n=1\\n\neq m}}^\infty\frac{1}{mn}\int_0^\alpha(\varphi-\pi)\cos(m-n)\varphi\,\partial\varphi = A_{21} + A_{22}, \tag{5.83}$$

$$A_{21} = \frac{1}{4}[(\varphi-\pi)^2 - \pi^2]\zeta_2(1), \tag{5.84}$$

$$A_{21} = \frac{1}{4}(\alpha-\pi)\sum_{m=1}^\infty\sum_{\substack{n\neq m\\n=1}}^\infty\frac{\sin(m-n)\alpha}{mn(m-n)} + \frac{1}{2}\sum_{m=1}^\infty\sum_{\substack{n\neq m\\n=1}}^\infty\frac{\cos(m-n)\alpha}{mn(m-n)^2} - \frac{1}{2}\sum_{m=1}^\infty\sum_{\substack{n\neq m\\n=1}}^\infty\frac{1}{mn(m-n)^2}. \tag{5.85}$$

It may further be written as

$$\sum_{m=1}^\infty\sum_{\substack{n\neq m\\n=1}}^\infty a_{m,n} = \sum_{m=2}^\infty\sum_{n=1}^{m-1}a_{m,n} + \sum_{m=1}^\infty\sum_{n=m+1}^\infty a_{m,n} = \sum_{m=1}^\infty\sum_{n=1}^\infty(a_{m,n} + a_{n,m}), \tag{5.86}$$

where

$$a_{m,n}=\frac{\sin(m-n)a}{mn(m-n)},\quad a_{m,n}=\frac{\cos(m-n)a}{mn(m-n)^2},\quad a_{m,n}=\frac{1}{mn(m-n)^2}\implies \tag{5.87}$$

$$a_{m,n}=a_{n,m}\implies \sum_{m=1}^{\infty}\sum_{\substack{n\neq m\\ n=1}}^{\infty}\frac{\sin(m-n)a}{mn(m-n)}=2\sum_{m=1}^{\infty}\sum_{\substack{n\neq m\\ n=1}}^{\infty}\frac{\sin(n-m)a}{mn(n-m)}$$

$$=2\sum_{k=1}^{\infty}\frac{\sin ka}{k}\sum_{m=1}^{\infty}\frac{1}{m(m+k)}=2\sum_{k=1}^{\infty}\frac{\sin ka}{k^2}\sum_{m=1}^{\infty}\frac{1}{m}.$$

We conclude (by analogy with the established relations in (5.87)) that

$$\sum_{m=1}^{\infty}\sum_{\substack{n\neq m\\ n=1}}^{\infty}=\frac{\cos(m-n)a}{mn(m-n)^2}=2\sum_{k=1}^{\infty}\frac{\cos ka}{k^3}\sum_{m=1}^{k}\frac{1}{m}, \tag{5.88}$$

$$\sum_{m=1}^{\infty}\sum_{\substack{n\neq m\\ n=1}}^{\infty}=\frac{1}{mn(m-n)^2}=2G_{(3,1)}(1). \tag{5.89}$$

From the above formulas and from (4.A56), (4.A57), and (5.80) it follows that

$$\begin{aligned}&\sum_{m=1}^{\infty}\sum_{n=1}^{\infty}\sum_{p=1}^{\infty}\frac{\cos(m+n+p)a}{mnp(m+n+p)}\\ &=3(a-\pi)\sum_{m=1}^{\infty}\frac{\sin ma}{m^2}\sum_{n=1}^{m}\frac{1}{n}+3\sum_{m=1}^{\infty}\frac{\cos ma}{m^3}\sum_{n=1}^{m}\frac{1}{n}-\frac{3}{2}(a-\pi)\sum_{m=1}^{\infty}\frac{\sin ma}{m^3}\\ &\quad-\frac{3}{2}\sum_{m=1}^{\infty}\frac{\cos ma}{m^4}-\frac{1}{32}(a-\pi)^4+\frac{3}{8}(a-\pi)^2\zeta_2(1)+\frac{15}{16}\zeta_4(1).\end{aligned} \tag{5.90}$$

We will prove the following two formulas:

$$\sum_{m=1}^{k-2}\frac{1}{m(k-m)}\sum_{n=1}^{k-m-1}\frac{1}{n}=\frac{3}{k}\sum_{m=1}^{k}\frac{1}{m}\sum_{n=1}^{m}\frac{1}{n}-\frac{3}{k^2}\sum_{m=1}^{k}\frac{1}{m}-\frac{3}{k}\sum_{m=1}^{k}\frac{1}{m^2}+\frac{3}{k^3}, \tag{5.91}$$

$$\begin{aligned}\sum_{m=1}^{\infty}\sum_{n=1}^{\infty}\sum_{p=1}^{\infty}\frac{f(m+n+p)}{mnp(m+n+p)}&=6\sum_{m=1}^{\infty}\frac{f(m)}{m^2}\sum_{n=1}^{m}\frac{1}{n}\sum_{p=1}^{n}\frac{1}{p}-6\sum_{m=1}^{\infty}\frac{f(m)}{m^3}\sum_{n=1}^{m}\frac{1}{n}\\ &\quad-6\sum_{m=1}^{\infty}\frac{f(m)}{m^2}\sum_{n=1}^{m}\frac{1}{n^2}+6\sum_{m=1}^{\infty}\frac{f(m)}{m^4}\quad(|f(k)|\leqslant 1).\end{aligned} \tag{5.92}$$

Proof. Denote

$$H_k=\sum_{m=1}^{k}\frac{1}{m},\quad s_k=k\sum_{m=1}^{k-2}\frac{1}{m(k-m)}\sum_{n=1}^{k-m-1}\frac{1}{n}. \tag{5.93}$$

Then

$$s_k = k\sum_{m=1}^{k-2}\frac{1}{m(k-m)}H_{k-m-1} = \sum_{m=1}^{k-2}\frac{1}{m}H_{k-m-1} + \sum_{m=1}^{k-2}\frac{1}{k-m}H_{k-m-1}$$

$$= \sum_{m=1}^{k-2}\frac{1}{m}H_{k-m-1} + \sum_{m=2}^{k-1}\frac{1}{m}H_{m-1}, \tag{5.94}$$

$$s_{k-1} = \sum_{m=1}^{k-3}\frac{1}{m}H_{k-m-2} + \sum_{m=2}^{k-2}\frac{1}{m}H_{m-1}, \tag{5.95}$$

$$s_k - s_{k-1} = \frac{1}{k-2} + \sum_{m=1}^{k-3}\frac{1}{m(k-m-1)} + \frac{1}{k-1}H_{k-2} = \frac{3}{k-1}H_{k-2} \implies$$

$$s_k = s_3 + \sum_{r=4}^{k}\frac{1}{r-1}H_{r-2} = \frac{3}{2} + 3\sum_{s=3}^{k-1}\frac{1}{s}\sum_{t=1}^{s-1}\frac{1}{t} \quad (k \geqslant 3), \tag{5.96}$$

and by the equations thus defined it follows that

$$s_k = 3\sum_{s=1}^{k}\frac{1}{s}\sum_{t=1}^{s}\frac{1}{t} - \frac{3}{k}\sum_{s=1}^{k}\frac{1}{s} - 3\sum_{s=1}^{k}\frac{1}{s^2} + \frac{3}{k^2}. \tag{5.97}$$

The right-hand side of this equation is equal to zero for $k = 1, 2$, i. e., $s_1 = s_2 = 0$. If $s = \frac{1}{k}s$, we obtain (5.91) from (5.97). It follows that

$$\sum_{m=1}^{\infty}\sum_{n=1}^{\infty}\sum_{p=1}^{\infty}\frac{f(m+n+p)}{mnp(m+n+p)}$$

$$= \sum_{k=3}^{\infty}\frac{f(k)}{k}\sum_{m=1}^{k-2}\frac{1}{m}\sum_{n=1}^{k-m-1}\frac{1}{n(k-m-n)}$$

$$= 2\sum_{k=3}^{\infty}\frac{f(k)}{k}\sum_{m=1}^{k-2}\frac{1}{m(k-m)}\sum_{n=1}^{k-m-1}\frac{1}{n} = 2\sum_{k=3}^{\infty}\frac{f(k)}{k^2}s_k = 2\sum_{k=1}^{\infty}\frac{f(k)}{k^2}s_k. \tag{5.98}$$

Formula (5.92) follows from the above and from (5.91) and (5.97). □

By formula (5.92) for $f(m) = \cos ma$ and by (5.90) we obtain

$$2\sum_{m=1}^{\infty}\frac{\cos ma}{m^2}\sum_{n=1}^{m}\frac{1}{n}\sum_{p=1}^{n}\frac{1}{p}$$

$$= 3\sum_{m=1}^{\infty}\frac{\cos ma}{m^3}\sum_{n=1}^{m}\frac{1}{n} + 2\sum_{m=1}^{\infty}\frac{\cos ma}{m^2}\sum_{n=1}^{m}\frac{1}{n^2}$$

$$+ (a-\pi)\sum_{m=1}^{\infty}\frac{\sin ma}{m^2}\sum_{n=1}^{m}\frac{1}{n} - \frac{5}{2}\sum_{m=1}^{\infty}\frac{\cos ma}{m^4} - \frac{1}{2}(a-\pi)\sum_{m=1}^{\infty}\frac{\sin ma}{m^3}$$

$$- \frac{(a-\pi)^4}{96} + \frac{1}{8}(a-\pi)^2\zeta_2(1) + \frac{5}{16}\zeta_4(1), \quad 0 \le a \le \pi. \tag{5.99}$$

This for $\alpha = \pi$, together with (4.A57) and (4.A58), leads to

$$\sum_{m=1}^{\infty} \frac{(-1)^m}{m^2} \sum_{n=1}^{m} \frac{1}{n} \sum_{p=1}^{n} \frac{1}{p} = -\frac{31}{64}\zeta_4(1) + \frac{1}{4}G_{(2,2)}(-1). \tag{5.100}$$

Integrating (5.99) along α and including the values for $\alpha = \pi$, we get

$$\begin{aligned}
2\sum_{m=1}^{\infty} \frac{\sin m\alpha}{m^3} \sum_{n=1}^{m} \frac{1}{n} \sum_{p=1}^{n} \frac{1}{p}
&= 4\sum_{m=1}^{\infty} \frac{\sin m\alpha}{m^4} \sum_{n=1}^{m} \frac{1}{n} + 2\sum_{m=1}^{\infty} \frac{\sin m\alpha}{m^3} \sum_{n=1}^{m} \frac{1}{n^2} \\
&\quad - 3\sum_{m=1}^{\infty} \frac{\sin m\alpha}{m^5} - (\alpha-\pi)\sum_{m=1}^{\infty} \frac{\cos m\alpha}{m^3} \sum_{n=1}^{m} \frac{1}{n} + \frac{1}{2}(\alpha-\pi)\sum_{m=1}^{\infty} \frac{\cos m\alpha}{m^4} \\
&\quad - \frac{1}{4}\frac{(\alpha-\pi)^5}{5!} + \frac{1}{4}\frac{(\alpha-\pi)^3}{3!}\zeta_2(1) + \frac{5}{16}(\alpha-\pi)\zeta_4(1),
\end{aligned} \tag{5.101}$$

and integrating the obtained relation along α, we obtain

$$\begin{aligned}
2\sum_{m=1}^{\infty} \frac{\cos m\alpha}{m^4} \sum_{n=1}^{m} \frac{1}{n} \sum_{p=1}^{n} \frac{1}{p}
&= 5\sum_{m=1}^{\infty} \frac{\cos m\alpha}{m^5} \sum_{n=1}^{m} \frac{1}{n} + 2\sum_{m=1}^{\infty} \frac{\cos m\alpha}{m^4} \sum_{n=1}^{m} \frac{1}{n^2} \\
&\quad - \frac{7}{2}\sum_{m=1}^{\infty} \frac{\cos m\alpha}{m^6} + (\alpha-\pi)\sum_{m=1}^{\infty} \frac{\sin m\alpha}{m^4} \sum_{n=1}^{m} \frac{1}{n} - \frac{1}{2}(\alpha-\pi)\sum_{m=1}^{\infty} \frac{\sin m\alpha}{m^5} \\
&\quad + \frac{1}{4}\frac{(\alpha-\pi)^6}{6!} - \frac{1}{4}\frac{(\alpha-\pi)^4}{4!}\zeta_2(1) - \frac{5}{16}\frac{(\alpha-\pi)^2}{2!}\zeta_4(1) + C_1.
\end{aligned} \tag{5.102}$$

Continuing to integrate (5.102) along α, we have

$$\begin{aligned}
2\sum_{m=1}^{\infty} \frac{\sin m\alpha}{m^5} \sum_{n=1}^{m} \frac{1}{n} \sum_{p=1}^{n} \frac{1}{p}
&= 6\sum_{m=1}^{\infty} \frac{\sin m\alpha}{m^6} \sum_{n=1}^{m} \frac{1}{n} + 2\sum_{m=1}^{\infty} \frac{\sin m\alpha}{m^5} \sum_{n=1}^{m} \frac{1}{n^2} \\
&\quad - 4\sum_{m=1}^{\infty} \frac{\sin m\alpha}{m^7} - (\alpha-\pi)\sum_{m=1}^{\infty} \frac{\cos m\alpha}{m^5} \sum_{n=1}^{m} \frac{1}{n} + \frac{1}{2}(\alpha-\pi)\sum_{m=1}^{\infty} \frac{\cos m\alpha}{m^6} \\
&\quad + \frac{1}{4}\frac{(\alpha-\pi)^7}{7!} - \frac{1}{4}\frac{(\alpha-\pi)^5}{5!}\zeta_2(1) - \frac{5}{16}\frac{(\alpha-\pi)^3}{3!}\zeta_4(1) + C_1\alpha + C_2.
\end{aligned} \tag{5.103}$$

From (5.103) for $\alpha = \pi$ it follows that $C_1 = -\frac{1}{\pi}C_2$, and for $\alpha = 0$, by (4.A56) and (4.A57) we have

$$C_2 = -\pi\left[\frac{133}{64}\zeta_6(1) - \frac{1}{2}\zeta_3^2(1)\right], \quad C_1 = \frac{133}{64}\zeta_6(1) - \frac{1}{2}\zeta_3^2(1). \tag{5.104}$$

Similarly from (5.102), where C_1 is specified in (5.104), for $\alpha = 0$, by (4.A56)and (4.A57) we obtain

$$\sum_{m=1}^{\infty}\frac{1}{m^4}\sum_{n=1}^{m}\frac{1}{n}\sum_{p=1}^{n}\frac{1}{p} = \frac{89}{48}\zeta_6(1) - \frac{1}{2}\zeta_3^2(1). \tag{5.105}$$

We will prove the following formula:

$$\sum_{r=1}^{\infty}\frac{1}{r(r+s)}\sum_{t=1}^{r+s}\frac{1}{t} = 2\sum_{r=1}^{\infty}\frac{1}{r(r+s)}\sum_{t=1}^{r-1}\frac{1}{t} = \frac{2}{s}\sum_{p=1}^{s}\frac{1}{p}\sum_{q=1}^{p}\frac{1}{q}. \tag{5.106}$$

Proof. Since

$$\sum_{m=1}^{\infty}\sum_{n=1}^{\infty}\frac{1}{mn(m+n+s)} = \sum_{m=1}^{\infty}\frac{1}{m}\sum_{n=1}^{r+s}\frac{1}{n(m+n+s)} = \sum_{m=1}^{\infty}\frac{1}{m(m+s)}\sum_{n=1}^{m+s}\frac{1}{n}, \tag{5.107}$$

$$\sum_{m=1}^{\infty}\sum_{n=1}^{\infty}\frac{1}{mn(m+n+s)} = \sum_{r=2}^{\infty}\frac{1}{r+s}\sum_{n=1}^{r-1}\frac{1}{n(r-n)} = 2\sum_{r=2}^{\infty}\frac{1}{r(r+s)}\sum_{n=1}^{r-1}\frac{1}{n}, \tag{5.108}$$

for $\psi(x) = \frac{\Gamma'(x)}{\Gamma(x)}$, we have

$$\psi(x) = -\gamma + \sum_{s=0}^{\infty}\left(\frac{1}{s+1} - \frac{1}{x+s}\right) \quad (x > 0). \tag{5.109}$$

The following formula is given in the book by *Nielsen*, page 52:

$$(\psi(x) + \gamma)^2 = \psi'(x) - \zeta_2(1) - 2\xi(x), \tag{5.110}$$

where

$$\xi(x) = \sum_{s=1}^{\infty}\left(\frac{1}{x+s} - \frac{1}{s+1}\right)\sum_{t=1}^{s}\frac{1}{t}. \tag{5.111}$$

For $x = r+1$ (where r is positive integer number), from (5.109) it follows that

$$\psi(r+1) + \gamma = \sum_{s=1}^{r}\frac{1}{s}, \tag{5.112}$$

$$\psi'(r+1) = \zeta_2(1) - \sum_{s=1}^{r}\frac{1}{s^2}. \tag{5.113}$$

From (5.111) we have

$$\xi(r+1) = -r\sum_{s=2}^{\infty}\frac{1}{s(s+r)}\sum_{t=1}^{s-1}\frac{1}{t}, \tag{5.114}$$

and from (5.110)

$$\left(\sum_{s=1}^{r}\frac{1}{s}\right)^2 = -\sum_{s=1}^{r}\frac{1}{s^2} + 2r\sum_{s=2}^{\infty}\frac{1}{s(s+r)}\sum_{t=1}^{s-1}\frac{1}{t}. \tag{5.115}$$

By (5.115) and (5.75) we obtain the right equality in (5.106). □

We will also prove the following formulas:

$$\sum_{m=1}^{\infty}\sum_{n=1}^{\infty}\sum_{p=1}^{\infty}\frac{(-1)^m}{mnp(m+n+p)} = -\frac{31}{32}\zeta_4(1) + \frac{1}{2}G_{(2,2)}(-1), \tag{5.116}$$

$$\sum_{m=1}^{\infty}\sum_{n=1}^{\infty}\sum_{p=1}^{\infty}\frac{(-1)^{m+n}}{mnp(m+n+p)} = \frac{3}{8}\zeta_4(1), \tag{5.117}$$

$$\sum_{m=1}^{\infty}\sum_{n=1}^{\infty}\sum_{p=1}^{\infty}\frac{(-1)^{m+n+p}}{mnp(m+n+p)} = -\frac{39}{32}\zeta_4(1) - \frac{3}{2}G_{(2,2)}(-1). \tag{5.118}$$

Proof. Formula (5.118) follows from (5.90) for $\alpha = \pi$, and from (4.A58) (Appendix to Chapter 4). Let us denote

$$M = \int_0^1 \frac{1}{x}\log^3(1-x)\,\partial x \overset{(5.64)}{\underset{(5.69)}{=}} -6\zeta_4(1), \tag{5.119}$$

$$N = \int_0^1 \frac{1}{x}\log^3(1+x)\,\partial x = -\sum_{m=1}^{\infty}\sum_{n=1}^{\infty}\sum_{p=1}^{\infty}\frac{(-1)^{m+n+p}}{mnp(m+n+p)} \overset{(5.118)}{=} \frac{39}{32}\zeta_4(1) + \frac{3}{2}G_{(2,2)}(-1), \tag{5.120}$$

$$A = \int_0^1 \frac{1}{x}\log^2(1-x)\log(1+x)\,\partial x = -\sum_{m=1}^{\infty}\sum_{n=1}^{\infty}\sum_{p=1}^{\infty}\frac{(-1)^m}{mnp(m+n+p)}, \tag{5.121}$$

$$B = \int_0^1 \frac{1}{x}\log(1-x)\log^2(1+x)\,\partial x = -\sum_{m=1}^{\infty}\sum_{n=1}^{\infty}\sum_{p=1}^{\infty}\frac{(-1)^{m+n}}{mnp(m+n+p)}, \tag{5.122}$$

$$\begin{aligned} M &= \int_0^1 \frac{1}{x}\log^3(1-x)\,\partial x = 2\int_0^1 \frac{1}{t}\log^3(1-t^2)\,\partial t \\ &= 2\int_0^1 \frac{1}{t}[\log(1-t)+\log(1+t)]^3\,\partial t = 2(M+3A+3B+N). \end{aligned} \tag{5.123}$$

By these relations it follows that:

$$A + B = -\frac{1}{6}(M + 2N). \tag{5.124}$$

It further follows that

$$\begin{aligned} M &= \int_0^1 \frac{1}{x}\log^3(1-x)\,\partial x = \left| x = \frac{2t}{1+t} \right| = \int_0^1 \frac{1}{t(1+t)}\log^3\frac{1-t}{1+t}\,\partial t \\ &= \int_0^1 \frac{1}{t}[\log(1-t) - \log(1+t)]^3\,\partial t - \int_0^1 \frac{1}{1+t}\log^3\frac{1-t}{1+t}\,\partial t = M - 3A + 3B - N - P, \end{aligned} \tag{5.125}$$

$$P = \int_0^1 \frac{1}{1+t}\log^3\frac{1-t}{1+t}\,\partial t = \int_0^1 \frac{\log^3 s}{1+s}\,\partial s = 6\zeta_4(-1), \tag{5.126}$$

and

$$A - B = -\frac{1}{3}(N + P). \tag{5.127}$$

By these formulas we determine A and B and the sums required, thus proving formulas (5.116) and (5.117). □

We will further prove the following formulas:

$$\sum_{m=1}^{\infty}\frac{1}{m^2}\sum_{n=1}^{m}\frac{1}{n}\sum_{p=1}^{n}\frac{(-1)^p}{p} = -\frac{39}{64}\zeta_4(1) - \frac{7}{4}\zeta_3(1)\log 2 - \frac{3}{4}\zeta_2(1)\log^2 2 - \frac{3}{4}G_{(2,2)}(-1), \tag{5.128}$$

$$\sum_{m=1}^{\infty}\frac{1}{m^2}\sum_{n=1}^{m}\frac{(-1)^n}{n}\sum_{p=1}^{n}\frac{1}{p} = -\frac{29}{16}\zeta_4(1) - \frac{7}{8}\zeta_3(1)\log 2 + \frac{3}{4}\zeta_2(1)\log^2 2 - \frac{3}{4}G_{(2,2)}(-1), \tag{5.129}$$

$$\sum_{m=1}^{\infty}\frac{1}{m^2}\sum_{n=1}^{m}\frac{(-1)^n}{n}\sum_{p=1}^{n}\frac{(-1)^p}{p} = \frac{55}{64}\zeta_4(1) - \frac{7}{8}\zeta_3(1)\log 2 + \frac{3}{2}\zeta_2(1)\log^2 2 - \frac{1}{4}G_{(2,2)}(-1), \tag{5.130}$$

$$\sum_{m=1}^{\infty}\frac{(-1)^m}{m^2}\sum_{n=1}^{m}\frac{1}{n}\sum_{p=1}^{n}\frac{(-1)^p}{p} = \frac{3}{16}\zeta_4(1) + \frac{3}{4}\zeta_2(1)\log^2 2, \tag{5.131}$$

$$\sum_{m=1}^{\infty}\frac{(-1)^m}{m^2}\sum_{n=1}^{m}\frac{(-1)^n}{n}\sum_{p=1}^{n}\frac{4}{p} = \frac{3}{64}\zeta_4(1) + \frac{21}{8}\zeta_3(1)\log 2 - \frac{3}{4}\zeta_2(1)\log^2 2 + \frac{3}{4}G_{(2,2)}(-1), \tag{5.132}$$

$$\sum_{m=1}^{\infty}\frac{(-1)^m}{m^2}\sum_{n=1}^{m}\frac{(-1)^n}{n}\sum_{p=1}^{n}\frac{(-1)^p}{p} = \frac{5}{16}\zeta_4(1) + \frac{7}{8}\zeta_3(1)\log 2 - \frac{3}{2}\zeta_2(1)\log^2 2 + G_{(2,2)}(-1), \tag{5.133}$$

as well as formulas (5.76) and (5.100) already mentioned.

Proof. Let the sums in the above formulas be denoted by y_j, where $j = 1$ for (5.76), $j = 5$ for (5.100), and $j = 2, 3, 4, 6, 7, 8$, respectively, for the sums in (5.128)–(5.133). Let

$$S_1 = \sum_{m=1}^{\infty}\sum_{n=1}^{\infty}\sum_{p=1}^{\infty} \frac{(-1)^m}{mnp(m+n+p)} \quad \Longrightarrow \quad S_1 = \sum_{m=1}^{\infty}\sum_{n=1}^{\infty} \frac{1}{np} \sum_{p=1}^{\infty} \frac{(-1)^m}{m(m+n+p)}. \tag{5.134}$$

It further follows that

$$\sum_{r=1}^{\infty} \frac{z^{r+s}}{r(r+s)} = \frac{1}{s}\left\{(z^s - 1)\zeta_1(z) + \sum_{r=1}^{s} \frac{z^r}{r}\right\} \quad (|z| \leqslant 1,\ s \in \mathbb{N}). \tag{5.135}$$

Passing to the limit as $z \to -1$, by *Abel's* theorem we have

$$\sum_{r=1}^{\infty} \frac{(-1)^s}{r(r+s)} = \frac{1}{s}[1 - (-1)^s]\zeta_1(-1) + \frac{(-1)^s}{s} \sum_{r=1}^{s} \frac{(-1)^r}{r} \quad (s \in \mathbb{N}). \tag{5.136}$$

By the above it follows that

$$S_1 = \zeta_1(-1)\sum_{n=1}^{\infty}\sum_{p=1}^{\infty} \frac{1}{np(n+p)} - \zeta_1(-1)\sum_{n=1}^{\infty}\sum_{p=1}^{\infty} \frac{(-1)^{n+p}}{np(n+p)} + \sum_{n=1}^{\infty}\sum_{p=1}^{\infty} \frac{(-1)^{n+p}}{np(n+p)} \sum_{m=1}^{n+p} \frac{(-1)^m}{m}. \tag{5.137}$$

Denoting by D_1, D_2, D_3, respectively, the sums on the right-hand side of the equality, we have

$$D_1 = 2\zeta_3(1) \tag{5.138}$$

(which is in fact the formula already given as (5.71)) and

$$\begin{aligned} D_2 &= \sum_{k=2}^{\infty} \frac{(-1)^k}{k} \sum_{n=1}^{k-1} \frac{1}{n(k-n)} = 2\sum_{k=2}^{\infty} \frac{(-1)^k}{k^2} \sum_{n=1}^{k-1} \frac{1}{n} \\ &= 2(G_{(2,1)}(-1) - \zeta_3(-1)) \overset{(4.127)}{=} \frac{1}{4}\zeta_3(1). \end{aligned} \tag{5.139}$$

We can now conclude that

$$S_1 = -\frac{7}{4}\zeta_3(1)\log 2 + \sum_{s=1}^{\infty} \frac{(-1)^s}{s} \sum_{r=1}^{\infty} \frac{(-1)^r}{r(r+s)} \sum_{t=1}^{r+s} \frac{(-1)^t}{t}. \tag{5.140}$$

□

We will also prove the equality

$$\sum_{r=1}^{\infty}\frac{(-1)^r}{r(r+s)}\sum_{t=1}^{r+s}\frac{(-1)^t}{t}=\left(\frac{1}{s}-\frac{(-1)^s}{s}\right)\left(\frac{1}{2}\zeta_2(1)+\frac{1}{2}\log^2 2\right)$$
$$+\frac{1}{s}\sum_{r=1}^{s}\frac{(-1)^r}{r}\sum_{t=1}^{r}\frac{1}{t}+\frac{(-1)^s}{s}\sum_{r=1}^{s}\frac{(-1)^r}{r}\sum_{t=1}^{r}\frac{(-1)^t}{t}\quad (s\in\mathbb{N}). \tag{5.141}$$

Proof. Formula (5.141) can be verified as follows:

$$\sum_{r=1}^{\infty}\frac{z^{r+s}}{r(r+s)}\sum_{t=1}^{r+s}\frac{(-1)^t}{t}=\frac{z^s}{s}\sum_{r=1}^{\infty}\frac{z^r}{r}\sum_{t=1}^{r+s}\frac{(-1)^t}{t}-\frac{1}{s}\sum_{r=s+1}^{\infty}\frac{z^r}{r}\sum_{t=1}^{r}\frac{(-1)^t}{t}$$
$$=\frac{z^s}{s}\sum_{r=1}^{\infty}\frac{z^r}{r}\sum_{t=1}^{r+s}\frac{(-1)^t}{t}+\frac{z^s}{s}\sum_{r=1}^{\infty}\frac{z^r}{r}\sum_{t=r+1}^{r+s}\frac{(-1)^t}{t}$$
$$-\frac{1}{s}\sum_{r=1}^{\infty}\frac{z^r}{r}\sum_{t=1}^{r}\frac{(-1)^t}{t}+\frac{1}{s}\sum_{r=1}^{\infty}\frac{z^r}{r}\sum_{t=1}^{r}\frac{(-1)^t}{t}, \tag{5.142}$$

$$\frac{z^s}{s}\sum_{r=1}^{\infty}\frac{z^r}{r}\sum_{t=r+1}^{r+s}\frac{(-1)^t}{t}=\frac{z^s}{s}\sum_{r=1}^{\infty}(-1)^r\frac{z^r}{r}\sum_{t=1}^{s}\frac{(-1)^t}{t+}$$
$$=\frac{z^s}{s}\sum_{t=1}^{s}(-1)^t\sum_{r=1}^{\infty}\frac{(-1)^r z^r}{r(t+r)}\quad (|z|<1). \tag{5.143}$$

Passing to the limit as $z\to -1$, we obtain formula (5.141). □

From (5.140), (5.116), (5.141), and (4.A82) it follows that

$$y_4+y_7=\frac{29}{32}\zeta_4(1)+\frac{7}{4}\zeta_3(1)\log 2+\frac{3}{4}\zeta_2(1)\log^2 2+\frac{1}{2}G_{(2,2)}(-1). \tag{5.144}$$

Further, by (5.140) we obtain

$$S_1+\frac{7}{4}\zeta_3(1)\log 2=\sum_{n=1}^{\infty}\sum_{p=1}^{\infty}\frac{(-1)^{n+p}}{np(n+p)}\sum_{m=1}^{n+p}\frac{(-1)^m}{m}=\sum_{k=1}^{n+p}\frac{(-1)^k}{k}\sum_{n=1}^{k-1}\frac{1}{n(k-n)}\sum_{m=1}^{k}\frac{(-1)^m}{m}$$
$$=2\sum_{k=2}^{\infty}\frac{(-1)^k}{k^2}\sum_{n=1}^{k-1}\frac{1}{n}\sum_{m=1}^{k}\frac{(-1)^m}{m}=2\sum_{k=1}^{\infty}\frac{(-1)^k}{k^2}\sum_{m=1}^{k}\frac{(-1)^m}{m}\sum_{n=1}^{k}\frac{1}{n}-2F_{(3,1)}(-1). \tag{5.145}$$

By this relation, (5.116), and (4.A94) it follows that

$$\sum_{k=1}^{\infty}\frac{(-1)^k}{k^2}\sum_{m=1}^{k}\frac{(-1)^m}{m}\sum_{n=1}^{k}\frac{1}{n}=-\frac{37}{64}\zeta_4(1)+\frac{21}{8}\zeta_3(1)\log 2+\frac{3}{4}G_{(2,2)}(-1), \tag{5.146}$$

$$y_6=\sum_{m=1}^{\infty}\frac{(-1)^m}{m^2}\sum_{n=1}^{m}\frac{1}{n}\sum_{p=1}^{n}\frac{(-1)^p}{p}=\sum_{m=1}^{\infty}\frac{(-1)^m}{m^2}\sum_{p=1}^{m}\frac{(-1)^p}{p}\sum_{n=p}^{m}\frac{1}{n}$$

$$= \sum_{m=1}^{\infty} \frac{(-1)^m}{m^2} \sum_{p=1}^{m} \frac{(-1)^p}{p} \sum_{n=1}^{m} \frac{1}{n} - y_7 + F_{(2,2)}(-1). \tag{5.147}$$

From (5.146) and (4.132) we obtain

$$y_6 + y_7 = \frac{15}{64}\zeta_4(1) + \frac{21}{8}\zeta_3(1)\log 2 + \frac{3}{4}G_{(2,2)}(-1), \tag{5.148}$$

where we accept that

$$S_2 = \sum_{m=1}^{\infty}\sum_{n=1}^{\infty}\sum_{p=1}^{\infty} \frac{(-1)^{m+n}}{mnp(m+n+p)}, \tag{5.149}$$

$$\begin{aligned} S_2 &= \sum_{m=1}^{\infty}\sum_{n=1}^{\infty} \frac{(-1)^{m+n}}{mn} \sum_{p=1}^{\infty} \frac{1}{p(m+n+p)} \\ &= \sum_{m=1}^{\infty}\sum_{n=1}^{\infty} \frac{(-1)^{m+n}}{mn(m+n)} \sum_{p=1}^{\infty} \frac{1}{p} = \sum_{s=1}^{\infty} \frac{(-1)^s}{s} \sum_{r=1}^{\infty} \frac{(-1)^r}{r(r+s)} \sum_{t=1}^{r+s} \frac{1}{t}. \end{aligned} \tag{5.150}$$

We will prove that

$$\begin{aligned} \sum_{r=1}^{\infty} \frac{(-1)^r}{r(r+s)} \sum_{t=1}^{r+s} \frac{1}{t} &= \left(\frac{1}{s} - \frac{(-1)^s}{s}\right)\left(-\frac{1}{2}\zeta_2(1) + \frac{1}{2}\log^2 2\right) \\ &\quad + \left(\frac{1}{s}\sum_{r=1}^{s} \frac{(-1)^r}{r} - \frac{1}{s}\sum_{r=1}^{s} \frac{1}{r}\right)\log 2 + \frac{(-1)^s}{s}\sum_{r=1}^{s} \frac{(-1)^r}{r}\sum_{t=1}^{r} \frac{1}{t} \\ &\quad + \frac{1}{s}\sum_{r=1}^{s} \frac{(-1)^r}{r}\sum_{t=1}^{r} \frac{(-1)^t}{t} \qquad (s \in \mathbb{N}). \end{aligned} \tag{5.151}$$

Proof. We have

$$\begin{aligned} \sum_{r=1}^{\infty} \frac{z^{r+s}}{r(r+s)} \sum_{t=1}^{r+s} \frac{1}{t} &= \frac{z^s}{s}\sum_{r=1}^{\infty} \frac{z^r}{r}\sum_{t=1}^{r+s} \frac{1}{t} - \frac{1}{s}\sum_{r=s+1}^{\infty} \frac{z^r}{r}\sum_{t=1}^{r} \frac{1}{t} \\ &= \frac{z^s}{s}G_{(1,1)}(z) + \frac{z^s}{s}\sum_{r=1}^{\infty} \frac{z^r}{r}\sum_{t=1}^{r+s} \frac{1}{t} - \frac{1}{s}\left\{G_{(1,1)}(z) - \sum_{r=1}^{\infty} \frac{z^r}{r}\sum_{t=1}^{r} \frac{1}{t}\right\}, \end{aligned} \tag{5.152}$$

$$\frac{z^s}{s}\sum_{r=1}^{\infty} \frac{z^r}{r}\sum_{t=r+1}^{r+s} \frac{1}{t} = \frac{z^s}{s}\sum_{r=1}^{\infty} \frac{z^r}{r}\sum_{t=1}^{s} \frac{1}{t+r} = \frac{z^s}{s}\sum_{t=1}^{s}\sum_{r=1}^{\infty} \frac{z^r}{r(t+r)} \qquad (|z| < 1). \tag{5.153}$$

Passing to the limit as $z \to -1$, by (5.136) we get (5.151). □

From the last derived relation we obtain the expression for S_2:

$$S_2 = \zeta_2(-1)G_{(1,1)}(-1) + \zeta_1(-1)G_{(2,1)}(-1) - \zeta_1(-1)F_{(2,1)}(-1) + y_8 - \zeta_2(1)G_{(1,1)}(-1) + y_3, \tag{5.154}$$

from which by (5.117), (4.A46), (4.127), (4.129), and (4.A56) we obtain

$$y_3+y_8=-\frac{3}{2}\zeta_4(1)-\frac{3}{4}\zeta_2(1)\log^2 2. \tag{5.155}$$

Further, we can write

$$\begin{aligned}
S_2&=\sum_{k=3}^{\infty}\frac{1}{k}\sum_{m=1}^{k-2}\frac{(-1)^m}{m}\sum_{n=1}^{k-m-1}\frac{(-1)^n}{n(k-m-n)}\\
&=\sum_{m=1}^{\infty}\frac{(-1)^m}{m}\sum_{k=m+2}^{\infty}\frac{1}{k(k-m)}\sum_{n=1}^{k-m-1}(-1)^n\left(\frac{1}{n}+\frac{1}{k-m-n}\right)\\
&=\sum_{m=1}^{\infty}\frac{(-1)^m}{m}\sum_{k=m+2}^{\infty}\frac{1}{k(k-m)}\sum_{n=1}^{k-m-1}\frac{(-1)^n}{n}+\sum_{m=1}^{\infty}\frac{1}{m}\sum_{k=m+2}^{\infty}\frac{(-1)^k}{k(k-m)}\sum_{n=1}^{k-m-1}\frac{(-1)^n}{n}\\
&=\sum_{m=1}^{\infty}\frac{(-1)^m}{m}\sum_{r=2}^{\infty}\frac{1}{r(r+m)}\sum_{n=1}^{r-1}\frac{(-1)^n}{n}+\sum_{m=1}^{\infty}\frac{(-1)^m}{m}\sum_{r=2}^{\infty}\frac{(-1)^r}{r(r+m)}\sum_{n=1}^{r-1}\frac{(-1)^n}{n}=S_{21}+S_{22},
\end{aligned} \tag{5.156}$$

$$\begin{aligned}
&\sum_{r=2}^{\infty}\frac{1}{r(r+s)}\sum_{m=1}^{r-1}\frac{(-1)^m}{m}\\
&\quad=\sum_{n=1}^{\infty}\frac{(-1)^n}{n}\sum_{r=n+1}^{\infty}\frac{1}{r(r+s)}\\
&\quad=\sum_{n=1}^{\infty}\frac{(-1)^n}{n}\left(\sum_{r=1}^{\infty}\frac{1}{r(r+s)}-\sum_{r=1}^{n}\frac{1}{r(r+s)}\right)\\
&\quad=\sum_{n=1}^{\infty}\frac{(-1)^n}{n}\left(\frac{1}{s}\sum_{r=1}^{s}\frac{1}{r}-\frac{1}{s}\sum_{r=1}^{s}\frac{1}{r}+\frac{1}{s}\sum_{r=1}^{s}\frac{1}{r+s}\right)\\
&\quad=\frac{1}{s}\sum_{n=1}^{\infty}\frac{(-1)^n}{n}\sum_{r=n+1}^{n+s}\frac{1}{r}=\frac{1}{s}\sum_{r=1}^{s}\sum_{n=1}^{\infty}\frac{(-1)^n}{n(n+r)}\\
&\overset{(5.136)}{=}\frac{1}{s}\zeta_1(-1)\sum_{r=1}^{s}\frac{1}{r}-\frac{1}{s}\zeta_1(-1)\sum_{r=1}^{s}\frac{(-1)^r}{r}+\frac{1}{s}\sum_{r=1}^{s}\frac{(-1)^r}{r}\sum_{t=1}^{r}\frac{(-1)^t}{t}.
\end{aligned} \tag{5.157}$$

It follows that

$$S_{21}=\zeta_1(-1)G_{(2,1)}(-1)-\zeta_1(-1)F_{(2,1)}(-1)+y_8, \tag{5.158}$$

which, together with formulas (4.127) and (4.129), results in

$$S_{21}=\frac{3}{2}\zeta_2(1)\log^2 2+y_8. \tag{5.159}$$

The following expression also provides the connection with S_{22}:

$$\left|\sum_{r=2}^{\infty}\frac{(-1)^r}{r(r+m)}\sum_{n=1}^{r-1}\frac{(-1)^n}{n}\right|\leqslant\sum_{r=2}^{\infty}\frac{1}{r(r+m)}\sum_{n=1}^{r-1}\frac{1}{n}$$

$$= \sum_{n=1}^{\infty} \frac{1}{n} \sum_{r=n+1}^{\infty} \frac{1}{r(r+m)} = \frac{1}{m} \sum_{n=1}^{\infty} \frac{1}{n} \sum_{r=n+1}^{m+n} \frac{1}{r} = \frac{1}{m} \sum_{n=1}^{\infty} \frac{1}{n} \sum_{r=1}^{m} \frac{1}{n+r}$$
$$= \frac{1}{m} \sum_{r=1}^{m} \frac{1}{n} \sum_{n=1}^{\infty} \frac{1}{n(n+r)} = \frac{1}{m} \sum_{r=1}^{m} \frac{1}{r} \sum_{n=1}^{r} \frac{1}{n}, \tag{5.160}$$

and so the right equality in (5.106) is proved. Then for $|z| \leqslant 1$, we have

$$\sum_{m=1}^{\infty} \left| \frac{z^m}{m} \sum_{r=2}^{\infty} \frac{(-1)^r}{r(r+m)} \sum_{n=1}^{r-1} \frac{z^n}{n} \right| \leqslant \sum_{m=1}^{\infty} \frac{1}{m} \sum_{r=2}^{\infty} \frac{1}{r(r+m)} \sum_{n=1}^{r-1} \frac{1}{n} = \sum_{m=1}^{\infty} \frac{1}{m^2} \sum_{n=1}^{m} \frac{1}{n} \sum_{p=1}^{n} \frac{1}{p}. \tag{5.161}$$

As it is already shown that the last series converges, the expression to calculate S_{22} can be obtained by applying *Abel*'s theorem.

$$P(z) = \sum_{m=1}^{\infty} \frac{z^m}{m} \sum_{r=2}^{\infty} \frac{(-1)^r}{r(r+m)} \sum_{n=1}^{r-1} \frac{z^n}{n} = \sum_{m=1}^{\infty} \frac{z^m}{m} \sum_{n=1}^{\infty} \frac{z^n}{n} \sum_{r=n+1}^{r-1} \frac{(-1)^r}{r(r+m)}$$
$$\overset{(5.136)}{=} \sum_{m=1}^{\infty} \frac{z^m}{m} \sum_{n=1}^{\infty} \frac{z^n}{n} \left\{ \left[\frac{1}{m} - \frac{(-1)^m}{m} \right] \zeta_1(-1) - \frac{1}{m} \sum_{r=1}^{n} \frac{(-1)^r}{r} + \frac{(-1)^m}{m} \sum_{r=1}^{m+n} \frac{(-1)^r}{r} \right\}$$
$$= [\zeta_2(z) - \zeta_2(-z)] \zeta_1(z) \zeta_1(-1) - \zeta_2(z) \sum_{m=1}^{\infty} \frac{z^m}{m} \sum_{n=1}^{m} \frac{(-1)^m}{n} + Q(z), \tag{5.162}$$

$$Q(z) = \sum_{m=1}^{\infty} \frac{(-1)^m z^m}{m^2} \sum_{n=1}^{\infty} \frac{z^n}{n} \sum_{r=1}^{m+n} \frac{(-1)^r}{r} = \sum_{k=2}^{\infty} z^k \sum_{m=1}^{k-1} \frac{(-1)^m}{m^2(k-m)} \sum_{r=1}^{k} \frac{(-1)^r}{r}$$
$$= \sum_{k=2}^{\infty} \frac{z^k}{k} \sum_{m=1}^{k-1} \frac{(-1)^m}{m^2} \sum_{r=1}^{k} \frac{(-1)^r}{r} + \sum_{k=2}^{\infty} \frac{z^k}{k^2} \sum_{m=1}^{k-1} (-1)^m \left(\frac{1}{m} + \frac{1}{k-m} \right) \sum_{r=1}^{k} \frac{(-1)^r}{r}$$
$$= \sum_{m=1}^{\infty} \frac{(-1)^m}{m^2} \sum_{k=m+1}^{\infty} \frac{z^k}{k} \sum_{r=1}^{k} \frac{(-1)^r}{r} + \sum_{k=2}^{\infty} \frac{z^k}{k^2} \sum_{m=1}^{k-1} \frac{(-1)^m}{m} \sum_{r=1}^{k} \frac{(-1)^r}{r}$$
$$+ \sum_{k=2}^{\infty} \frac{(-1)^k z^k}{k^2} \sum_{m=1}^{k-1} \frac{(-1)^m}{m} \sum_{r=1}^{k} \frac{(-1)^r}{r}. \tag{5.163}$$

Similarly to formula (5.75), we have that

$$\left(\sum_{r=1}^{k} \frac{(-1)^r}{r} \right)^2 = 2 \sum_{r=1}^{k} \frac{(-1)^r}{r} \sum_{s=1}^{r} \frac{(-1)^s}{s} - \sum_{r=1}^{k} \frac{1}{r^2}, \tag{5.164}$$

from which it follows that

$$Q(z) = \zeta_2(-1) \sum_{k=1}^{\infty} \frac{z^k}{k} \sum_{r=1}^{k} \frac{(-1)^r}{r} - \sum_{m=1}^{\infty} \frac{(-1)^m}{m^2} \sum_{k=1}^{m} \frac{z^k}{k} \sum_{r=1}^{k} \frac{(-1)^r}{r}$$
$$+ 2 \sum_{k=1}^{\infty} \frac{z^k}{k^2} \sum_{m=1}^{k} \frac{(-1)^m}{m} \sum_{r=1}^{k} \frac{(-1)^r}{r} - \sum_{k=1}^{\infty} \frac{(-1)^k z^k}{k^3} \sum_{r=1}^{k} \frac{(-1)^r}{r}$$

$$- \sum_{k=1}^{\infty} \frac{z^k}{k^2} \sum_{r=1}^{k} \frac{1}{r^2} + 2 \sum_{k=1}^{\infty} \frac{(-1)^k z^k}{k^2} \sum_{m=1}^{k} \frac{(-1)^m}{m} \sum_{r=1}^{m} \frac{(-1)^r}{r}$$

$$- \sum_{k=1}^{\infty} \frac{z^k}{k^3} \sum_{r=1}^{k} \frac{(-1)^r}{r} - \sum_{k=1}^{\infty} \frac{(-1)^k z^k}{k^2} \sum_{r=1}^{k} \frac{1}{r^2}. \tag{5.165}$$

The expressions $Q(z)$ and $P(z)$ are calculated in the region $|z| < 1$, and passing to the limit as $z \to -1$, we get

$$S_{22} = [\zeta_2(-1) - \zeta_2(1)]\zeta_1^2 + y_8 - H_{(3,1)}(-1) - G_{(2,2)}(-1) + 2y_4 - F_{(3,1)}(-1) - G_{(2,2)}(1). \tag{5.166}$$

From here, by the derived equations in (4.134) and (4.A57) it follows that

$$S_{22} = -\frac{63}{32}\zeta_4(1) - \frac{3}{2}\zeta_2(1)\log^2 2 - \frac{3}{2}G_{(2,2)}(-1) + 2y_4 + y_8, \tag{5.167}$$

which, together with (5.159), gives

$$S_2 = 2y_4 + 2y_8 - \frac{63}{32}\zeta_4(1) - \frac{3}{2}G_{(2,2)}(-1). \tag{5.168}$$

Therefore by (5.117) we have

$$y_4 + y_8 = \frac{75}{64}\zeta_4(1) + \frac{3}{4}G_{(2,2)}(-1). \tag{5.169}$$

Let

$$S_3 = \sum_{m=1}^{\infty}\sum_{n=1}^{\infty}\sum_{p=1}^{\infty} \frac{(-1)^{m+n+p}}{mnp(m+n+p)}, \tag{5.170}$$

$$\Longrightarrow \quad S_3 = \sum_{m=1}^{\infty}\sum_{n=1}^{\infty}\sum_{p=1}^{\infty} \frac{(-1)^{m+n}}{mn(m+n)} \sum_{p=1}^{\infty} \frac{(-1)^p}{p(m+n+p)}. \tag{5.171}$$

As in the calculation of S_1, by (5.136) we have

$$S_3 = \frac{7}{4}\zeta_3(1)\log 2 + \sum_{m=1}^{\infty}\sum_{n=1}^{\infty} \frac{1}{mn(m+n)} \sum_{p=1}^{m+n} \frac{(-1)^p}{p}, \tag{5.172}$$

$$\sum_{m=1}^{\infty}\sum_{n=1}^{\infty} \frac{1}{mn(m+n)} \sum_{p=1}^{m+n} \frac{(-1)^p}{p} = \sum_{k=2}^{\infty} \frac{1}{k} \sum_{n=1}^{k-1} \frac{1}{n(k-n)} \sum_{p=1}^{k} \frac{(-1)^p}{p}$$

$$= 2\sum_{k=1}^{\infty} \frac{1}{k^2} \sum_{m=1}^{k} \frac{(-1)^m}{m} \sum_{n=1}^{k} \frac{1}{n} - 2H_{(3,1)}(-1). \tag{5.173}$$

By the established relation, (5.118), and (4.134) it follows that

$$\sum_{k=1}^{\infty} \frac{1}{k^2} \sum_{m=1}^{k} \frac{(-1)^m}{m} \sum_{n=1}^{k} \frac{1}{n} = -\frac{19}{64}\zeta_4(1) - \frac{21}{8}\zeta_3(1)\log 2 - \frac{3}{4}G_{(2,2)}(-1). \tag{5.174}$$

In addition, we can conclude that

$$
\begin{aligned}
y_3 &= \sum_{m=1}^{\infty} \frac{1}{m^2} \sum_{n=1}^{m} \frac{(-1)^n}{n} \sum_{p=1}^{n} \frac{1}{p} = \sum_{m=1}^{\infty} \frac{1}{m^2} \sum_{p=1}^{m} \frac{1}{p} \sum_{n=p}^{m} \frac{(-1)^n}{n} \\
&= \sum_{m=1}^{\infty} \frac{1}{m^2} \sum_{n=1}^{m} \frac{(-1)^n}{n} \sum_{p=1}^{m} \frac{1}{p} - y_2 + H_{(2,2)}(-1).
\end{aligned} \tag{5.175}
$$

From (5.175), (5.174), and (4.139) we obtain

$$
y_2 + y_3 = -\frac{155}{64}\zeta_4(1) - \frac{21}{8}\zeta_3(1)\log 2 - \frac{7}{4}G_{(2,2)}(-1). \tag{5.176}
$$

It follows that

$$
\begin{aligned}
S_3 &= \sum_{k=3}^{\infty} \frac{(-1)^k}{k} \sum_{m=1}^{k-2} \frac{1}{m} \sum_{n=1}^{k-m-1} \frac{1}{n(k-m-n)} \\
&= \sum_{m=1}^{\infty} \frac{1}{m} \sum_{k=m+2}^{\infty} \frac{(-1)^k}{k} \sum_{n=1}^{k-m-1} \frac{1}{n(k-m-n)} \\
&= 2\sum_{m=1}^{\infty} \frac{1}{m} \sum_{k=m+2}^{\infty} \frac{(-1)^k}{k(k-m)} \sum_{n=1}^{k-m-1} \frac{1}{n} = 2\sum_{m=1}^{\infty} \frac{(-1)^m}{m} \sum_{r=2}^{\infty} \frac{(-1)^r}{r(r+m)} \sum_{n=1}^{r-1} \frac{1}{n},
\end{aligned} \tag{5.177}
$$

$$
\lambda_m(z) = m\sum_{r=2}^{\infty} \frac{z^{r+m}}{r(r+m)} \sum_{n=1}^{r-1} \frac{1}{n} \quad (|z| \leqslant 1), \tag{5.178}
$$

$$
\begin{aligned}
\lambda_m(z) - \lambda_{m-1}(z) &= \sum_{r=2}^{\infty} z^{r+m}\left(\frac{1}{r} - \frac{1}{r+m}\right)\sum_{n=1}^{r-1}\frac{1}{n} - \sum_{r=1}^{\infty} z^{r+m}\left(\frac{1}{r+1} - \frac{1}{r+m}\right)\sum_{n=1}^{r}\frac{1}{n} \\
&= -\sum_{r=1}^{\infty} z^{r+m}\left(\frac{1}{r+1} - \frac{1}{r+m}\right)\frac{1}{r} + \sum_{r=2}^{\infty} z^{r+m}\left(\frac{1}{r} - \frac{1}{r+1}\right)\sum_{n=1}^{r-1}\frac{1}{n} \\
&= -z^{m-1}\sum_{r=1}^{\infty}\frac{z^{r+1}}{r(r+1)} + \sum_{r=1}^{\infty}\frac{z^{r+m}}{r(r+m)} + z^m\left[G_{(1,1)}(z) - \zeta_2(z)\right] \\
&\quad - z^{m-1}\left\{G_{(1,1)}(z) - z - (\zeta_2(z) - z) - \sum_{r=1}^{\infty}\frac{z^{r+1}}{r(r+1)}\right\} \\
&= (z^m - z^{m-1})\left[G_{(1,1)}(z) - \zeta_2(z)\right] + \sum_{r=1}^{\infty}\frac{z^{r+m}}{r(r+m)} \\
&= (z^m - z^{m-1})\left[G_{(1,1)}(z) - \zeta_2(z)\right] + \frac{1}{m}\left\{(z^m - 1)\zeta_1(z) + \sum_{r=1}^{m}\frac{z^r}{r}\right\},
\end{aligned} \tag{5.179}
$$

$$
\lambda_m(z) = \sum_{s=1}^{m}(\lambda_s(z) - \lambda_{s-1}(z)) = (z^m - 1)\left[G_{(1,1)}(z) - \zeta_2(z)\right]
$$

$$+\left(\sum_{r=1}^{m}\frac{z^r}{r}-\sum_{r=1}^{m}\frac{1}{r}\right)\zeta_1(z)+\sum_{r=1}^{m}\frac{1}{r}\sum_{s=1}^{m}\frac{z^s}{s}. \tag{5.180}$$

Passing to the limit as $z \to -1$, we get $\lambda_m(-1)$ and then

$$S_3 = \left[-\zeta_2(1)+\zeta_2(-1)\right]\log^2 2 + 2\zeta_1(-1)H_{(2,1)}(-1) - 2\zeta_1(-1)G_{(2,1)}(1) + 2y_2. \tag{5.181}$$

Formula (5.128) follows from the established equality in the last relation, and from (4.127), (4.132), and (5.118). Formula (5.129) follows from (5.128) and (5.176), and by (5.129) and (5.155) we obtain (5.133). Formula (5.130) follows from (5.133) and (5.169), and by (5.130) and (5.144) we obtain (5.132). From (5.132) and (5.148) we arrive at (5.131). So all formulas (5.128)–(5.133) are proved.

Further, we can specify the following formulas:

$$\sum_{m=1}^{\infty}\frac{(-1)^m}{(2m-1)^{2k-1}}\sum_{n=1}^{2m-1}\frac{1}{n} = -k\beta_{2k}(-1) - \frac{\pi}{2^{2k+1}}\zeta_{2k-1}(-1) + \sum_{r=1}^{k-1}\beta_{2k}(-1)\zeta_{2k+1-2r}(1), \tag{5.182}$$

$$\begin{aligned}\sum_{m=1}^{\infty}\frac{(-1)^m}{(2m-1)^{2k-1}}\sum_{n=1}^{2m-1}\frac{1}{n} &= -k\beta_{2k}(-1) - \frac{\pi}{2^{2k+1}}\zeta_{2k-1}(-1)\\ &\quad - \frac{1}{2}\sum_{s=1}^{k-1}(-1)^s\left(\frac{\pi}{2}\right)^{2s-1}\frac{E_{2s-2}}{(2s-2)!}\zeta_{2k+1-2s}(1).\end{aligned} \tag{5.183}$$

Proof. Differentiating (5.57) along α, for $\alpha = \frac{\pi}{2}$, we have

$$\sum_{m=1}^{\infty}\frac{\sin m\frac{\pi}{2}}{m^p} = \beta_p(-1), \quad \sum_{m=1}^{\infty}\frac{\cos m\frac{\pi}{2}}{m^p} = \frac{1}{2^p}\zeta_p(-1), \tag{5.184}$$

and so we obtain (5.182), and (5.183) follows from (5.49) and (5.182). □

We will also prove the following formulas:

$$\sum_{m=1}^{\infty}\sum_{n=1}^{\infty}\sum_{p=1}^{\infty}\frac{1}{mnp(m+n+p)^2} = 12\zeta_5(1) - 6\zeta_2(1)\zeta_3(1), \tag{5.185}$$

$$\sum_{m=1}^{\infty}\sum_{n=1}^{\infty}\sum_{p=1}^{\infty}\frac{1}{mnp(m+n+p)^3} = \frac{69}{8}\zeta_6(1) - 6\zeta_3^2(1), \tag{5.186}$$

$$\sum_{m=1}^{\infty}\frac{1}{m^3}\sum_{n=1}^{\infty}\frac{1}{n}\sum_{p=1}^{\infty}\frac{1}{p} = -\frac{1}{2}\zeta_5(1) + \zeta_2(1)\zeta_3(1), \tag{5.187}$$

$$\sum_{m=1}^{\infty}\frac{1}{m^2}\sum_{n=1}^{\infty}\frac{1}{n^2}\sum_{p=1}^{\infty}\frac{1}{p} = 2\zeta_5(1), \tag{5.188}$$

as well as the already given formula (5.72).

Proof. Formula (5.185) is stated in [189] and is the result of the already mentioned *Mordell's* study [66]. The proof we will give here is different from that in [66], and as such provides the possibility of deriving the formulas (5.187) and (5.188). We can conclude that

$$\begin{aligned}
&\sum_{m=1}^{\infty}\sum_{n=1}^{\infty}\sum_{p=1}^{\infty}\frac{1}{mnp(m+n+p)^2}\\
&\quad=\sum_{m=1}^{\infty}\sum_{n=1}^{\infty}\frac{1}{mn(m+n)}\sum_{p=1}^{\infty}\frac{1}{m+n+p}\left(\frac{1}{p}-\frac{1}{m+n+p}\right)\\
&\quad=\sum_{m=1}^{\infty}\sum_{n=1}^{\infty}\frac{1}{mn(m+n)^2}\sum_{p=1}^{m+n}\frac{1}{p}-\sum_{m=1}^{\infty}\sum_{n=1}^{\infty}\frac{1}{mn(m+n)}\sum_{p=m+n+1}^{\infty}\frac{1}{p^2}\\
&\quad=\sum_{k=2}^{\infty}\frac{1}{k^2}\sum_{n=1}^{k-1}\frac{1}{n(k-n)}\sum_{p=1}^{k}\frac{1}{p}+\sum_{k=2}^{\infty}\frac{1}{k}\sum_{n=1}^{k-1}\frac{1}{n(k-n)}\sum_{p=1}^{k}\frac{1}{p^2}-2\zeta_2(1)\zeta_3(1)\\
&\quad=P+Q-2\zeta_2(1)\zeta_3(1), \qquad (5.189)\\
Q&=2\sum_{k=2}^{\infty}\frac{1}{k^2}\sum_{n=1}^{k-1}\frac{1}{n(k-n)}\sum_{p=1}^{k}\frac{1}{p^2}=2\sum_{p=1}^{\infty}\frac{1}{p^2}\sum_{k=p}^{\infty}\frac{1}{k^2}\sum_{n=1}^{k}\frac{1}{n}-2G_{(3,2)}(1)\\
&=2G_{(4,1)}(1)+4\zeta_2(1)\zeta_3(1)-2G_{(3,2)}(1)-2\sum_{k=1}^{\infty}\frac{1}{k^2}\sum_{n=1}^{k}\frac{1}{n^2}\sum_{p=1}^{n}\frac{1}{p}, \qquad (5.190)\\
P&=2\sum_{k=2}^{\infty}\frac{1}{k^3}\sum_{n=1}^{k-1}\frac{1}{n}\sum_{p=1}^{k}\frac{1}{p}=2\sum_{k=2}^{\infty}\frac{1}{k^3}\left(\sum_{n=1}^{k}\frac{1}{n}\right)^2-2G_{(4,1)}\\
&\overset{(5.75)}{=}\sum_{k=1}^{\infty}\frac{1}{k^3}\sum_{n=1}^{k}\frac{1}{n}\sum_{p=1}^{n}\frac{1}{p}-2G_{(3,2)}-2G_{(4,1)}. \qquad (5.191)
\end{aligned}$$

It follows that

$$\begin{aligned}
&\sum_{m=1}^{\infty}\sum_{n=1}^{\infty}\sum_{p=1}^{\infty}\frac{1}{mnp(m+n+p)^2}\\
&\quad=4\sum_{k=1}^{\infty}\frac{1}{k^3}\sum_{n=1}^{k}\frac{1}{n}\sum_{p=1}^{n}\frac{1}{p}-2\sum_{k=1}^{\infty}\frac{1}{k^2}\sum_{n=1}^{k}\frac{1}{n^2}\sum_{p=1}^{n}\frac{1}{p}-4G_{(3,2)}+2\zeta_2(1)\zeta_3(1). \qquad (5.192)
\end{aligned}$$

On the other hand,

$$\begin{aligned}
P&=\sum_{n=1}^{\infty}\frac{1}{n}\sum_{k=n+1}^{\infty}\frac{1}{k^2(k-n)}\sum_{p=1}^{k}\frac{1}{p}\\
&=\sum_{n=1}^{\infty}\frac{1}{n^2}\sum_{k=n+1}^{\infty}\frac{1}{k(k-n)}\sum_{p=1}^{k}\frac{1}{p}-\sum_{n=1}^{\infty}\frac{1}{n^2}\sum_{k=n+1}^{\infty}\frac{1}{k^2}\sum_{p=1}^{k}\frac{1}{p}
\end{aligned}$$

$$= \sum_{n=1}^{\infty} \frac{1}{n^2} \sum_{r=1}^{\infty} \frac{1}{r(r+n)} \sum_{p=1}^{r+n} \frac{1}{p} - \zeta_2(1) G_{(2,1)}(1) + \sum_{k=1}^{\infty} \frac{1}{k^2} \sum_{n=1}^{k} \frac{1}{n^2} \sum_{p=1}^{n} \frac{1}{p}$$

$$\overset{(4.A83)}{=} 2 \sum_{k=1}^{\infty} \frac{1}{k^3} \sum_{n=1}^{k} \frac{1}{n} \sum_{p=1}^{n} \frac{1}{p} + \sum_{k=1}^{\infty} \frac{1}{k^2} \sum_{n=1}^{k} \frac{1}{n^2} \sum_{p=1}^{n} \frac{1}{p} - 2\zeta_2(1)\zeta_3(1). \tag{5.193}$$

From (5.191) and (5.193) we have

$$2 \sum_{k=1}^{\infty} \frac{1}{k^3} \sum_{n=1}^{k} \frac{1}{n} \sum_{p=1}^{n} \frac{1}{p} - \sum_{k=1}^{\infty} \frac{1}{k^2} \sum_{n=1}^{k} \frac{1}{n^2} \sum_{p=1}^{n} \frac{1}{p} = -3\zeta_5(1) + 2\zeta_2(1)\zeta_3(1). \tag{5.194}$$

Using (4.86), we obtain

$$G_{(4,1)}(1) = 3\zeta_5(1) - \zeta_2(1)\zeta_3(1), \tag{5.195}$$

$$G_{(3,2)}(1) = -\frac{9}{2}\zeta_5(1) + 3\zeta_2(1)\zeta_3(1). \tag{5.196}$$

Formula (5.185) follows from (5.192), (5.194), and (5.196). Further, by (5.92) for $f(k) = \frac{1}{k}$, we have

$$\sum_{m=1}^{\infty} \sum_{n=1}^{\infty} \sum_{p=1}^{\infty} \frac{1}{mnp(m+n+p)^2} = 6 \sum_{k=1}^{\infty} \frac{1}{k^3} \sum_{n=1}^{k} \frac{1}{n} \sum_{p=1}^{n} \frac{1}{p} - 6G_{(4,1)}(1) - 6G_{(3,2)}(1) + 6\zeta_5(1). \tag{5.197}$$

By (5.185), (5.195), (5.196), and (5.197) we obtain (5.187). Formula (5.188) follows from (5.187) and (5.194). Further, for $f(k) = \frac{1}{k^2}$, from (5.92) it follows that

$$\sum_{m=1}^{\infty} \sum_{n=1}^{\infty} \sum_{p=1}^{\infty} \frac{1}{mnp(m+n+p)^3} = 6 \left\{ \sum_{k=1}^{\infty} \frac{1}{k^4} \sum_{n=1}^{k} \frac{1}{n} \sum_{p=1}^{n} \frac{1}{p} - G_{(5,1)}(1) - G_{(4,2)}(1) + \zeta_6(1) \right\}, \tag{5.198}$$

Formula (5.186) is derived from (5.198), (5.105), and (4.A57). □

At the end of this chapter, we will prove the convergence of the following series:

$$\sum_{k=1}^{\infty} \frac{z^m}{m^\lambda} \sum_{k_1=1}^{m} \frac{z^{k_1}}{k_1^{\lambda_1}} \sum_{k_2=1}^{k_1} \frac{z^{k_2}}{k_2^{\lambda_2}} \cdots \sum_{k_r=1}^{k_{r-1}} \frac{z^{k_r}}{k_r^{\lambda_1}} \quad (|z| \leqslant 1,\ \lambda \geqslant 2,\ \lambda_1 \geqslant 1,\ \lambda_2 \geqslant 1, \dots, \lambda_r \geqslant 1). \tag{5.199}$$

Proof. We have

$$\left| \frac{z^m}{m^\lambda} \sum_{k_1=1}^{m} \frac{z^{k_1}}{k_1^{\lambda_1}} \sum_{k_2=1}^{k_1} \frac{z^{k_2}}{k_2^{\lambda_2}} \sum_{k_3=1}^{k_2} \frac{z^{k_3}}{k_3^{\lambda_3}} \cdots \sum_{k_r=1}^{k_{r-1}} \frac{z^{k_r}}{k_r^{\lambda_1}} \right|$$

$$\leqslant \frac{1}{m^2} \sum_{k_1=1}^{m} \frac{1}{k_1} \sum_{k_2=1}^{k_1} \frac{1}{k_2} \cdots \sum_{k_r=1}^{k_r=1} \frac{1}{k_r} = \frac{1}{m^2} \lambda(r,m) = a_m, \tag{5.200}$$

$$
\begin{aligned}
\lim_{m\to\infty} m\left(\frac{a_m}{a_{m+1}}-1\right) &= \lim_{m\to\infty} m\left(\frac{\lambda(r,m)}{\lambda(r,m+1)}\cdot\frac{(m+1)^2}{m^2}-1\right)\\
&= \lim_{m\to\infty} m\left(\frac{(m+1)^2}{m^2}\cdot\frac{\lambda(r,m+1)-\frac{1}{m+1}\lambda(r-1,m+1)}{\lambda(r,m+1)}-1\right)\\
&= \lim_{m\to\infty} m\left\{m\left[\left(1+\frac{1}{m}\right)^2-1\right]-\frac{m}{m+1}\,\frac{\lambda(r-1,m+1)}{\lambda(r,m+1)}\right\}\\
&= 2-\lim_{m\to\infty}\frac{\lambda(r-1,m+1)}{\lambda(r,m+1)} = 2-\lim_{m\to\infty}\frac{\lambda(r-1,m+1)-\lambda(r-1,m)}{\lambda(r,m+1)-\lambda(r,m)}\\
&= 2-\lim_{m\to\infty}\frac{\frac{1}{m+1}\lambda(r-2,m+1)}{\frac{1}{m+1}\lambda(r-1,m+1)} = 2-\lim_{m\to\infty}\frac{\lambda(r-2,m+1)}{\lambda(r-1,m+1)}=\cdots\\
&= 2-\lim_{m\to\infty}\frac{\sum_{k_1=1}^{m+1}\frac{1}{k_1}}{\sum_{k_1=1}^{m+1}\frac{1}{k_1}\sum_{k_2=1}^{k_1}\frac{1}{k_2}} = 2-\lim_{m\to\infty}\frac{1}{\sum_{k_1=1}^{m+1}\frac{1}{k_1}} = 2.
\end{aligned}
\tag{5.201}
$$

The *Stolz* theorem was used to calculate the limit. By the obtained result, according to the *Raabe* criterion, the series $\sum_{m=1}^{\infty} a_m$ converges, and according to the comparison criterion, the first series also converges. □

By the formulas provided in Chapters 4 and 5 we conclude that the multiple summation of the series is directly related to the summations given in (5.199), so the term **multiple summation** can also be applied to them.

5.2 Formulas for *Mordell*'s sum and the sums similar to it

Mordell [66] considered the following expressions:

$$
\sum_{r=1}^{\infty}\sum_{s=1}^{\infty}\frac{1}{r^n s^n (r+s)^n} = S_n \quad (n\in\mathbb{N}). \tag{5.202}
$$

For $n = 2k$, the result for S_{2k} is given in [189], whereas for $n = 2k+1$, the result for the above-stated double sum is offered in this chapter; see formula (5.257).

Combining even and odd cases, we get the unique formula

$$
S_n = \frac{4}{2(-1)^n+1}\cdot\sum_{0\leqslant 2t\leqslant n}\binom{2n-1-2t}{n-1}\zeta_{2t}(1)\zeta_{3n-2t}(1) \quad (n\in\mathbb{N}), \tag{5.203}
$$

provided that $\zeta_0(1) = \zeta(0) = -\frac{1}{2}$. The formula for S_{2k+1} is obtained by applying the results of Chapter 4.

Tornheim [49] proves (without obtaining the expression for S_{2k+1}) that S_{2k+1} is a polynomial of $\zeta_2(1), \zeta_3(1), \ldots, \zeta_{6k+3}(1)$ with rational coefficients. Based on the result obtained for S_{2k+1}, we can conclude that S_{2k+1} is a polynomial of degree two with respect to $\zeta_2(1)$,

$\zeta_4(1), \ldots, \zeta_{2k}(1), \zeta_{4k+3}(1), \zeta_{4k+5}(1), \ldots, \zeta_{6k+3}(1), \zeta_0(1)$, with integer coefficients. In [189] the authors deal with the problem of calculating the following sums:

$$T_n = \sum_{r=1}^{\infty}\sum_{s=1}^{\infty} \frac{(-1)^r}{r^n s^n (r+s)^n}, \quad V_n = \sum_{r=1}^{\infty}\sum_{s=1}^{\infty} \frac{(-1)^{r+s}}{r^n s^n (r+s)^n}. \tag{5.204}$$

These sums are calculated for $n = 2k+1$. The sums given are a particular case of the sums of the form:

$$S_{(m,n,p)} = \sum_{r=1}^{\infty}\sum_{s=1}^{\infty} \frac{1}{r^m s^n (r+s)^p}, \quad T_{(m,n,p)} = \sum_{r=1}^{\infty}\sum_{s=1}^{\infty} \frac{(-1)^r}{r^n s^n (r+s)^p},$$
$$V_{(m,n,p)} = \sum_{r=1}^{\infty}\sum_{s=1}^{\infty} \frac{(-1)^{r+s}}{r^n s^n (r+s)^p} \quad (m,n,p \in \mathbb{N}), \tag{5.205}$$

where

$$S_{(n,n,n)} = S_n, \quad T_{(n,n,n)} = T_n, \quad V_{(n,n,n)} = V_n. \tag{5.206}$$

In this section, we calculate the following sums: $S_{(m,n,p)}$, $T_{(m,n,p)}$, $V_{(m,n,p)}$ when $m+n+p = 2k+1$. Among the expressions S_n, T_n, V_n for $n = 2k+1$, only S_1, T_1, V_1 were defined, whereas for $n \geqslant 3$, none of them was calculated.

If

$$S_{(m,n,p)} = \sum_{r=1}^{\infty}\sum_{s=1}^{\infty} \frac{1}{r^m s^n (r+s)^p}, \quad S(m,m,m) = S_m, \tag{5.207}$$

$$T_{(m,n,p)} = \sum_{r=1}^{\infty}\sum_{s=1}^{\infty} \frac{(-1)^r}{r^m s^n (r+s)^p}, \quad T(m,m,m) = T_m, (m,n,p \in \mathbb{N}), \tag{5.208}$$

$$V_{(m,n,p)} = \sum_{r=1}^{\infty}\sum_{s=1}^{\infty} \frac{(-1)^{r+s}}{r^m s^n (r+s)^p}, \quad V(m,m,m) = V_m. \tag{5.209}$$

Then

$$S_{(m,n,p)} = \sum_{k=2}^{\infty} \frac{1}{k^p} \sum_{r=1}^{k-1} \frac{1}{r^m (k-r)^n}. \tag{5.210}$$

According to (4.13), we know that

$$\frac{1}{r^m (k-r)^n} = \sum_{i=0}^{m-1} \binom{n+i-1}{n-1} \frac{1}{r^{m-i} k^{n+i}} + \sum_{j=0}^{n-1} \binom{m+j-1}{m-1} \frac{1}{r^{m+j} (k-r)^{n-j}}. \tag{5.211}$$

By the above it follows that

$$S_{(m,n,p)} = \sum_{k=2}^{\infty} \frac{1}{k^p} \sum_{r=1}^{k-1}\sum_{i=0}^{m-1} \binom{n-1+i}{n-1} \frac{1}{r^{m-i} k^{n+i}} + \sum_{k=2}^{\infty} \frac{1}{k^p} \sum_{r=1}^{k-1}\sum_{j=0}^{n-1} \binom{m-1+j}{m-1} \frac{1}{k^{m+j} (k-r)^{n-j}}$$

$$= \sum_{i=0}^{m-1} \binom{n-1+i}{n-1} \sum_{k=2}^{\infty} \frac{1}{k^{n+p+i}} \sum_{r=1}^{k-1} \frac{1}{r^{m-i}} + \sum_{j=0}^{n-1} \binom{m-1+j}{m-1} \sum_{k=2}^{\infty} \frac{1}{k^{n+p+j}} \sum_{r=1}^{k-1} \frac{1}{r^{n-j}}$$

$$= \sum_{i=0}^{m-1} \binom{n-1+i}{n-1} G_{(n+p+i,m-i)}(1) - \zeta_{m+n+p}(1) \cdot \sum_{i=0}^{m-1} \binom{n-1+i}{n-1}$$
$$+ \sum_{j=0}^{n-1} \binom{m-1+j}{m-1} G_{(m+p+j,\, n-j)}(1) - \zeta_{m+n+p}(1) \cdot \sum_{j=0}^{n-1} \binom{m-1+j}{m-1}. \tag{5.212}$$

Since

$$\sum_{i=0}^{m-1} \binom{n-1+i}{n-1} = \binom{m+n-1}{m-1}, \quad \sum_{j=0}^{n-1} \binom{m-1+j}{n-1} = \binom{m+n-1}{n-1}, \tag{5.213}$$

we get the following formulas:

$$S_{(m,n,p)} = \sum_{i=0}^{m-1} \binom{n-1+i}{n-1} G_{(n+p+i,m-i)}(1)$$
$$+ \sum_{j=0}^{n-1} \binom{m-1+j}{m-1} G_{(m+p+j,\, n-j)}(1) - \binom{m+n}{m} \zeta_{m+n+p}(1), \tag{5.214}$$

$$S_{(m,n,p)} = \sum_{i=1}^{m} \binom{m+n-1-i}{n-1} G_{(m+n+p-i,i)}(1)$$
$$+ \sum_{j=1}^{n} \binom{m+n-1-j}{m-1} G_{(m+n+p-j,j)}(1) - \binom{m+n}{m} \zeta_{m+n+p}(1). \tag{5.215}$$

In particular, for $m = n = p$, we have:

$$S_m = 2 \sum_{j=1}^{m} \binom{2m-1-j}{m-1} G_{(3m-j,j)}(1) - \binom{2m}{m} \zeta_{3m}(1). \tag{5.216}$$

By analogy we also obtain

$$V_{(m,n,p)} = \sum_{i=1}^{m} \binom{m+n-1-i}{n-1} G_{(m+n+p-i,i)}(-1)$$
$$+ \sum_{j=1}^{n} \binom{m+n-1-j}{m-1} G_{(m+n+p-j,\, j)}(-1) - \binom{m+n}{m} \zeta_{m+n+p}(-1), \tag{5.217}$$

$$V_m = 2 \sum_{j=1}^{m} \binom{2m-1-j}{m-1} G_{(3m-j,j)}(-1) - \binom{2m}{m} \zeta_{3m}(-1), \tag{5.218}$$

$$T_{(m,n,p)} = \sum_{i=1}^{m} \binom{m+n-1-i}{n-1} H_{(m+n+p-i,i)}(-1) + \sum_{j=1}^{n} \binom{m+n-1-j}{m-1} F_{(m+n+p-j,j)}(-1)$$

$$-\binom{m+n-1}{n}\zeta_{m+n+p}(-1)-\binom{m+n-1}{m}\zeta_{m+n+p}(1), \tag{5.219}$$

$$T_m=\sum_{i=1}^{m}\binom{2m-1-i}{m-1}H_{(3m-i,i)}(-1)+\sum_{j=1}^{m}\binom{2m-1-j}{m-1}F_{(3m-j,j)}(-1)$$
$$-\binom{2m-1}{m}\zeta_{3m}(-1)-\binom{2m-1}{m}\zeta_{3m}(-1). \tag{5.220}$$

Further, let $m+n+p=2k+1$. By (5.215) it follows that

$$2S(m,n,p)=A_1+A_2-2\binom{m+n}{m}\zeta_{2k+1}(1), \tag{5.221}$$

$$A_1=A_1(m,n,p)=2\sum_{i=1}^{m}\binom{m+n-1-i}{n-1}2G_{(2k+1-i,i)}(1), \tag{5.222}$$

$$A_2=A_2(m,n,p)=\sum_{j=1}^{n}\binom{m+n-1-j}{m-1}2G_{(2k+1-j,j)}(1)=A_1(m,n,p). \tag{5.223}$$

By (4.84) with $p=2k+1-i$ we have

$$A_1=\sum_{i=1}^{m}\binom{m+n-1-i}{n-1}\bigg\{\bigg[1+(-1)^{i+1}\binom{2k+1}{i}\bigg]\zeta_{2k+1}(1)$$
$$+2(-1)^i\sum_{2\leqslant 2t\leqslant 2k-i}\binom{2k-2t}{i-1}\zeta_{2t}(1)\zeta_{2k+1-2t}(1)$$
$$+2(-1)^i\sum_{2\leqslant 2t\leqslant i}\binom{2k-2t}{2k-i}\zeta_{2t}(1)\zeta_{2k+1-2t}(1)\bigg\}, \tag{5.224}$$

$$A_{11}=\zeta_{2k+1}(1)\sum_{i=1}^{m}\binom{m+n-1-i}{n-1}=\zeta_{2k+1}(1)\sum_{j=0}^{m-1}\binom{n-1+j}{j}=\zeta_{2k+1}(1)\binom{m+n-1}{n}, \tag{5.225}$$

$$A_{12}=-\zeta_{2k+1}(1)\sum_{i=1}^{m}(-1)^i\binom{m+n-1-i}{n-1}\binom{2k+1}{i}=-\zeta_{2k+1}(1)A_{12}', \tag{5.226}$$

$$A_{12}'=\sum_{i=1}^{m}(-1)^i\binom{m+n-1-i}{n-1}\bigg[\binom{2k}{i-1}+\binom{2k}{i}\bigg]$$
$$=-\sum_{i=0}^{m-1}(-1)^i\binom{m+n-2-i}{n-1}\binom{2k}{i}+\sum_{i=1}^{m}(-1)^i\binom{m+n-1-i}{n-1}\binom{2k}{i}$$
$$=-\binom{m+n-2}{n-1}+\sum_{i=1}^{m}(-1)^i\binom{m+n-2-i}{n-2}\binom{2k}{i}$$
$$=\cdots=-\binom{m+n-2}{n-1}-\binom{m+n-3}{n-1}-\cdots-\binom{m}{1}$$
$$+\sum_{i=1}^{m}(-1)^i\binom{m+n-1-i-(n-1)}{n-1-(n-1)}\binom{2k+1-(n-1)}{i}$$

$$= -\sum_{j=1}^{n-1}\binom{m-1+j}{j} + \sum_{i=1}^{m}(-1)^i\binom{m+p+1}{i}$$

$$= -\sum_{j=0}^{n-1}\binom{m-1+j}{j} + \sum_{i=0}^{m}(-1)^i\binom{m+p+1}{i}$$

$$= -\binom{m+n-1}{m} + \sum_{i=0}^{m}(-1)^i\binom{m+p+1}{i}. \tag{5.227}$$

By the already known formula

$$\sum_{r=0}^{m}(-1)^r\binom{n}{r} = \begin{cases}\delta_{o,n}, & m \geqslant n,\\ (-1)^m\binom{n-1}{m}, & m \leqslant n-1,\end{cases} \tag{5.228}$$

we have

$$\sum_{r=0}^{m}(-1)^i\binom{m+p+1}{i} = (-1)^m\binom{m+p}{m}, \tag{5.229}$$

i. e.,

$$A_{12} = \left[\binom{m+n-1}{m} + (-1)^{m+1}\binom{m+p}{m}\right]\zeta_{2k+1}(1), \tag{5.230}$$

$$A_{11} + A_{12} = \left[\binom{m+n}{m} + (-1)^{m+1}\binom{m+p}{m}\right]\zeta_{2k+1}(1). \tag{5.231}$$

We can further obtain

$$A_{13} = 2\sum_{i=1}^{m}(-1)^i\binom{m+n-1-i}{n-1}\sum_{2\leqslant 2t\leqslant 2k-i}\binom{2k-2t}{i-1}\zeta_{2t}(1)\zeta_{2k+1-2t}(1), \tag{5.232}$$

$$\begin{aligned}A_{13} &= 2\sum_{2\leqslant 2t\leqslant 2k-m}\zeta_{2t}(1)\zeta_{2k+1-2t}(1)\sum_{1\leqslant i\leqslant m}^{m}(-1)^i\binom{m+n-1-i}{n-1}\binom{2k-2t}{i-1}\\ &\quad + 2\sum_{n+p\leqslant 2t\leqslant 2k-2}\zeta_{2t}(1)\zeta_{2k+1-2t}(1)\sum_{1\leqslant i\leqslant 2k-2t}^{m}(-1)^i\binom{m+n-1-i}{n-1}\binom{2k-2t}{i-1} = A_{13}^{(1)} + A_{13}^{(2)}.\end{aligned} \tag{5.233}$$

First, let $2 \leqslant 2t \leqslant 2k - m = n + p - 1$. Under this assumption, we obtain

$$\begin{aligned}A'_{13} &= \sum_{1\leqslant i\leqslant m}(-1)^i\binom{m+n-1-i}{n-1}\binom{2k-2t}{i-1}\\ &= -\sum_{i=0}^{m-1}(-1)^i\binom{m+n-2-i}{n-1}\binom{2k-2t}{i},\end{aligned} \tag{5.234}$$

$$\sum_{i=0}^{m-1}(-1)^i\binom{m+n-2-i}{n-1}\binom{2k-2t}{i}$$

$$= \sum_{i=0}^{m-1} (-1)^i \binom{m+n-2-1-i}{n-1-1} \binom{2k-2t-1}{i} = \cdots$$

$$= \cdots = \sum_{i=0}^{m-1} (-1)^i \binom{m+n-2-r-i}{n-1-r} \binom{2k-2t-r}{i}. \qquad (5.235)$$

We can conclude the following:

1° $n = 1 \implies 2k - m - 1 < p, 2t \leqslant 2k - m \leqslant p, m - 1 \leqslant 2k - 2t - 1,$

$$-A'_{13} = \sum_{i=0}^{m-1} (-1)^i \binom{2k-2t}{i} = (-1)^{m-1} \binom{2k-2t-1}{m-1} = (-1)^{m-1} \binom{m+p-1-2t}{m-1}. \qquad (5.236)$$

2° $n \geqslant 2 \implies p \leqslant 2k - m - 1,$

α) $n - 1 < 2k - 2t, 2t \leqslant m + p - 1, r = n - 1,$

$$-A'_{13} = \sum_{i=0}^{m-1} (-1)^i \binom{m+p-2t}{i}. \qquad (5.237)$$

$\alpha 1$) $m - 1 \leqslant m + p - 2t - 1, 2t \leqslant p,$

$$-A'_{13} = (-1)^{m-1} \binom{m+p-1-2t}{m-1}. \qquad (5.238)$$

$\alpha 2$) $m - 1 \leqslant m + p - 2t, 2t \geqslant p + 1,$

$$-A'_{13} = \delta_{o,m+p-2t} = 0. \qquad (5.239)$$

β) $2k - 2t \leqslant n - 1, 2t \geqslant m + p, r = 2k - 2t - m + 1,$

$$-A'_{13} \sum_{i=0}^{m-1} (-1)^i \binom{2t+m-p-2-i}{2t-p-1} \binom{m-1}{i}$$

$$= \sum_{i=0}^{m-1-r} (-1)^i \binom{2t+m-p-2-r-i}{2t-p-1-r} \binom{m-1-r}{i} = \binom{2t-1-p}{m-1}. \qquad (5.240)$$

However, for $m + p - 1 - 2t < 0$, i. e., for $2t \geqslant m + p$, we have

$$(-1)^m \binom{m+p-1-2t}{m-1} = -\binom{2t-p-1}{m-1}. \qquad (5.241)$$

From this equality it follows that

$$A_{13}^{(1)} = 2(-1)^m \sum_{2 \leqslant 2t \leqslant p} \binom{m+p-1-2t}{m-1} \zeta_{2t}(1)\zeta_{2k+1-2t}(1)$$

$$+ 2(-1)^m \sum_{m+p \leqslant 2t \leqslant n+p-1} \binom{m+p-1-2t}{m-1} \zeta_{2t}(1)\zeta_{2k+1-2t}(1). \qquad (5.242)$$

Now let $n+p \leqslant 2t \leqslant 2k-1$. Under this condition, we have

$$A''_{13} = -\sum_{i=0}^{2k-2t-1}(-1)^i\binom{m+n-2-i}{n-1}\binom{2k-2t}{i} = -A^*_{13} + \binom{2t-p-1}{n-1}, \tag{5.243}$$

$$\begin{aligned} A^*_{13} &= \sum_{i=0}^{2k-2t}(-1)^i\binom{m+n-2-i}{n-1}\binom{2k-2t}{i} \\ &= \sum_{i=0}^{2k-2t-r}(-1)^i\binom{m+n-2-r-i}{n-1-r}\binom{2k-2t-r}{i}, \end{aligned} \tag{5.244}$$

1° $n=1, 2t \leqslant 2k-1 = m+p-1$,

$$A^*_{13} = \sum_{i=0}^{2k-2t}(-1)^i\binom{2k-2t}{i} = \delta_{0,2k-2t} = \delta_{0,m+p-2t} = 0. \tag{5.245}$$

2° $n \geqslant 2$,

α) $n-1 < 2k-2t,\ 2t \leqslant m+p-1,\ r = n-1$,

$$A^*_{13} = \sum_{i=0}^{2k-2t-(n-1)}(-1)^i\binom{2k-2t-(n-1)}{i} = \delta_{0,2k-2t-(n-1)} = \delta_{0,m+p-2t} = 0. \tag{5.246}$$

β) $2k-2t \leqslant n-1,\ 2t \geqslant m+p,\ r = 2k-2t$,

$$A^*_{13} = \binom{2t-p-1}{m-1} \overset{(15')}{=} (-1)^{m-1}\binom{m+p-1-2t}{m-1}. \tag{5.247}$$

In a similar way, we can conclude that

$$\binom{2t-p-1}{n-1} = (-1)^{n-1}\binom{n+p-1-2t}{n-1} \quad (2t \geqslant n+p), \tag{5.248}$$

so that

$$\begin{aligned} A^{(2)}_{13} &= 2(-1)^m \sum_{q\leqslant 2t\leqslant 2k-1}\binom{m+p-1-2t}{m-1}\zeta_{2t}(1)\zeta_{2k+1-2t}(1) \\ &\quad - 2(-1)^n \sum_{n+p\leqslant 2t\leqslant 2k-1}\binom{n+p-1-2t}{n-1}\zeta_{2t}(1)\zeta_{2k+1-2t}(1) \\ &\quad (q = \max\{m+p, n+p\}). \end{aligned} \tag{5.249}$$

From (5.233), (5.242), and (5.249) we obtain

$$\begin{aligned} A_{13} &= 2(-1)^m \sum_{2\leqslant 2t\leqslant p}\binom{m+p-1-2t}{m-1}\zeta_{2t}(1)\zeta_{2k+1-2t}(1) \\ &\quad + 2(-1)^m \sum_{m+p\leqslant 2t\leqslant 2k-1}\binom{m+p-1-2t}{m-1}\cdot\zeta_{2t}(1)\zeta_{2k+1-2t}(1) \end{aligned}$$

$$-2(-1)^n \sum_{n+p\leqslant 2t\leqslant 2k-1} \binom{n+p-1-2t}{n-1} \zeta_{2t}(1)\zeta_{2k+1-2t}(1). \tag{5.250}$$

We can further conclude that

$$A_{14} = 2\sum_{i=1}^{m}(-1)^i \binom{m+p-1-i}{n-1} \sum_{2\leqslant 2t\leqslant i} \binom{2k-2t}{2k-i} \zeta_{2t}(1)\zeta_{2k+1-2t}(1), \tag{5.251}$$

$$A_{14} = 2\sum_{2\leqslant 2t\leqslant m} \zeta_{2t}(1)\zeta_{2k+1-2t}(1) \sum_{2t\leqslant i\leqslant m} (-1)^i \binom{m+n-1-i}{n-1}\binom{2k-2t}{2k-i}. \tag{5.252}$$

For $2 \leqslant 2t \leqslant m$, we have

$$\begin{aligned}
A'_{14} &= \sum_{i=2t}^{m}(-1)^i \binom{m+n-1-i}{n-1}\binom{2k-2t}{2k-i} = \cdots \\
&= (-1)^r \sum_{i=2t+r}^{m+r} (-1)^i \binom{m+n-1-r}{n-1-r}\binom{2k-2t-r}{2k-i} \\
&= (-1)^{n-1} \sum_{i=2t+n-1}^{m+n-1} (-1)^i \binom{2k-2t-n+1}{2k-i} = \sum_{j=0}^{m-2t} (-1)^j \binom{2k-2t-n+1}{j} \\
&= \sum_{j=0}^{m-2t} (-1)^j \binom{m+p-2t}{j} = (-1)^m \binom{m+p-1-2t}{p-1}.
\end{aligned} \tag{5.253}$$

By this relation it follows that

$$A_{14} = 2(-1)^m \sum_{2\leqslant 2t\leqslant m} \binom{m+p-1-2t}{p-1} \zeta_{2t}(1)\zeta_{2k+1-2t}(1). \tag{5.254}$$

From formulas (5.231), (5.250), and (5.254) we obtain

$$\begin{aligned}
A_1(m,n,p) = &\left[\binom{m+n}{m} + (-1)^m \binom{m+p}{m}\right] \zeta_{2k+1}(1) \\
&+ 2(-1)^m \sum_{2\leqslant 2t\leqslant m} \binom{m+p-1-2t}{m-1} \zeta_{2t}(1)\zeta_{2k+1-2t}(1) \\
&+ 2(-1)^m \sum_{2\leqslant 2t\leqslant m} \binom{m+p-1-2t}{p-1} \zeta_{2t}(1)\zeta_{2k+1-2t}(1) \\
&+ 2(-1)^m \sum_{m+p\leqslant 2t\leqslant 2k-1} \binom{m+p-1-2t}{m-1} \zeta_{2t}(1)\zeta_{2k+1-2t}(1) \\
&- 2(-1)^n \sum_{n+p\leqslant 2t\leqslant 2k-1} \binom{n+p-1-2t}{n-1} \zeta_{2t}(1)\zeta_{2k+1-2t}(1).
\end{aligned} \tag{5.255}$$

From (5.221), (5.223), and (5.255) for $m + n + p = 2k + 1$, we get

$$\begin{aligned} S_{(m,n,p)} &= \sum_{r=1}^{\infty}\sum_{s=1}^{\infty} \frac{1}{r^m s^n (r+s)^p} \\ &= \frac{1}{2}\left\{(-1)^{m+1}\binom{m+p}{m} + (-1)^{n+1}\binom{n+p}{n}\right\}\zeta_{2k+1}(1) \\ &+ (-1)^m \sum_{2\leqslant 2t\leqslant m} \binom{m+p-1-2t}{p-1}\zeta_{2t}(1)\zeta_{2k+1-2t}(1) \\ &+ (-1)^n \sum_{2\leqslant 2t\leqslant n} \binom{m+p-1-2t}{p-1}\zeta_{2t}(1)\zeta_{2k+1-2t}(1) \\ &+ (-1)^m \sum_{2\leqslant 2t\leqslant p} \binom{m+p-1-2t}{p-1}\zeta_{2t}(1)\zeta_{2k+1-2t}(1) \\ &+ (-1)^n \sum_{2\leqslant 2t\leqslant p} \binom{n+p-1-2t}{n-1}\zeta_{2t}(1)\zeta_{2k+1-2t}(1). \end{aligned} \tag{5.256}$$

For $m = n = p$ and $3m = 2k + 1$,

$$\begin{aligned} S_m &= \sum_{r=1}^{\infty}\sum_{s=1}^{\infty} \frac{1}{r^m s^n (r+s)^m} \\ &= \binom{2m}{m}\zeta_{3m}(1) - 4\sum_{2\leqslant 2t\leqslant m} \binom{2m-1-2t}{m-1}\zeta_{2t}(1)\zeta_{3m-2t}(1) \quad (m = 2q-1, q \in \mathbb{N}). \end{aligned} \tag{5.257}$$

In the paper [189] already mentioned, the following formula is also given:

$$S_{2m} = \frac{8m}{3((2m)!)^2} \sum_{0\leqslant r\leqslant m} \binom{2m}{2r}(2r)!(4m-2r-1)!\zeta_{2r}(1)\zeta_{6m-2r}(1). \tag{5.258}$$

Combining the even and odd cases, we have that

$$S_m = \frac{4}{2(-1)^m + 1} \sum_{0\leqslant 2t\leqslant m} \binom{2m-1-2t}{m-1}\zeta_{2t}(1)\zeta_{3m-2r}(1) \quad (m \in \mathbb{N}). \tag{5.259}$$

In particular, from (5.257) we obtain

$$S_1 = 2\zeta_3(1). \tag{5.260}$$

This result was also obtained by *Euler* (see the comment related to (5.71)). In addition, the following is also true:

$$S_3 = 20\zeta_9(1) - 12\zeta_2(1)\zeta_7(1), \tag{5.261}$$

$$S_5 = 252\zeta_{15}(1) - 140\zeta_2(1)\zeta_{13}(1) - 20\zeta_4(1)\zeta_{11}(1). \tag{5.262}$$

We come to the position that we can derive a formula for $V(m,n,p)$ in the case $m+n+p=2k+1$:

$$2V(m,n,p) = B_1 + B_2 - 2\binom{m+n}{m}\zeta_{2k+1}(-1), \tag{5.263}$$

$$B_1 = B_1(m,n,p) = \sum_{i=1}^{m}\binom{m+n-1-i}{n-1}2G_{(2k+1-i,i)}(-1), \tag{5.264}$$

$$B_2 = B_2(m,n,p) = \sum_{j=1}^{n}\binom{m+n-1-j}{m-1}2G_{(2k+1-j,j)}(-1) = B_1(n,m,p). \tag{5.265}$$

By (4.86) with $p = 2k+1-i$ we have

$$\begin{aligned} B_1 = \sum_{i=1}^{m}\binom{m+n-1-i}{n-1}&\left\{\left[1+(-1)^{i+1}\binom{2k}{i}\right]\zeta_{2k+1}(-1) + (-1)^{i+1}\binom{2k}{i-1}\zeta_{2k+1}(1)\right. \\ &+ 2(-1)^i \sum_{2\leqslant 2t\leqslant 2k-i}\binom{2k-2t}{i-1}\zeta_{2t}(-1)\zeta_{2k+1-2t}(1) \\ &\left.+ 2(-1)^i \sum_{2\leqslant 2t\leqslant i}\binom{2k-2t}{2k-i}\zeta_{2t}(-1)\zeta_{2k+1-2t}(-1)\right\}. \end{aligned} \tag{5.266}$$

By the derived relation for A_{11} we obtain the formulas

$$B_{11} = \binom{m+n-1}{n}\zeta_{2k+1}(-1), \tag{5.267}$$

$$B_{121} = -\zeta_{2k+1}(-1)\sum_{i=1}^{m}\binom{m+n-1-i}{n-1}\binom{2k}{i}. \tag{5.268}$$

Similarly to determining the expression for A'_{12} and A_{12}, we can obtain the formulas

$$B_{121} = \left[\binom{m+n-1}{m} + (-1)^{m+1}\binom{m+p-1}{m}\right]\zeta_{2k+1}(-1), \tag{5.269}$$

$$\begin{aligned} B_{122} &= -\zeta_{2k+1}(1)\sum_{i=1}^{m}(-1)^i\binom{m+n-1-i}{n-1}\binom{2k}{i-1} - \sum_{i=1}^{m}(-1)^i\binom{m+n-1-i}{n-1}\binom{2k}{i-1} \\ &= \sum_{i=0}^{m-1}(-1)^i\binom{m+n-2-i}{n-1}\binom{2k}{i} = \cdots \\ &= \sum_{i=0}^{m-1}(-1)^i\binom{m+n-2-r-i}{n-1-r}\binom{2k-r}{i} = |r=n-1| \\ &= |r=n-1| = \sum_{i=0}^{m-1}(-1)^i\binom{m+p}{i} = (-1)^{m-1}\binom{m+p-1}{p}. \end{aligned} \tag{5.270}$$

It follows that

$$B_{11}+B_{12}=\left[\binom{m+n}{m}+(-1)^{m+1}\binom{m+p-1}{m}\right]\zeta_{2k+1}(-1)+(-1)^{m+1}\binom{m+p-1}{m}\zeta_{2k+1}(-1). \tag{5.271}$$

The double sums in the expressions for A_{13} and B_{13}, have the same limits, the same coefficients, and the only difference is that in the sum A_{13}, we have the product $\zeta_{2t}(1)\zeta_{2k+1-2t}(1)$, and in B_{13}, the product $\zeta_{2t}(-1)\zeta_{2k+1-2t}(1)$. According to (5.250) for A_{13}, we get

$$\begin{aligned}B_{13}&=2(-1)^m\sum_{2\leqslant 2t\leqslant p}\binom{m+p-1-2t}{m-1}\zeta_{2t}(-1)\zeta_{2k+1-2t}(1)\\&+2(-1)^m\sum_{m+p\leqslant 2t\leqslant 2k-1}\binom{m+p-1-2t}{m-1}\zeta_{2t}(-1)\zeta_{2k+1-2t}(1)\\&-2(-1)^n\sum_{n+p\leqslant 2t\leqslant 2k-1}\binom{n+p-1-2t}{n-1}\zeta_{2t}(-1)\zeta_{2k+1-2t}(1).\end{aligned} \tag{5.272}$$

Comparing the expressions A_{14} and B_{14}, we conclude that the only difference is that in A_{14}, we have the product $\zeta_{2t}(1)\zeta_{2k+1-2t}(1)$, and in B_{14}, the product $\zeta_{2t}(-1)\zeta_{2k+1-2t}(-1)$. Therefore by (5.254) it follows that

$$B_{14}=2(-1)^m\sum_{2\leqslant 2t\leqslant m}\binom{m+p-1-2t}{p-1}\zeta_{2t}(-1)\zeta_{2k+1-2t}(-1). \tag{5.273}$$

From formulas (5.266) and (5.271)–(5.273) it follows that

$$\begin{aligned}B_1(m,n,p)&=\left[\binom{m+n}{m}+(-1)^{m+1}\binom{m+p-1}{m}\right]\zeta_{2k+1}(-1)\\&+(-1)^{m+1}\binom{m+p-1}{p}\zeta_{2k+1}(1)\\&+2(-1)^m\sum_{2\leqslant 2t\leqslant p}\binom{m+p-1-2t}{m-1}\zeta_{2t}(-1)\zeta_{2k+1-2t}(1)\\&+2(-1)^m\sum_{m+p\leqslant 2t\leqslant 2k-1}\binom{m+p-1-2t}{m-1}\zeta_{2t}(-1)\zeta_{2k+1-2t}(1)\\&-2(-1)^n\sum_{n+p\leqslant 2t\leqslant 2k-1}\binom{n+p-1-2t}{n-1}\zeta_{2t}(-1)\zeta_{2k+1-2t}(1)\\&+2(-1)^m\sum_{2\leqslant 2t\leqslant m}\binom{m+p-1-2t}{p-1}\zeta_{2t}(-1)\zeta_{2k+1-2t}(-1),\end{aligned} \tag{5.274}$$

and by (5.274), (5.263), and (5.265) we arrive at

$$V(m,n,p)=\sum_{r=1}^{\infty}\sum_{s=1}^{\infty}\frac{(-1)^{r=s}}{r^m s^n (r+s)^p}$$

$$
\begin{aligned}
&= \frac{1}{2}\left\{\left[(-1)^{m+1}\binom{m+p-1}{m} + (-1)^{n+1}\binom{n+p-1}{n}\right]\zeta_{2k+1}(-1)\right.\\
&\quad \left.+\left[(-1)^{m+1}\binom{m+p-1}{p} + (-1)^{n+1}\binom{n+p-1}{n}\right]\zeta_{2k+1}(1)\right\}\\
&\quad + (-1)^m \sum_{2\leqslant 2t\leqslant p}\binom{m+p-1-2t}{m-1}\zeta_{2t}(-1)\zeta_{2k+1-2t}(1)\\
&\quad + (-1)^n \sum_{2\leqslant 2t\leqslant p}\binom{n+p-1-2t}{n-1}\zeta_{2t}(-1)\zeta_{2k+1-2t}(1)\\
&\quad + (-1)^m \sum_{2\leqslant 2t\leqslant m}\binom{n+p-1-2t}{p-1}\zeta_{2t}(-1)\zeta_{2k+1-2t}(-1)\\
&\quad + (-1)^n \sum_{2\leqslant 2t\leqslant n}\binom{n+p-1-2t}{p-1}\zeta_{2t}(-1)\zeta_{2k+1-2t}(-1) \quad (m+n+p=2k+1).
\end{aligned} \tag{5.275}
$$

For $m = n = p$ and $3m = 2k+1$,

$$
\begin{aligned}
V_m = \sum_{r=1}^{\infty}\sum_{s=1}^{\infty}\frac{(-1)^{r+s}}{r^m s^m (r+s)^m} &= \binom{2m-1}{m}\zeta_{3m}(-1) + \binom{2m-1}{m}\zeta_{3m}(1)\\
&- 2\sum_{2\leqslant 2t\leqslant m}\binom{2m-1-2t}{m-1}\zeta_{2t}(-1)\zeta_{3m-2t}(1)\\
&- 2\sum_{2\leqslant 2t\leqslant m}\binom{2m-1-2t}{m-1}\zeta_{2t}(-1)\zeta_{3m-2t}(-1) \quad (m=2q-1,\ m\in\mathbb{N}).
\end{aligned} \tag{5.276}
$$

It follows that

$$
V_1 = \zeta_3(-1) + \zeta_3(1) = \frac{1}{4}\zeta_3(1), \tag{5.277}
$$

which is a well-known result. It follows that

$$
V_3 = 10\zeta_9(-1) + 10\zeta_9(1) - 6\zeta_2(-1)\zeta_7(-1) - 6\zeta_2(-1)\zeta_7(1), \tag{5.278}
$$

$$
V_3 = \frac{1}{128}\{5\zeta_9(1) + 6\zeta_2(1)\zeta_7(1)\}, \tag{5.279}
$$

$$
\begin{aligned}
V_5 = 126\{\zeta_{15}(-1) + \zeta_{15}(1)\} &- 2\left\{\binom{7}{3}\zeta_2(-1)\zeta_{13}(1) + \binom{5}{1}\zeta_4(-1)\zeta_{11}(1)\right\}\\
&- 2\left\{\binom{7}{3}\zeta_2(-1)\zeta_{13}(-1) + \binom{5}{1}\zeta_4(-1)\zeta_{11}(-1)\right\},
\end{aligned} \tag{5.280}
$$

$$
V_5 = \frac{7}{8192}\{9\zeta_{15}(1) + 10\zeta_2(1)\zeta_{13}(1) + 10\zeta_4(1)\zeta_{11}(1)\}. \tag{5.281}
$$

We will also derive the formula for $T(m,n,p)$ in the case where $m+n+p = 2k+1$:

$$
2T(m,n,p) = C_1 + C_2 - 2\binom{m+n-1}{n}\zeta_{2k+1}(-1) - 2\binom{m+n-1}{m}\zeta_{2k+1}(1), \tag{5.282}
$$

$$C_1 = \sum_{i=1}^{m} \binom{m+n-1-i}{n-1} 2H_{(2k+1-i,i)}(-1),$$
$$C_2 = \sum_{j=1}^{n} \binom{m+n-1-j}{m-1} 2F_{(2k+1-j,j)}(-1). \tag{5.283}$$

According to (4.88), in the case where $p = 2k + 1 - i$, we have

$$\begin{aligned} C_1 = \sum_{i=1}^{m} \binom{m+n-1-i}{n-1} \Bigg\{ & \left[1 + (-1)^{i+1}\binom{2k}{i-1}\right] \zeta_{2k+1}(-1) \\ & + (-1)^{i+1}\binom{2k}{i}\zeta_{2k+1}(1) + 2\zeta_{2k+1-i}(1)\zeta_i(-1) \\ & + 2(-1)^i \sum_{2\leqslant 2t\leqslant i-1} \binom{2k-2t}{2k-i} \zeta_{2t}(-1)\zeta_{2k+1-2t}(1) \\ & + 2(-1)^i \sum_{2\leqslant 2t\leqslant 2k+1-i} \binom{2k-2t}{i-1} \zeta_{2t}(-1)\zeta_{2k+1-2t}(-1) \Bigg\}, \end{aligned} \tag{5.284}$$

$$C_{11} = \binom{m+n-1}{n} \zeta_{2k+1}(-1) \tag{5.285}$$

(obtained on the basis of the already derived relation for B_{11}). Among other things, it follows that

$$C_{121} = (-1)^{m+1}\binom{m+n-1}{p} \zeta_{2k+1}(-1) \tag{5.286}$$

(obtained on the basis of B_{122}),

$$C_{122} = \left\{\binom{m+n-1}{m} + (-1)^{m+1}\binom{m+p-1}{m}\right\} \zeta_{2k+1}(1) \tag{5.287}$$

(obtained on the basis of B_{121}),

$$\begin{aligned} C_{123} &= 2\sum_{i=1}^{m} \binom{m+n-1-i}{n-1} \zeta_i(-1)\zeta_{2k+1-i}(1) \\ &= 2\sum_{2\leqslant 2t\leqslant m} \binom{m+n-1-2t}{n-1} \zeta_{2t}(-1)\zeta_{2k+1-2t}(1) \\ &\quad + 2\sum_{n+p\leqslant 2t\leqslant 2k}^{m} \binom{2t-p-1}{n-1} \zeta_{2t}(1)\zeta_{2k+1-2t}(-1) \overset{(5.241)}{\Longrightarrow} \end{aligned} \tag{5.288}$$

$$\begin{aligned} C_{123} &= 2\sum_{2\leqslant 2t\leqslant m} \binom{m+n-1-2t}{n-1} \zeta_{2t}(-1)\zeta_{2k+1-2t}(1) \\ &\quad - 2(-1)^n \sum_{n+p\leqslant 2t\leqslant 2k} \binom{n+p-1-2t}{n-1} \zeta_{2t}(1)\zeta_{2k+1-2t}(-1), \end{aligned} \tag{5.289}$$

$$
\begin{aligned}
C_{13} &= 2\sum_{i=1}(-1)^i\binom{m+n-1-i}{n-1}\cdot\Bigg\{\sum_{2\leqslant 2t\leqslant i}\binom{2k-2t}{2k-i}\zeta_{2t}(-1)\zeta_{2k+1-2t}(1)\\
&\quad-\frac{1+(-1)^i}{2}\zeta_i(-1)\zeta_{2k+1-i}(1)\Bigg\}\\
&\overset{(5.254)}{=} 2(-1)^m\sum_{2\leqslant 2t\leqslant m}\binom{m+p-1-2t}{p-1}\zeta_{2t}(-1)\zeta_{2k+1-2t}(1)\\
&\quad-2\sum_{2\leqslant 2t\leqslant m}\binom{m+n-1-2t}{n-1}\zeta_{2t}(-1)\zeta_{2k+1-2t}(1),
\end{aligned}
\tag{5.290}
$$

$$
\begin{aligned}
C_{14} &= 2\sum_{i=1}^{m}(-1)^i\binom{m+n-1-i}{n-1}\Bigg\{\sum_{2\leqslant 2t\leqslant 2k-i}\binom{2k-2t}{i-1}\zeta_{2t}(-1)\zeta_{2k+1-2t}(-1)\\
&\quad+\frac{1-(-1)^i}{2}\zeta_i(-1)\zeta_{2k+1-i}(-1)\Bigg\}\\
&\underset{(5.241)}{\overset{(5.250)}{=}} 2(-1)^m\sum_{2\leqslant 2t\leqslant p}\binom{m+p-1-2t}{m-1}\zeta_{2t}(-1)\zeta_{2k+1-2t}(-1)\\
&\quad+2(-1)^m\sum_{m+p\leqslant 2t\leqslant 2k-1}\binom{m+p-1-2t}{m-1}\zeta_{2t}(-1)\zeta_{2k+1-i}(-1)\\
&\quad-2(-1)^n\sum_{n+p\leqslant 2t\leqslant 2k-1}\binom{n+p-1-2t}{n-1}\zeta_{2t}(-1)\zeta_{2k+1-i}(-1)\\
&\quad+2(-1)^n\sum_{n+p\leqslant 2t\leqslant 2k}\binom{n+p-1-2t}{n-1}\zeta_{2t}(-1)\zeta_{2k+1-i}(-1).
\end{aligned}
\tag{5.291}
$$

Since

$$
\begin{aligned}
2(-1)^n\binom{n+p-1-2k}{n-1} &= 2(-1)^n\binom{-m}{n-1} = -2\binom{m+n-2}{n-1} = -2\binom{n+m-2}{m-1}\\
&= 2(-1)^m\binom{-n}{m-1} = 2(-1)^m\binom{m+p-1-2k}{m-1},
\end{aligned}
\tag{5.292}
$$

it follows that

$$
\begin{aligned}
C_{14} &= 2(-1)^m\sum_{2\leqslant 2t\leqslant p}\binom{m+p-1-2t}{m-1}\zeta_{2t}(-1)\zeta_{2k+1-2t}(-1)\\
&\quad+2(-1)^m\sum_{m+p\leqslant 2t\leqslant 2k}\binom{m+p-1-2t}{m-1}\zeta_{2t}(-1)\zeta_{2k+1-2t}(-1).
\end{aligned}
\tag{5.293}
$$

From (5.284)–(5.287), (5.289)–(5.290), and (5.293) we have

$$
\begin{aligned}
C_1 &= \left\{\binom{m+n-1}{n}+(-1)^{m+1}\binom{m+p-1}{p}\right\}\zeta_{2k+1}(-1)\\
&\quad+\left\{\binom{m+n-1}{m}+(-1)^{m+1}\binom{m+p-1}{m}\right\}\zeta_{2k+1}(1)
\end{aligned}
$$

$$+ 2(-1)^m \sum_{2\leqslant 2t\leqslant m} \binom{m+p-1-2t}{p-1} \zeta_{2t}(-1)\zeta_{2k+1-2t}(-1)$$
$$- 2(-1)^n \sum_{n+p\leqslant 2t\leqslant 2k} \binom{n+p-1-2t}{n-1} \zeta_{2t}(1)\zeta_{2k+1-2t}(-1)$$
$$+ 2(-1)^m \sum_{2\leqslant 2t\leqslant p} \binom{m+p-1-2t}{m-1} \zeta_{2t}(-1)\zeta_{2k+1-2t}(-1)$$
$$- 2(-1)^m \sum_{m+p\leqslant 2t\leqslant 2k} \binom{m+p-1-2t}{m-1} \zeta_{2t}(-1)\zeta_{2k+1-2t}(-1). \tag{5.294}$$

By (5.283) and (4.84), for $p = 2k+1-i$, we obtain

$$C_2 = \sum_{i=1}^{n} \binom{m+n-1-i}{m-1} \Big\{ \zeta_{2k+1}(1) + (-1)^{i+1} \binom{2k+1}{i} \zeta_{2k+1}(-1)$$
$$+ [1-(-1)^i]\zeta_i(-1)\zeta_{2k+1-i}(-1)$$
$$+ 2(-1)^i \sum_{2\leqslant 2t\leqslant 2k+1-i} \binom{2k-2t}{i-1} \zeta_{2t}(1)\zeta_{2k+1-2t}(-1)$$
$$+ 2(-1)^i \sum_{2\leqslant 2t\leqslant i} \binom{2k-2t}{2k-i} \zeta_{2t}(1)\zeta_{2k+1-2t}(-1) \Big\}, \tag{5.295}$$
$$C_{21} = \binom{m+n-1}{m} \zeta_{2k+1}(1), \tag{5.296}$$
$$C_{221} = -\zeta_{2k+1}(-1) \cdot C'_{221}. \tag{5.297}$$

The expression C'_{221} is derived from the expression A'_{12} after the replacements $m \to n$ and $n \to m$. It follows that

$$C_{221} = \left\{ \binom{m+n-1}{n} + (-1)^{n+1} \binom{n+p}{n} \zeta_{2k+1}(1) \right\} \zeta_{2k+1}(-1), \tag{5.298}$$
$$C_{222} = 2 \sum_{1\leqslant 2t+1\leqslant n} \binom{m+n-2-2t}{m-1} \zeta_{2t+1}(-1)\zeta_{2k-2t}(1)$$
$$= 2 \sum_{m+p\leqslant 2t\leqslant 2k} \binom{2t-p-1}{m-1} \zeta_{2t}(-1)\zeta_{2k+1-2t}(-1) \overset{(5.241)}{\Longrightarrow}$$
$$C_{222} = -2(-1)^m \sum_{m+p\leqslant 2t\leqslant 2k} \binom{m+n-1-2t}{m-1} \zeta_{2t}(-1)\zeta_{2k-2t}(-1). \tag{5.299}$$

If in the expression for C_{14}, we replace $m \to n$ and $n \to m$, then by (5.293) we obtain

$$C_{23} = 2(-1)^n \sum_{2\leqslant 2t\leqslant p} \binom{n+p-1-2t}{n-1} \zeta_{2t}(1)\zeta_{2k+1-2t}(-1)$$
$$+ 2(-1)^n \sum_{n+p\leqslant 2t\leqslant 2k} \binom{n+p-1-2t}{n-1} \zeta_{2t}(1)\zeta_{2k+1-2t}(-1), \tag{5.300}$$

and if in the expression for A_{14}, we replace $m \to n, n \to m$, and $\zeta_{2k+1-2t}(1) \to \zeta_{2k+1-2t}(-1)$, then by (5.254) we get

$$C_{24} = 2(-1)^n \sum_{2\leqslant 2t\leqslant n} \binom{n+p-1-2t}{p-1} \zeta_{2t}(1)\zeta_{2k-2t}(-1). \tag{5.301}$$

From (5.295)–(5.301) it follows that

$$\begin{aligned} C_2 = &\left\{\binom{m+n-1}{n} + (-1)^{n+1}\binom{n+p}{n}\right\}\zeta_{2k+1}(-1) \\ &+ \binom{m+n-1}{m}\zeta_{2k+1}(1) \\ &+ 2(-1)^n \sum_{2\leqslant 2t\leqslant p} \binom{n+p-1-2t}{n-1}\zeta_{2t}(1)\zeta_{2k+1-2t}(-1) \\ &+ 2(-1)^n \sum_{2\leqslant 2t\leqslant n} \binom{n+p-1-2t}{p-1}\zeta_{2t}(1)\zeta_{2k+1-2t}(-1) \\ &+ 2(-1)^n \sum_{n+p\leqslant 2t\leqslant 2k} \binom{n+p-1-2t}{n-1}\zeta_{2t}(1)\zeta_{2k+1-2t}(-1) \\ &- 2(-1)^m \sum_{m+p\leqslant 2t\leqslant 2k} \binom{m+p-1-2t}{m-1}\zeta_{2t}(-1)\zeta_{2k+1-2t}(-1). \end{aligned} \tag{5.302}$$

By (5.282), (5.294), and (5.301) we obtain

$$\begin{aligned} T(m,n,p) &= \sum_{r=1}^{\infty}\sum_{s=1}^{\infty} \frac{(-1)^r}{r^m s^n (r+s)^p} \\ &= \frac{1}{2}\left\{\left[(-1)^{m+1}\binom{m+p-1}{p} + (-1)^{n+1}\binom{n+p}{n}\right]\zeta_{2k+1}(-1)\right. \\ &\quad \left. + (-1)^{m+1}\binom{m+p-1}{m}\zeta_{2k+1}(1)\right\} \\ &\quad + (-1)^m \sum_{2\leqslant 2t\leqslant m} \binom{m+p-1-2t}{p-1}\zeta_{2t}(1)\zeta_{2k-2t}(-1) \\ &\quad + (-1)^n \sum_{2\leqslant 2t\leqslant n} \binom{n+p-1-2t}{p-1}\zeta_{2t}(1)\zeta_{2k-2t}(-1) \\ &\quad + (-1)^n \sum_{2\leqslant 2t\leqslant p} \binom{n+p-1-2t}{n-1}\zeta_{2t}(1)\zeta_{2k-2t}(-1) \\ &\quad + (-1)^m \sum_{2\leqslant 2t\leqslant p} \binom{m+p-1-2t}{n-1}\zeta_{2t}(-1)\zeta_{2k-2t}(-1) \quad (m+n+p=2k+1). \end{aligned} \tag{5.303}$$

For $m = n = p$ and $3m = 2k + 1$,

$$\begin{aligned} T_m = \sum_{r=1}^{\infty}\sum_{s=1}^{\infty} \frac{(-1)^r}{r^m s^n (r+s)^p} &= \frac{1}{2}\binom{2m-1}{m}\{3\zeta_{3m}(-1) + \zeta_{3m}(1)\} \\ &- \sum_{2\leqslant 2t\leqslant m} \binom{2m-1-2t}{m-1}\zeta_{2t}(-1)\zeta_{3m-2t}(1) \\ &- \sum_{2\leqslant 2t\leqslant m} \binom{2m-1-2t}{m-1}\zeta_{2t}(-1)\zeta_{3m-2t}(-1) \\ &- 2\sum_{2\leqslant 2t\leqslant m} \binom{2m-1-2t}{m-1}\zeta_{2t}(1)\zeta_{3m-2t}(-1) \quad (m = 2q-1, q \in \mathbb{N}). \end{aligned} \tag{5.304}$$

It directly follows that

$$T_1 = -\frac{5}{8}\zeta_3(1), \tag{5.305}$$

which is also a known result. It follows that

$$T_3 = 15\zeta_9(-1) + 5\zeta_9(1) - 3\zeta_2(-1)\zeta_7(1) - 3\zeta_2(-1)\zeta_7(-1) - 6\zeta_2(1)\zeta_7(-1), \tag{5.306}$$

$$T_3 = \frac{1}{256}\{-2545\zeta_9(1) + 1518\zeta_2(1)\zeta_7(1)\}. \tag{5.307}$$

Concerning the convergence of the given series, for $m, n, p \in \mathbb{N}$ and $\lambda, \mu \in \{0, 1\}$, we can conclude the following:

$$\sum_{r=1}^{\infty}\left|\sum_{s=1}^{\infty} \frac{(-1)^{\lambda r+\mu s}}{r^m s^n (r+s)^p}\right| \leqslant \sum_{r=1}^{\infty}\frac{1}{r}\sum_{s=1}^{\infty}\frac{1}{s(r+s)} = \sum_{r=1}^{\infty}\frac{1}{r^2}\sum_{s=1}^{\infty}\frac{1}{s} = 2\zeta_3(1), \tag{5.308}$$

and therefore all the series listed above absolutely converge. From $\zeta_0(-1) = -\frac{1}{2}$, $\zeta_0(1) = -\frac{1}{2}$, and formulas (5.257), (5.276), and (5.304) it follows that

$$S_m = -4\sum_{0\leqslant 2t\leqslant m}\binom{2m-1-2t}{m-1}\zeta_{2t}(1)\zeta_{3m-2t}(1), \tag{5.309}$$

$$V_m = \frac{1}{2^{3m-3}}\sum_{0\leqslant 2t\leqslant m}(2^{2t-1}-1)\binom{2m-1-2t}{m-1}\zeta_{2t}(1)\zeta_{3m-2t}(1), \tag{5.310}$$

$$T_m = \frac{1}{2^{3m-3}}\sum_{0\leqslant 2t\leqslant m}(2^{3m-1}-2^{2t-1}-1)\binom{2m-1-2t}{m-1}\zeta_{2t}(1)\zeta_{3m-2t}(1) \tag{5.311}$$

for $m = 2q - 1$ and $q \in \mathbb{N}$.

6 Formulas for the basic 4th-order double sums

6.1 Formulas for the basic 4th-order double sums

In this chapter, we consider the sums of the form $\omega_j(p,q)$ (with the corresponding signs already given in Chapter 4) when $p+q=4$. There are 40 such sums in total, from which formulas were derived for 16 sums with $p=3$ and $p=2$ and 8 sums with $p=1$. We derive 26 formulas, containing π, $\log 2$, $\zeta_2(1)$, $\zeta_3(1)$, $\zeta_4(1)$, $\beta_2(-1)=G$ (*Catalan*'s constant), and $\beta_4(-1)$. The remaining 9 sums are expressed in terms of $\omega_5(2,2)$. Since $\beta_4(-1)$ appears in one group of the formulas, and $G_{(2,2)}(-1)$ in the second group, the question arises regarding their relationship. Only some of the formulas presented here are listed by *Sitaramachandrarao* [196], but based on the results in [43], we obtain a formula from which, by a simple manipulation, we can conclude that $G_{(2,2)}(-1)\sim -0.78438$.

The following formulas define the sums $\omega_j(p,q)$ if $p+q=4$, $j=1,2,\dots,16$:

$$\omega_1(2,2)=G_{(2,2)}(1)=\sum_{m=1}^{\infty}\frac{1}{m^2}\sum_{n=1}^{m}\frac{1}{n^2}=\frac{7}{4}\zeta_4(1), \tag{6.1}$$

$$\omega_1(3,1)=G_{(3,1)}(1)=\sum_{m=1}^{\infty}\frac{1}{m^3}\sum_{n=1}^{m}\frac{1}{n}=\frac{5}{4}\zeta_4(1), \tag{6.2}$$

$$\omega_2(2,2)=H_{(2,2)}(-1)=\sum_{m=1}^{\infty}\frac{1}{m^2}\sum_{n=1}^{m}\frac{(-1)^n}{n^2}=-\frac{17}{8}\zeta_4(1)-G_{(2,2)}(-1), \tag{6.3}$$

$$\omega_2(3,1)=H_{(3,1)}(-1)=\sum_{m=1}^{\infty}\frac{1}{m^3}\sum_{n=1}^{m}\frac{(-1)^n}{n}=-\frac{7}{4}\zeta_3(1)\log 2+\frac{5}{16}\zeta_4(1), \tag{6.4}$$

$$\omega_3(2,2)=\sum_{m=1}^{\infty}\frac{1}{m^2}\sum_{n=1}^{m}\frac{(-1)^n}{(2n-1)^2}=\frac{49}{16}\zeta_4(1)+2G_{(2,2)}(-1), \tag{6.5}$$

$$\omega_3(3,1)=\sum_{m=1}^{\infty}\frac{1}{m^3}\sum_{n=1}^{m}\frac{1}{2n-1}=-\frac{1}{4}\zeta_4(1)-2G_{(2,2)}(-1), \tag{6.6}$$

$$\omega_4(2,2)=\sum_{m=1}^{\infty}\frac{1}{m^2}\sum_{n=1}^{m}\frac{(-1)^n}{(2n-1)^2}=-6\beta_4(1)-\frac{3}{2}G\zeta_2(1)+\frac{7}{4}\pi\zeta_3(1), \tag{6.7}$$

$$\omega_4(3,1)=\sum_{m=1}^{\infty}\frac{1}{m^2}\sum_{n=1}^{m}\frac{(-1)^n}{2n-1}=-4\beta_4(1)+G\zeta_2(1)-\frac{7}{4}\pi\zeta_3(1), \tag{6.8}$$

$$\omega_5(1,3)=G_{(1,3)}(-1)=\sum_{m=1}^{\infty}\frac{(-1)^m}{m}\sum_{n=1}^{m}\frac{1}{n^3}=\frac{3}{4}\zeta_3(1)\log 2-\frac{19}{16}\zeta_4(1), \tag{6.9}$$

$$\omega_5(2,2)=G_{(2,2)}(-1)=\sum_{m=1}^{\infty}\frac{(-1)^m}{m^2}\sum_{n=1}^{m}\frac{1}{n^2}, \tag{6.10}$$

$$\omega_5(3,1)=G_{(3,1)}(-1)=\sum_{m=1}^{\infty}\frac{(-1)^m}{m^3}\sum_{n=1}^{m}\frac{1}{n}=-\frac{37}{32}\zeta_4(1)-\frac{1}{2}G_{(2,2)}(-1), \tag{6.11}$$

https://doi.org/10.1515/9783112233276-006

$$\omega_6(1,3) = F_{(1,3)}(-1) = \sum_{m=1}^{\infty} \frac{(-1)^m}{m} \sum_{n=1}^{m} \frac{(-1)^n}{n^3} = \frac{35}{32}\zeta_4(1) - \zeta_3(1)\log 2 - \frac{1}{2}G_{(2,2)}(-1), \quad (6.12)$$

$$\omega_6(2,2) = F_{(2,2)}(-1) = \sum_{m=1}^{\infty} \frac{(-1)^m}{m^2} \sum_{n=1}^{m} \frac{(-1)^n}{n^2} = \frac{13}{16}\zeta_4(1), \quad (6.13)$$

$$\omega_6(3,1) = F_{(3,1)}(-1) = \sum_{m=1}^{\infty} \frac{(-1)^m}{m^3} \sum_{n=1}^{m} \frac{(-1)^n}{n} = -\frac{3}{32}\zeta_4(1) + \frac{7}{4}\zeta_3(1)\log 2 + \frac{1}{2}G_{(2,2)}(-1), \quad (6.14)$$

$$\omega_7(1,3) = \sum_{m=1}^{\infty} \frac{(-1)^m}{m} \sum_{n=1}^{m} \frac{1}{(2n-1)^3} = G^2 - \frac{45}{32}\zeta_4(1), \quad (6.15)$$

$$\omega_7(2,2) = \sum_{m=1}^{\infty} \frac{(-1)^m}{m^2} \sum_{n=1}^{m} \frac{1}{(2n-1)^2}, \quad (6.16)$$

$$\omega_7(3,1) = \sum_{m=1}^{\infty} \frac{(-1)^m}{m^3} \sum_{n=1}^{m} \frac{1}{2n-1} = -2G^2 - \omega_7(2,2), \quad (6.17)$$

$$\omega_8(1,3) = \sum_{m=1}^{\infty} \frac{(-1)^m}{m} \sum_{n=1}^{m} \frac{(-1)^n}{(2n-1)^3} = 3\beta_4(-1) - \frac{3}{2}G\zeta_2(1), \quad (6.18)$$

$$\omega_8(2,2) = \sum_{m=1}^{\infty} \frac{(-1)^m}{m^2} \sum_{n=1}^{m} \frac{(-1)^n}{(2n-1)^2} = -6\beta_4(-1) + \frac{9}{2}G\zeta_2(1), \quad (6.19)$$

$$\omega_8(3,1) = \sum_{m=1}^{\infty} \frac{(-1)^m}{m^3} \sum_{n=1}^{m} \frac{(-1)^n}{2n-1} = 4\beta_4(-1) - 2G\zeta_2(1), \quad (6.20)$$

$$\omega_9(2,2) = \sum_{m=1}^{\infty} \frac{1}{(2m-1)^2} \sum_{n=1}^{m} \frac{1}{n^2} = -\frac{19}{16}\zeta_4(1) - 2G_{(2,2)}(-1) + 4\zeta_2(1) - 8\log 2, \quad (6.21)$$

$$\omega_9(3,1) = \sum_{m=1}^{\infty} \frac{1}{(2m-1)^3} \sum_{n=1}^{m} \frac{1}{n} = \frac{45}{32}\zeta_4(1) - \frac{7}{4}\zeta_3(1)\log 2 + \frac{7}{4}\zeta_3(1) - \frac{3}{2}\zeta_2(1) + 2\log 2, \quad (6.22)$$

$$\omega_{10}(2,2) = \sum_{m=1}^{\infty} \frac{1}{(2m-1)^2} \sum_{n=1}^{m} \frac{(-1)^n}{n^2} = -\frac{15}{16}\zeta_4(1) - \omega_7(2,2) - \frac{1}{2}\zeta_2(1) - 4G + 2\pi - 4\log 2, \quad (6.23)$$

$$\omega_{10}(3,1) = \sum_{m=1}^{\infty} \frac{1}{(2m-1)^3} \sum_{n=1}^{m} \frac{(-1)^n}{n} = \frac{45}{32}\zeta_4(1) - \frac{7}{8}\zeta_3(1)\log 2 - G^2 - \frac{\pi^3}{16} + 2G - \frac{\pi}{2} + \log 2, \quad (6.24)$$

$$\omega_{11}(2,2) = \sum_{m=1}^{\infty} \frac{1}{(2m-1)^2} \sum_{n=1}^{m} \frac{1}{(2n-1)^2} = \frac{75}{64}\zeta_4(1), \quad (6.25)$$

$$\omega_{11}(3,1) = \sum_{m=1}^{\infty} \frac{1}{(2m-1)^3} \sum_{n=1}^{m} \frac{1}{2n-1} = \frac{1}{2}\zeta_4(1) + \frac{7}{8}\zeta_3(1)\log + \frac{1}{4}G_{(2,2)}(-1), \quad (6.26)$$

$$\omega_{12}(2,2) = \sum_{m=1}^{\infty} \frac{1}{(2m-1)^2} \sum_{n=1}^{m} \frac{(-1)^n}{(2n-1)^2} = -\frac{1}{2}\beta_4(-1) - \frac{3}{4}G\zeta_2(1) + \frac{7}{64}\pi\zeta_3(1), \quad (6.27)$$

$$\omega_{12}(3,1) = \sum_{m=1}^{\infty} \frac{1}{(2m-1)^3} \sum_{n=1}^{m} \frac{(-1)^n}{2n-1} = -\frac{1}{2}\beta_4(-1) - \frac{7}{128}\pi\zeta_3(1) - \frac{1}{64}\pi^3 \log 2, \tag{6.28}$$

$$\begin{aligned}\omega_{13}(1,3) &= \sum_{m=1}^{\infty} \frac{(-1)^m}{2m-1} \sum_{n=1}^{m} \frac{1}{n^3} \\ &= -4\beta_4(-1) - G\zeta_2(1) + \frac{3}{2}\pi\zeta_3(1) + \frac{3}{4}\zeta_3(1) + \zeta_2(1) - 2\pi + 4\log 2,\end{aligned} \tag{6.29}$$

$$\begin{aligned}\omega_{13}(2,2) &= \sum_{m=1}^{\infty} \frac{(-1)^m}{(2m-1)^2} \sum_{n=1}^{m} \frac{1}{n^2} \\ &= 6\beta_4(-1) + \frac{1}{2}G\zeta_2(1) - \frac{7}{4}\pi\zeta_3(1) - 4G - \frac{1}{2}\zeta_2(1) + 2\pi - 4\log 2,\end{aligned} \tag{6.30}$$

$$\begin{aligned}\omega_{13}(3,1) &= \sum_{m=1}^{\infty} \frac{(-1)^m}{(2m-1)^3} \sum_{n=1}^{m} \frac{1}{n} \\ &= -3\beta_4(-1) + \frac{1}{32}\pi^3 \log 2 + \frac{7}{16}\pi\zeta_3(1) - \frac{1}{16}\pi^3 + 2G - \frac{1}{2}\pi + \log 2,\end{aligned} \tag{6.31}$$

$$\omega_{14}(1,3) = \sum_{m=1}^{\infty} \frac{(-1)^m}{2m-1} \sum_{n=1}^{m} \frac{1}{n^3} = -4\beta_4(-1) + 2G\zeta_2(1) + \frac{3}{16}\pi\zeta_3(1) - \zeta_3(1) - 2\zeta_2(1) + 8\log 2, \tag{6.32}$$

$$\omega_{14}(2,2) = \sum_{m=1}^{\infty} \frac{(-1)^m}{(2m-1)^2} \sum_{n=1}^{m} \frac{(-1)^n}{n^2} = 6\beta_4(-1) - 4G\zeta_2(1) + 4\zeta_2(1) - 8\log 2, \tag{6.33}$$

$$\begin{aligned}\omega_{14}(3,1) &= \sum_{m=1}^{\infty} \frac{(-1)^m}{(2m-1)^3} \sum_{n=1}^{m} \frac{(-1)^n}{n} \\ &= -3\beta_4(-1) + \frac{3}{2}G\zeta_2(1) + \frac{1}{32}\pi^3 \log 2 + \frac{7}{4}\zeta_3(1) - \frac{3}{2}\zeta_2(1) + 2\log 2,\end{aligned} \tag{6.34}$$

$$\omega_{15}(1,3) = \sum_{m=1}^{\infty} \frac{(-1)^m}{2m-1} \sum_{n=1}^{m} \frac{1}{(2n-1)^3} = -\frac{1}{2}\beta_4(-1) - \frac{21}{128}\pi\zeta_3(1) + \frac{1}{64}\pi^3 \log 2, \tag{6.35}$$

$$\omega_{15}(2,2) = \sum_{m=1}^{\infty} \frac{(-1)^m}{(2m-1)^2} \sum_{n=1}^{m} \frac{1}{(2n-1)^2} = -\frac{1}{2}\beta_4(-1) - \frac{7}{64}\pi\zeta_3(1), \tag{6.36}$$

$$\omega_{15}(3,1) = \sum_{m=1}^{\infty} \frac{(-1)^m}{(2m-1)^3} \sum_{n=1}^{m} \frac{1}{2n-1} = -\frac{1}{2}\beta_4(-1) + \frac{7}{128}\pi\zeta_3(1) - \frac{1}{64}\pi^3 \log 2, \tag{6.37}$$

$$\omega_{16}(1,3) = \sum_{m=1}^{\infty} \frac{(-1)^m}{2m-1} \sum_{n=1}^{m} \frac{(-1)^n}{(2n-1)^3} = \frac{75}{64}\zeta_4(1) - \frac{1}{4}G^2 - \frac{7}{16}\zeta_3(1)\log 2 - \frac{1}{8}\omega_7(2,2), \tag{6.38}$$

$$\omega_{16}(2,2) = \sum_{m=1}^{\infty} \frac{(-1)^m}{(2m-1)^2} \sum_{n=1}^{m} \frac{(-1)^n}{(2n-1)^2} = \frac{15}{32}\zeta_4(1) + \frac{1}{2}G^2, \tag{6.39}$$

$$\omega_{16}(3,1) = \sum_{m=1}^{\infty} \frac{(-1)^m}{(2m-1)^3} \sum_{n=1}^{m} \frac{(-1)^n}{2n-1} = \frac{15}{32}\zeta_4(1) + \frac{1}{4}G^2 + \frac{7}{16}\zeta_3(1)\log 2 + \frac{1}{8}\omega_7(2,2). \tag{6.40}$$

Proof. Formula (6.1) is the already given formula (4.A57); (6.2) is (4.A57); (6.3) is the same as (4.A95); (6.4) is (4.91); and (6.7) follows from (4.127) for $p = q = 2$.

Further analysis of the already obtained results leads to the conclusion that (6.8) follows from (4.127) for $p = 3$ and $q = 1$; (6.9) is identical to (4.A58); (6.11) is (4.A92); (6.12) is (4.A93); (6.13) is (4.A89); (6.14) is (4.A94); and (6.15) is (4.A55).

Formula (6.18) follows from (4.135) for $p = 1$ and $q = 3$; (6.19) follows from (4.135) for $p = q = 2$; (6.20) follows from (4.135) for $p = 3$ and $q = 1$; (6.27) follows from (4.127) for $p = 2q = 2$; (6.28) is a consequence of (4.127) for $p = 3$ and $q = 1$; (6.29) is derived from (4.130) for $p = 1$ and $q = 3$; (6.30) follows from (4.130) for $p = q = 2$; (6.31) follows from the already given formula (4.130) for $p = 3$ and $q = 1$; (6.30) is derived from (4.136) for $p = 1$ and $q = 3$; (6.33) follows from (4.136) for $p = q = 2$; (6.34) follows from (4.136) for $p = 3$ and $q = 1$; finally formulas (6.35), (6.36), and (6.37) can be derived from (4.129) for $p = 1$ and $q = 3$, $p = q = 2$, and $p = 3$ and $q = 1$, respectively.

We can derive formula (6.5) from the following relation:

$$\omega_1(2,2) + \omega_5(2,2) = \frac{1}{8}\omega_1(2,2) + \frac{1}{2}\omega_3(2,2) \implies \omega_3(2,2) = \frac{7}{4}\omega_1(2,2) + 2\omega_5(2,2). \tag{6.41}$$

Formula (6.5) follows from the last relation and (6.1). Since

$$\omega_1(3,1) + \omega_5(3,1) = \frac{1}{8}\omega_1(3,1) + \frac{1}{4}\omega_3(3,1) \implies \omega_3(3,1) = \frac{7}{2}\omega_1(3,1) + 4\omega_5(3,1), \tag{6.42}$$

(6.6) follows from (6.2) and (6.11). To prove (6.17), we will use the following functions:

$$Q_r = \sum_{k=1}^{\infty} \frac{z^k}{(2k-1)^r} \quad \text{(where } r \text{ is a nonnegative integer number)}, \quad |z| \leqslant 1, \tag{6.43}$$

$$Q_r = \frac{z}{1-z} \ (z \neq 1), \quad Q_r(-1) = -\beta_r(-1), \quad Q_1(-1) = -\frac{\pi}{4}, \quad Q_2(-1) = -G, \tag{6.44}$$

$$Q_3(-1) = -\frac{\pi^3}{32}, \quad Q_4(-1) = -\beta_4(-1), \tag{6.45}$$

$$Q_2^2(z) = \sum_{k=2}^{\infty} z^k \sum_{m=1}^{k-1} \frac{1}{(2m-1)^2(2k-2m-1)^2}, \tag{6.46}$$

$$\begin{aligned} \frac{1}{(2m-1)^2(2k-2m-1)^2} &= \frac{1}{4(k-1)^2}\frac{1}{(2m-1)^2} + \frac{1}{4(k-1)^2}\frac{1}{(2k-2m-1)^2} \\ &+ \frac{1}{4(k-1)^3}\frac{1}{2m-1} + \frac{1}{4(k-1)^3}\frac{1}{2k-2m-1}, \end{aligned} \tag{6.47}$$

$$Q_2^2(z) = \frac{z}{2}\sum_{k=1}^{\infty}\frac{z^k}{k^2}\sum_{n=1}^{k}\frac{1}{(2n-1)^2} + \frac{z}{2}\sum_{k=1}^{\infty}\frac{z^k}{k^3}\sum_{n=1}^{k}\frac{1}{2n-1}. \tag{6.48}$$

Taking $z = -1$ here, we get (6.17). We can also conclude that

$$\omega_9(2,2) + \omega_3(2,2) = \sum_{n=1}^{\infty}\frac{1}{n^2(2n-1)^2} + \sum_{m=1}^{\infty}\frac{1}{m^2}\sum_{n=1}^{\infty}\frac{1}{(2n-1)^2}, \tag{6.49}$$

$$\sum_{n=1}^{\infty}\frac{1}{n^2(2n-1)^2} = 4\zeta_2(1) - 8\log 2. \tag{6.50}$$

Formula (6.21) now follows from (6.5). In relation to (6.22), we can write

$$\omega_1(3,1) = \frac{1}{16}\omega_1(3,1) + \frac{1}{8}\omega_3(3,1) + \frac{1}{2}\sum_{m=2}^{\infty}\frac{(-1)^m}{(2m-1)^3}\sum_{n=1}^{m-1}\frac{1}{n} + \omega_{11}(3,1), \tag{6.51}$$

$$\omega_2(3,1) = \frac{1}{16}\omega_1(3,1) - \frac{1}{8}\omega_3(3,1) + \frac{1}{2}\sum_{m=2}^{\infty}\frac{1}{(2m-1)^3}\sum_{n=1}^{m-1}\frac{1}{n} - \omega_{11}(3,1). \tag{6.52}$$

By summation we get

$$\omega_9(3,1) = \frac{7}{8}\omega_1(3,1) + \omega_2(3,1) + \sum_{n=1}^{\infty}\frac{1}{n(2n-1)^3}, \tag{6.53}$$

$$\sum_{n=1}^{\infty}\frac{1}{n(2n-1)^3} = \frac{7}{4}\zeta_3(1) - \frac{3}{2}\zeta_2(1) + 2\log 2. \tag{6.54}$$

Formula (6.22) follows from the relation thus established and from (6.2) and (6.4). In the case of (6.23), we can conclude that

$$\omega_7(2,2) + \omega_{10}(2,2) = \sum_{n=1}^{\infty}\frac{(-1)^n}{n^2(2n-1)^2} + \sum_{m=1}^{\infty}\frac{(-1)^m}{m^2}\sum_{n=1}^{\infty}\frac{1}{(2n-1)^2}, \tag{6.55}$$

$$\sum_{n=1}^{\infty}\frac{(-1)^n}{n^2(2n-1)^2} = -\frac{1}{2}\zeta_2(1) - 4G + 2\pi - 4\log 2. \tag{6.56}$$

From the last two relations we obtain (6.23). For (6.24), it follows that

$$\omega_{10}(3,1) + \omega_7(1,3) = \sum_{n=1}^{\infty}\frac{(-1)^n}{n(2n-1)^3} + \sum_{m=1}^{\infty}\frac{1}{(2m-1)^3}\sum_{n=1}^{\infty}\frac{(-1)^n}{n}, \tag{6.57}$$

$$\sum_{n=1}^{\infty}\frac{(-1)^n}{n(2n-1)^3} = -\frac{\pi^3}{16} + 2G - \frac{\pi}{2} + \log 2. \tag{6.58}$$

Therefore from the relations thus established and (6.15) we obtain (6.24). Further, for (6.25), we have

$$2\omega_{11}(2,2) = \sum_{n=1}^{\infty}\frac{1}{(2n-1)^4} + \sum_{m=1}^{\infty}\frac{1}{(2m-1)^2}\sum_{n=1}^{\infty}\frac{1}{(2n-1)^2}, \tag{6.59}$$

from which (6.25) directly follows. In addition, with respect to (6.26), we can conclude that

$$\omega_1(3,1) - \omega_2(3,1) = \frac{1}{4}\omega_3(3,1) + 2\omega_{11}(3,1). \tag{6.60}$$

Formula (6.26) is derived from the above and from (6.2), (6.4), and (6.6). For (6.38) and (6.39), we can establish the following relations:

$$\omega_{16}(3,1)+\omega_{16}(1,3)=\sum_{n=1}^{\infty}\frac{1}{(2n-1)^4}+\sum_{m=1}^{\infty}\frac{(-1)^m}{(2m-1)^3}\sum_{n=1}^{\infty}\frac{(-1)^n}{2n-1}=\frac{105}{64}\zeta_4(1), \tag{6.61}$$

$$2\omega_{16}(2,2)=\sum_{n=1}^{\infty}\frac{1}{(2n-1)^4}+\sum_{m=1}^{\infty}\frac{(-1)^m}{(2m-1)^2}\sum_{n=1}^{\infty}\frac{(-1)^n}{(2n-1)^2}, \tag{6.62}$$

which lead to (6.39).

We have

$$Q_2(-z)\zeta_2(z)=\sum_{k=2}^{\infty}z^k\sum_{m=1}^{k-1}\frac{(-1)^m}{(2m-1)^2(k-m)^2}, \tag{6.63}$$

$$\begin{aligned}\frac{1}{(2m-1)^2(k-m)^2}&=\frac{4}{(2k-1)^2}\frac{1}{(2m-1)^2}+\frac{8}{(2k-1)^3}\frac{1}{2m-1}\\&\quad+\frac{1}{(2k-1)^2}\frac{1}{(k-m)^2}+\frac{4}{(2k-1)^3}\frac{1}{k-m},\end{aligned} \tag{6.64}$$

$$\begin{aligned}Q_2(-z)\zeta_2(z)&=4\sum_{k=1}^{\infty}\frac{z^k}{(2k-1)^2}\sum_{m=1}^{k}\frac{(-1)^m}{(2m-1)^2}\\&\quad+8\sum_{k=1}^{\infty}\frac{z^k}{(2k-1)^3}\sum_{m=1}^{k}\frac{(-1)^m}{2m-1}\\&\quad+\sum_{k=1}^{\infty}\frac{(-z)^m}{(2m-1)^2}\sum_{m=1}^{k}\frac{(-1)^m}{m^2}\\&\quad+4\sum_{k=1}^{\infty}\frac{(-z)^k}{(2k-1)^3}\sum_{m=1}^{k}\frac{(-1)^m}{m}-\sum_{k=1}^{\infty}\frac{z^k}{k^2(2k-1)^2}\\&\quad-4\sum_{k=1}^{\infty}\frac{z^k}{k(2k-1)}-12Q_4(-z),\end{aligned} \tag{6.65}$$

$$\sum_{k=1}^{\infty}\frac{z^k}{k^2(2k-1)^2}=\zeta_2(z)+4Q_2(z)-4\sum_{k=1}^{\infty}\frac{z^k}{k(2k-1)}, \tag{6.66}$$

$$\sum_{k=1}^{\infty}\frac{z^k}{k(2k-1)^3}=2Q_3(z)-2Q_2(z)+\sum_{k=1}^{\infty}\frac{z^k}{k(2k-1)}, \tag{6.67}$$

$$-\sum_{k=1}^{\infty}\frac{z^k}{k^2(2k-1)^2}-4\sum_{k=1}^{\infty}\frac{z^k}{k(2k-1)^3}=-8Q_3(z)+4Q_2(z)-\zeta_2(z). \tag{6.68}$$

For $z=-1$, from the formulas above it follows that

$$\begin{aligned}Q_2(1)\zeta_2(-1)&=4\omega_{16}(2,2)-12Q_4(1)+8\omega_{16}(3,1)+\omega_{10}(2,2)\\&\quad+4\omega_{10}(3,1)-8Q_3(-1)+4Q_2(-1)-\zeta_2(-1).\end{aligned} \tag{6.69}$$

Therefore from (6.39), (6.23), and (6.24) we derive (6.40). Formula (6.38) follows from the formulas used in the derivation of (6.38) and (6.40). □

6.2 Rational-linear transformation of multiple series

In this section, using the expressions $\zeta_k(\frac{1}{2}) = \sum_{n=1}^{\infty} \frac{1}{2^n n^k}$ $(k = 1, 2, 3)$, we determine a formula for $\zeta_4(\frac{1}{2})$. Various relations for the functions specified by multiple summation are presented, related to the *Spencer* relation. The relations for the functions given by integrals are also obtained, where these integrals are related to the *Riemann* zeta function. A relation between $G_{(2,2)}(-1)$ and $\zeta_4(\frac{1}{2})$ is also established. Several formulas are presented, from which we can try by analogy to deduce a possible form for $\zeta_4(\frac{1}{2})$ or for $G_{(2,2)}(-1)$. The results of one special form of rational-linear transformation in a multiple series are also given, which allows us (in some particular cases) to reduce the multiplicity of the summation by one. Finally, we give a recurrent relation by which the multiple sum $F_r(x)$ (formula (6.192)) can be determined, and the values for $F_r(x)$ $(r = 1, 2, \dots, 6)$ are presented. It is worth pointing out that the following functional dependence is also established:

$$G_{(2,2)}(-1) = \sum_{m=1}^{\infty} \frac{(-1)^m}{m^2} \sum_{n=1}^{m} \frac{1}{n^2} = \sum_{n=1}^{\infty} \frac{k_n}{n^2}, \quad k_n = \sum_{\substack{d \geqslant \sqrt{n} \\ d|n}} (-1)^d \quad (n \geqslant 1). \tag{6.70}$$

First of all, we list some formulas for multiple sums and integrals, related to the *Spencer* relation (formula (6.77)):

$$\zeta_2(x) + \zeta_2(1-x) = \zeta_2(1) - \log x \cdot \log(1-x) \quad (0 \leqslant x \leqslant 1), \tag{6.71}$$

$$\zeta_2\left(\frac{x}{x-1}\right) = -\zeta_2(x) - \frac{1}{2}\zeta_1^2(x) \quad (-1 \leqslant x \leqslant \tfrac{1}{2}), \tag{6.72}$$

$$\zeta_2\left(\frac{1}{x+1}\right) = \log x \cdot \log(1+x) + \zeta_2(-x) - \frac{1}{2}\log^2(1+x) + \zeta_2(1) \quad (0 \leqslant x \leqslant 1), \tag{6.73}$$

$$\zeta_2\left(\frac{1}{x-1}\right) = \log(-x) \cdot \log(1-x) + \zeta_2(x) - \frac{1}{2}\log^2(1-x) + \zeta_2(1) \quad (-1 \leqslant x \leqslant 0), \tag{6.74}$$

$$\zeta_2\left(\frac{x-1}{x}\right) = -\zeta_2(1-x) - \frac{1}{2}\log^2 x \quad (\tfrac{1}{2} \leqslant x \leqslant 1), \tag{6.75}$$

$$G_{(1,1)}(x) = \zeta_2(x) + \frac{1}{2}\zeta_1^2(x) \quad (-1 \leqslant x \leqslant 1), \tag{6.76}$$

$$\zeta_3\left(\frac{x}{x-1}\right) = -\zeta_3(x) - \zeta_3(x-1) + \zeta_2(1)\log(1-x) + \frac{1}{6}\log^3(1-x) - \frac{1}{2}\log x \cdot \log^2(1-x) + \zeta_3(1) \quad (0 \leqslant x \leqslant \tfrac{1}{2}), \tag{6.77}$$

$$G_{(2,1)}(x) = \zeta_3\left(\frac{x}{x-1}\right) + 2\zeta_3(x) - \frac{1}{6}\log^3(1-x) - \zeta_2(x)\log(1-x) \quad (-1 \leqslant x \leqslant \tfrac{1}{2}), \tag{6.78}$$

$$G_{(2,1)}(x) = \zeta_3(x) - \zeta_3(1-x) + [\zeta_2(1) - \zeta_2(x)]\log(1-x)$$
$$- \frac{1}{2}\log x \cdot \log^2(1-x) + \zeta_3(1) \quad (0 \leqslant x \leqslant 1), \tag{6.79}$$

$$2G_{(2,1)}(x) + G_{(1,2)}(x) = 3\zeta_3(x) + \zeta_1(x)\zeta_2(x) \quad (-1 \leqslant x < 1), \tag{6.80}$$

$$G_{(1,2)}(x) = \zeta_3(x) + 2\zeta_3(1-x) - 2\zeta_3(1) - [\zeta_2(1) - \zeta_2(1-x)]\log(1-x)$$
$$(0 \leqslant x < 1), \tag{6.81}$$

$$\lim_{x\uparrow 1}\{G_{(1,2)}(x) + \zeta_2(1)\log(1-x)\} = -\zeta_3(1), \tag{6.82}$$

$$\frac{1}{2}\sum_{m=1}^{\infty}\sum_{n=1}^{\infty}\frac{x^{m+n}}{mn(m+n)} = G_{(2,1)}(x) - \zeta_3(x), \tag{6.83}$$

$$\{G_{(2,1)}(x) - \zeta_3(x)\}' = \frac{1}{2}\frac{\log^2(1-x)}{x}, \tag{6.84}$$

$$\int_0^x \frac{\log(1-t)}{t}\,\partial t = -\zeta_2(x) \quad (-1 \leqslant x \leqslant 1), \tag{6.85}$$

$$\frac{1}{2}\int_0^x \frac{\log^2(1-t)}{t}\,\partial t = \zeta_2(1-x)\log(1-x) + \frac{1}{2}\log x \log^2(1-x)$$
$$- \zeta_3(1-x) + \zeta_3(1) \quad (0 \leqslant x \leqslant 1), \tag{6.86}$$

$$\frac{1}{2}\int_0^x \frac{\log^2(1-t)}{t}\,\partial t = \zeta_3\left(\frac{x}{x-1}\right) + \zeta_3(x) - \frac{1}{6}\log^3(1-x)$$
$$- \zeta_2(x)\log(1-x) \quad (-1 \leqslant x \leqslant \tfrac{1}{2}), \tag{6.87}$$

$$\frac{1}{2}\int_0^x \frac{\log^2(1-t)}{t}\,\partial t = -\zeta_3\left(\frac{1}{1-x}\right) - \frac{1}{2}\log(-x)\log^2(1-x) + \frac{1}{6}\log^3(1-x)$$
$$- \zeta_2(1)\log(1-x) - \zeta_2(x)\log(1-x) + \zeta_3(1) \quad (-1 \leqslant x \leqslant 0), \tag{6.88}$$

$$\frac{1}{6}\int_0^x \frac{\log^3(1-t)}{t}\,\partial t = \frac{1}{6}\log(x)\log^3(1-x) + \frac{1}{2}\zeta_2(1-x)\log^2(1-x)$$
$$- \zeta_3(1-x)\log(1-x) + \zeta_4(1-x) - \zeta_4(1) \quad (0 \leqslant x \leqslant 1), \tag{6.89}$$

$$\frac{1}{6}\int_0^x \frac{\log^3(1-t)}{t}\,\partial t = \frac{1}{2}G_{(2,2)}(x) + \zeta_4\left(\frac{x}{x-1}\right) + \zeta_3\left(\frac{x}{x-1}\right)\log(1-x) + \frac{1}{2}\zeta_4(x)$$
$$- \frac{1}{2}\zeta_2(x)\log^2(1-x) - \frac{1}{4}\zeta_2^2(x) - \frac{1}{8}\log^4(1-x) \quad (-1 \leqslant x \leqslant \tfrac{1}{2}), \tag{6.90}$$

$$\int_0^x \frac{\log(1+t)}{t}\,\partial t = -\zeta_2(-x), \tag{6.91}$$

$$\frac{1}{2}\int_0^x \frac{\log^2(1+t)}{t}\,\partial t = -\zeta_3\left(\frac{1}{1+x}\right) + \frac{1}{6}\log^3(1+x) - \frac{1}{2}\log x\log^2(1+x) - \zeta_2(1)\log(1+x)$$

$$- \zeta_2(1)\log(1+x) - \zeta_2(-x)\log(1+x) + \zeta_3(1) \quad (0 \leqslant x \leqslant 1), \quad (6.92)$$

$$\frac{1}{6}\int_0^x \frac{\log^3(1+t)}{t}\,\partial t = -\zeta_4\left(\frac{1}{1+x}\right) - \zeta_3\left(\frac{1}{1+x}\right)\log(1+x) + \frac{1}{8}\log^4(1+x)$$

$$- \frac{1}{3}\log x \cdot \log^3(1+x) - \zeta_2(-x)\log^2(1+x)$$

$$- \frac{1}{2}\zeta_2(1)\log^2(1+x) + \zeta_4(1) \quad (0 \leqslant x \leqslant 1), \quad (6.93)$$

$$G_{(1,3)}(x) = \zeta_4(x) - \frac{1}{2}\zeta_2^2(x) - \zeta_3(x)\log(1-x) \quad (-1 \leqslant x < 1), \quad (6.94)$$

$$G_{(2,2)}(x) + 2G_{(3,1)}(x) = 3\zeta_4(x) + \frac{1}{2}\zeta_2^2(x) \quad (-1 \leqslant x < 1), \quad (6.95)$$

$$G_{(2,2)}(x) = -2\zeta_4\left(\frac{x}{x-1}\right) - \zeta_4(x) + 2\zeta_4(1-x) - \zeta_2(1)\log^2(1-x)$$

$$- \frac{1}{12}\log^4(1-x) - 2\zeta_3(1)\log(1-x) + 2\zeta_3(x)\log(1-x)$$

$$+ \frac{1}{3}\log x\log^3(1-x) + \frac{1}{2}\zeta_2^2(x) - 2\zeta_4(1) \quad (0 \leqslant x \leqslant \tfrac{1}{2}), \quad (6.96)$$

$$G_{(3,1)}(x) = \zeta_4\left(\frac{x}{x-1}\right) + 2\zeta_4(x) - \zeta_4(1-x) + \frac{1}{2}\zeta_2(1)\log^2(1-x)$$

$$+ \frac{1}{24}\log^4(1-x) + \zeta_3(1)\log(1-x) - \zeta_3(x)\log(1-x)$$

$$- \frac{1}{6}\log x\log^3(1-x) + \zeta_4(1) \quad (0 \leqslant x \leqslant \tfrac{1}{2}). \quad (6.97)$$

We further prove formula (6.97), whereas the other formulas can be proven in a similar way.

Proof. Denoted by $A(x)$ the right-hand side of equality (6.97). Then

$$A'(x) = \frac{\zeta_3(\frac{x}{x-1})}{x(1-x)} + 2\frac{\zeta_3(x)}{x} + \frac{\zeta_3(1-x)}{1-x} - \zeta_2(1)\frac{\log(1-x)}{1-x} - \frac{1}{6}\frac{\log^3(1-x)}{1-x}$$

$$- \frac{\zeta_3(1)}{1-x} - \frac{\zeta_2(x)}{x}\log(1-x) + \frac{\zeta_3(x)}{1-x} - \frac{1}{6}\frac{\log^3(1-x)}{x} + \frac{1}{2}\frac{\log x \cdot \log^2(1-x)}{1-x}$$

$$\overset{(6.77)}{=} \left(\frac{1}{x} + \frac{1}{1-x}\right) \cdot \Big\{-\zeta_3(x) - \zeta_3(1-x) + \zeta_2(1)\log(1-x) + \frac{1}{6}\log^3(1-x)$$

$$- \frac{1}{2}\log x\log^2(1-x) + \zeta_3(1)\Big\}$$

$$+ 2\frac{\zeta_3(x)}{x} - \frac{\zeta_2(x)}{x}\log(1-x) - \frac{1}{6}\frac{\log^3(1-x)}{x} + \frac{\zeta_3(1-x)}{1-x} - \zeta_2(1)\frac{\log(1-x)}{1-x}$$

$$- \frac{1}{6}\frac{\log^3(1-x)}{1-x} - \frac{\zeta_3(1)}{1-x} + \frac{\zeta_3(x)}{1-x} + \frac{1}{2}\frac{\log x \cdot \log^2(1-x)}{1-x}$$

$$= \frac{1}{x}\left\{\zeta_3(x)-\zeta_3(1-x)+\zeta_2(1)\log(1-x)-\frac{1}{2}\log x\cdot\log^2(1-x)+\zeta_3(1)-\zeta_2(x)\log(1-x)\right\}$$

$$\overset{(6.79)}{=} \frac{1}{x}G_{(2,1)}(x) = G'_{(3,1)}(x) \implies A(x) = G_{(3,1)}(x) + C, \quad A(0) = 0 = G_{(3,1)}(0) + C = C. \tag{6.98}$$

This leads to the proof of formula (6.97). □

Some well-known formulas for $\zeta_k(\frac{1}{2}) = \sum_{n=1}^{\infty}\frac{1}{2^n n^k}$ $(k = 1,2,3)$ are

$$\begin{aligned} \zeta_1\left(\frac{1}{2}\right) &= \log 2, \\ \zeta_2\left(\frac{1}{2}\right) &= -\frac{1}{2}\log^2 2 + \frac{1}{2}\zeta_2(1), \\ \zeta_3\left(\frac{1}{2}\right) &= \frac{1}{6}\log^3 2 - \frac{1}{2}\zeta_2(1)\log 2 + \frac{7}{8}\zeta_3(1). \end{aligned} \tag{6.99}$$

The second formula in (6.99) is also given in the book by *Legendre* [6]. The notation $\zeta_1(\frac{1}{2})$ is due to *Legendre*, where $\zeta_k(x) = \sum_{n=1}^{\infty}\frac{x^n}{n^k}$, in contrast to the notation commonly used today, $Li_k(x) = \sum_{n=1}^{\infty}\frac{x^n}{n^k}$ $(|x| \leqslant 1)$. This function is called the polylogarithm, and for $k = 2$, the dilogarithm. In the monograph, the authors chose *Legendre*'s notation. The third formula in (6.99) follows from (6.77) for $x = \frac{1}{2}$, which can be found in [65]. Further, we have

$$G_{(1,1)}(-1) = -\zeta_2\left(\frac{1}{2}\right), \tag{6.100}$$

$$G_{(1,1)}\left(\frac{1}{2}\right) = -\zeta_2(-1), \tag{6.101}$$

$$G_{(1,2)}(-1) = -\zeta_3(1) + \frac{1}{2}\zeta_2(1)\log 2, \tag{6.102}$$

$$G_{(2,1)}(-1) = -\frac{5}{8}\zeta_3(1), \tag{6.103}$$

$$G_{(1,2)}\left(\frac{1}{2}\right) = -G_{(2,1)}(-1), \tag{6.104}$$

$$G_{(2,1)}\left(\frac{1}{2}\right) = -G_{(1,2)}(-1). \tag{6.105}$$

Equalities (5.31) and (6.101) are derived directly from (6.148) and (5.30); (6.103) and (6.102) are in fact (4.A84) and (4.A85); (6.104) and (6.105) follow from (6.79), (6.81), (6.99), (6.102), and (6.103). Further, we have

$$G_{(1,3)}\left(\frac{1}{2}\right) = \zeta_4\left(\frac{1}{2}\right) + \frac{1}{24}\log^4 2 - \frac{1}{4}\zeta_2(1)\log^2 2 + \frac{7}{8}\zeta_3(1)\log 2 - \frac{5}{16}\zeta_4(1), \tag{6.106}$$

$$G_{(2,2)}\left(\frac{1}{2}\right) = \zeta_4\left(\frac{1}{2}\right) + \frac{1}{24}\log^4 2 - \frac{1}{4}\zeta_2(1)\log^2 2 + \frac{1}{4}\zeta_3(1)\log 2 + \frac{1}{16}\zeta_4(1), \tag{6.107}$$

$$G_{(3,1)}\left(\frac{1}{2}\right) = \zeta_4\left(\frac{1}{2}\right) + \frac{1}{24}\log^4 2 - \frac{1}{8}\zeta_3(1)\log 2 + \frac{1}{8}\zeta_4(1). \tag{6.108}$$

We can conclude that (6.106) stems from (6.94) and (6.99); (6.107) follows from (6.96) and (6.99), whereas (6.108) is a consequence of (6.97) and (6.99). The following equations are also true:

$$\int_0^{\frac{1}{2}} \frac{\log(1-t)}{t}\,\partial t = -\zeta_2\left(\frac{1}{2}\right), \tag{6.109}$$

$$\frac{1}{2}\int_0^{\frac{1}{2}} \frac{\log^2(1-t)}{t}\,\partial t = -\frac{1}{6}\log^3 2 + \frac{1}{8}\zeta_3(1), \tag{6.110}$$

$$\frac{1}{6}\int_0^{\frac{1}{2}} \frac{\log^3(1-t)}{t}\,\partial t = \zeta_4\left(\frac{1}{2}\right) + \frac{1}{12}\log^4 2 - \frac{1}{4}\zeta_2(1)\log^2 2 + \frac{7}{8}\zeta_3(1)\log 2 - \zeta_4(1), \tag{6.111}$$

$$\int_0^{1} \frac{\log(1+t)}{t}\,\partial t = -\zeta_2(-1), \tag{6.112}$$

$$\frac{1}{2}\int_0^{1} \frac{\log^2(1+t)}{t}\,\partial t = \frac{1}{8}\zeta_3(1), \tag{6.113}$$

$$\frac{1}{6}\int_0^{1} \frac{\log^3(1+t)}{t}\,\partial t = -\zeta_4\left(\frac{1}{2}\right) - \frac{1}{24}\log^4 2 + \frac{1}{4}\zeta_2(1)\log^2 2 - \frac{7}{8}\zeta_3(1)\log 2 + \zeta_4(1), \tag{6.114}$$

$$\frac{1}{6}\int_0^{1} \frac{\log^3(1+t)}{t}\,\partial t = \frac{1}{4}G_{(2,2)}(-1) + \frac{13}{64}\zeta_4(1), \tag{6.115}$$

$$G_{(2,2)}(-1) = -4\zeta_4\left(\frac{1}{2}\right) - \frac{1}{6}\log^4 2 + \zeta_2(1)\log^2 2 - \frac{7}{2}\zeta_3(1)\log 2 + \frac{51}{16}\zeta_4(1). \tag{6.116}$$

From these relations we can conclude that (6.109) follows from (6.85), (6.110) from (6.86) and (6.99), (6.111) from (6.89) and (6.99), (6.112) from (6.91), (6.113) from (6.92) and (6.99), and (6.114) from (6.93) and (6.99).

On the basis of (6.90), from (6.99) for $x = -1$ and from (6.93) for $x = 1$ we can derive (6.116). Formula (6.115) follows from (6.114) and (6.116).

We also have the following relations:

$$G_{(1,3)}\left(\frac{1}{2}\right) = -\frac{1}{4}G_{(2,2)}(-1) + \frac{31}{64}\zeta_4(1), \tag{6.117}$$

$$G_{(2,2)}\left(\frac{1}{2}\right) = -\frac{1}{4}G_{(2,2)}(-1) - \frac{5}{8}\zeta_3(1)\log 2 + \frac{55}{64}\zeta_4(1), \tag{6.118}$$

$$G_{(3,1)}\left(\frac{1}{2}\right) = -\frac{1}{4}G_{(2,2)}(-1) + \frac{1}{4}\zeta_2(1)\log^2 2 - \zeta_3(1)\log 2 + \frac{59}{64}\zeta_4(1). \tag{6.119}$$

They follow from (6.106)–(6.108) and (6.116). Let

$$G_{(a,b,c)}(x) = \sum_{k=1}^{\infty}\frac{x^k}{k^a}\sum_{m=1}^{k}\frac{1}{m^b}\sum_{n=1}^{m}\frac{1}{n^c} \quad (a,b,c \in \mathbb{N};\ |x| \leqslant 1). \tag{6.120}$$

Before deriving proper formulas for the functions of this form, we will deal with the question of convergence of this series. To this end, we will prove that the series

$$G_{(\lambda,\lambda_1,\lambda_2,\dots,\lambda_r)}(z) = \sum_{k=1}^{\infty}\frac{x^k}{k^\lambda}\sum_{k_1=1}^{k}\frac{1}{k_1^{\lambda_1}}\sum_{k_2=1}^{k_1}\frac{1}{k_2^{\lambda_2}}\cdots\sum_{k_r=1}^{k_{r-1}}\frac{1}{k_r^{\lambda_r}} \quad (\lambda,\lambda_1,\lambda_2,\dots,\lambda_r \in \mathbb{N}) \tag{6.121}$$

has the following properties:

(1) Its radius of convergence is $R = 1$.
(2) For $\lambda \geqslant 2$, the series converges for $|z| \leqslant 1$.
(3) For $\lambda = \lambda_1 = \lambda_2 = \cdots = \lambda_r = 1$, at the point $z = -1$ the series converges.
(4) For $r = 2$, $(\lambda,\lambda_1,\lambda_2) = (1,2,1)$ and $(\lambda,\lambda_1,\lambda_2) = (1,1,2)$, at the point $z = -1$ the series converges.

Proof. Let

$$a_k = \frac{1}{k^\lambda}\sum_{k_1=1}^{k}\frac{1}{k_1^{\lambda_1}}\sum_{k_2=1}^{k}\frac{1}{k_2^{\lambda_2}}\cdots\sum_{k_r=1}^{k_{r-1}}\frac{1}{k_r^{\lambda_r}} = \frac{1}{k_r^\lambda}\mu(\lambda_1,\lambda_2,\dots,\lambda_r,k). \tag{6.122}$$

Then

$$\begin{aligned}
\lim_{k\to\infty} a_{k+1} &= \lim_{k\to\infty}\frac{\mu(\lambda_1,\lambda_2,\dots,\lambda_r,k+1)}{(k+1)^\lambda}\\
&= \lim_{k\to\infty}\frac{\mu(\lambda_1,\lambda_2,\dots,\lambda_r,k+1)-\mu(\lambda_1,\lambda_2,\dots,\lambda_r,k)}{(k+1)^\lambda - k^\lambda}\\
&= \lim_{k\to\infty}\frac{\frac{1}{(k+1)^{\lambda_1}}\mu(\lambda_2,\dots,\lambda_r,k+1)}{(k+1)^\lambda - k^\lambda} = \lim_{k\to\infty}\frac{(1+\frac{1}{k})^{\lambda-1}\mu(\lambda_2,\dots,\lambda_r,k+1)}{\frac{(1+\frac{1}{k})^\lambda-1}{\frac{1}{k}}(k+1)^{\lambda+\lambda_1-1}}\\
&= \frac{1}{\lambda}\lim_{k\to\infty}\frac{\mu(\lambda_2,\dots,\lambda_r,k+1)}{(k+1)^{\lambda+\lambda_1-1}} = \cdots\\
&= \frac{1}{\lambda}\frac{1}{\lambda+\lambda_1-1}\frac{1}{\lambda+\lambda_1+\lambda_2-2}\cdots\frac{1}{\lambda+\lambda_1+\cdots+\lambda_{r-2}-(r-2)}\\
&\quad\cdot\lim_{k\to\infty}\frac{\frac{1}{(k+1)^{\lambda_r}}}{(k+1)^{\lambda+\lambda_1+\cdots+\lambda_{r-1}-(r-1)} - k^{\lambda+\lambda_1+\cdots+\lambda_{r-1}-(r-1)}}\\
&= \frac{1}{\lambda}\frac{1}{\lambda+\lambda_1-1}\frac{1}{\lambda+\lambda_1+\lambda_2-2}\cdots\frac{1}{\lambda+\lambda_1+\cdots+\lambda_{r-2}-(r-2)}
\end{aligned}$$

$$\cdot \frac{1}{\lambda+\lambda_1+\cdots+\lambda_{r-1}-(r-1)} \lim_{k\to\infty} \frac{1}{(k+1)^{\lambda+\lambda_1+\cdots+\lambda_r-r-1}} = 0. \tag{6.123}$$

(1)

$$\begin{aligned}
\lim_{k\to\infty} \frac{a_k}{a_{k+1}} &= \lim_{k\to\infty} \left(1+\frac{1}{k}\right)^{\lambda} \cdot \frac{\mu(\lambda_1,\lambda_2,\ldots,\lambda_r,k)}{\mu(\lambda_1,\lambda_2,\ldots,\lambda_r,k+1)} \\
&= \lim_{k\to\infty} \frac{\mu(\lambda_1,\lambda_2,\ldots,\lambda_r,k)}{\mu(\lambda_1,\lambda_2,\ldots,\lambda_r,k+1)} \\
&= \lim_{k\to\infty} \frac{\mu(\lambda_1,\lambda_2,\ldots,\lambda_r,k)-\mu(\lambda_1,\lambda_2,\ldots,\lambda_r,k-1)}{\mu(\lambda_1,\lambda_2,\ldots,\lambda_r,k+1)-\mu(\lambda_1,\lambda_2,\ldots,\lambda_r,k)} \\
&= \lim_{k\to\infty} \frac{\frac{1}{k^{\lambda_1}}\mu(\lambda_1,\lambda_2,\ldots,\lambda_r,k)}{\frac{1}{(k+1)^{\lambda_1}}\mu(\lambda_1,\lambda_2,\ldots,\lambda_r,k+1)} \\
&= \lim_{k\to\infty} \frac{\mu(\lambda_2,\ldots,\lambda_r,k)}{\mu(\lambda_2,\ldots,\lambda_r,k+1)} = \cdots = \lim_{k\to\infty} \frac{\mu(\lambda_r,k)}{\mu(\lambda_r,k+1)} \\
&= \lim_{k\to\infty} \frac{\mu(\lambda_r,k)-\mu(\lambda_r,k-1)}{\mu(\lambda_r,k+1)-\mu(\lambda_r,k)} \\
&= \lim_{k\to\infty} \frac{\frac{1}{k^{\lambda_r}}}{\frac{1}{(k+1)^{\lambda_r}}} = 1 \Longrightarrow R = 1.
\end{aligned} \tag{6.124}$$

(2) The property follows from (6.71) and (5.199).

(3) It is already proved that $\lim_{k\to\infty} a_k = 0$. To apply the *Leibniz* criterion, we need to prove that $a_k > a_{k+1}$ $(k \geqslant 1)$:

$$\begin{aligned}
a_k > a_{k+1} &\Longleftrightarrow \frac{1}{k}\mu(r,k) > \frac{1}{k+1}\mu(r,k+1) \\
&\Longleftrightarrow \frac{1}{k}\left\{\mu(r,k+1) - \frac{1}{k+1}\mu(r-1,k+1)\right\} > \frac{1}{k+1}\mu(r,k+1) \\
&\Longleftrightarrow \mu(r,k+1) > \mu(r-1,k+1),
\end{aligned} \tag{6.125}$$

$$\mu(1,k+1) > 1 = \mu(0,k+1), (r=1), \tag{6.126}$$

$$r \geqslant 2, \quad \mu(r,k+1) = \sum_{k_1=1}^{k+1} \frac{1}{k_1} \cdots \sum_{k_{r-1}=1}^{k_{r-2}} \frac{1}{k_{r-1}} \sum_{k_r=1}^{k_{r-1}} \frac{1}{k_r} > \mu(r-1,k+1). \tag{6.127}$$

It follows that the last sum is greater than 1.

(4) The case $(\lambda,\lambda_1,\lambda_2) = (2,1,1)$ follows from (6.72).

$$(\lambda,\lambda_1,\lambda_2) = (1,2,1), \quad a_k = \frac{1}{k}\sum_{k_1=1}^{k} \frac{1}{k_1^2} \sum_{k_2=1}^{k_1} \frac{1}{k_2}. \tag{6.128}$$

$$a_k > a_{k+1} \Longrightarrow \sum_{k_1=1}^{k+1} \frac{1}{k_1^2} \sum_{k_2=1}^{k_1} \frac{1}{k_2} > \frac{1}{k+1} \sum_{k_1=1}^{k+1} \frac{1}{k_1}; \quad \sum_{k_2=1}^{k_1} \frac{1}{k_2} > \frac{k_1}{k+1}, \tag{6.129}$$

$$(\lambda, \lambda_1, \lambda_2) = (1,1,1), \quad a_k = \frac{1}{k} \sum_{k_1=1}^{k} \frac{1}{k_1} \sum_{k_2=1}^{k_1} \frac{1}{k_2^2}, \tag{6.130}$$

$$a_k > a_{k+1} \implies \sum_{k_1=1}^{k+1} \frac{1}{k_4} \sum_{k_2=1}^{k_1} \frac{1}{k_2^2} > \sum_{k_1=1}^{k+1} \frac{1}{k_1^2}; \quad \sum_{k_2=1}^{k_1} \frac{1}{k_2^2} > \frac{k_1}{k_1^2} = \frac{1}{k_1}. \tag{6.131}$$

□

In this way the issue of convergence of all the series presented in this chapter until formula (6.166) is solved. According to (5.91), we know that

$$\sum_{m=1}^{k-2} \sum_{n=1}^{k-m-1} \frac{1}{mn(k-m-n)} = \frac{3}{k} + \frac{6}{k} \sum_{r=3}^{k-1} \frac{1}{r} \sum_{s=1}^{r-1} \frac{1}{s} \quad (k \geqslant 3), \tag{6.132}$$

$$\frac{3}{k} + \frac{6}{k} \sum_{r=3}^{k-1} \frac{1}{r} \sum_{s=1}^{r-1} \frac{1}{s} = \frac{6}{k} \sum_{r=1}^{k} \frac{1}{r} \sum_{s=1}^{r} \frac{1}{s} - \frac{6}{k} \sum_{r=1}^{k} \frac{1}{r^2} - \frac{6}{k^2} \sum_{r=1}^{k} \frac{1}{r} + \frac{6}{k^3} = t_k \quad (k \geqslant 3),\ t_1 = t_2 = 0. \tag{6.133}$$

It follows that

$$\frac{1}{6} \zeta_1^3(x) = G_{(1,1,1)}(x) - G_{(1,2)}(x) - G_{(2,1)}(x) + \zeta_3(x) \quad (-1 \leqslant x < 1). \tag{6.134}$$

By the last relation and (6.78) and (6.80) we get

$$G_{(1,1,1)}(x) = -\zeta_3\left(\frac{x}{x-1}\right) \quad \left(-1 \leqslant x \leqslant \frac{1}{2}\right), \tag{6.135}$$

and further,

$$G_{(1,1,1)}(-1) = -\zeta_3\left(\frac{1}{2}\right), \tag{6.136}$$

$$G_{(1,1,1)}\left(\frac{1}{2}\right) = -\zeta_3(-1). \tag{6.137}$$

From the relations established in this way and from (6.135) it follows that

$$G_{(2,1,1)}(x) = -\zeta_4\left(\frac{x}{x-1}\right) - \zeta_3\left(\frac{x}{x-1}\right) \log(1-x) + \frac{1}{2} \zeta_2^2(x) + \frac{1}{2} \zeta_2(x) \log^2(1-x) + \frac{1}{8} \log^4(1-x) \quad \left(-1 \leqslant x \leqslant \tfrac{1}{2}\right). \tag{6.138}$$

On the other hand, (6.138) for $x = -1$, according to (6.99) and (6.106), leads to

$$G_{(2,1,1)}(-1) = -G_{(1,3)}\left(\frac{1}{2}\right). \tag{6.139}$$

From (6.138) for $x = \frac{1}{2}$ and (6.94) for $x = -1$ we obtain

$$G_{(2,1,1)}\left(\frac{1}{2}\right) = -G_{(1,3)}(-1). \tag{6.140}$$

Now we are in the position to prove the formula

$$\begin{aligned} G_{(1,2,1)}(x) &= 3\zeta_4\left(\frac{x}{x-1}\right) + 2\zeta_3\left(\frac{x}{x-1}\right)\log(1-x) + G_{(2,2)}(x) + 3\zeta_4(x) \\ &\quad - \zeta_2^2(x) - \frac{1}{2}\zeta_2(x)\log^2(1-x) - 2\zeta_3(x)\log(1-x) \\ &\quad - \frac{5}{24}\log^4(1-x) \quad \left(-1 \leqslant x \leqslant \frac{1}{2}\right). \end{aligned} \tag{6.141}$$

Proof. We have

$$\begin{aligned} G_{(1,2,1)}(x) &= \sum_{k=1}^{\infty}\frac{x^k}{k}\sum_{m=1}^{k}\frac{1}{m^2}\sum_{n=1}^{m}\frac{1}{n} = \sum_{m=1}^{\infty}\frac{1}{m^2}\sum_{n=1}^{m}\frac{1}{n}\sum_{k=m}^{\infty}\frac{x^k}{k} \\ &= G_{(3,1)}(x) + G_{(2,1)}(1)\zeta_1(x) - \sum_{m=1}^{\infty}\frac{1}{m^2}\sum_{n=1}^{m}\frac{1}{n}\sum_{k=m}^{\infty}\frac{x^k}{k}. \end{aligned} \tag{6.142}$$

From this we can conclude that

$$\begin{aligned} G'_{(1,2,1)}(x) &= \frac{1}{x}G_{(2,1)}(x) + \frac{G_{(2,1)}(1)}{1-x} - \frac{G_{(2,1)}(1)}{1-x} + \frac{G_{(2,1)}(x)}{1-x} \\ &= \left(\frac{1}{x} + \frac{1}{1-x}\right)G_{(2,1)}(x) \\ &\overset{(6.78)}{=} \left(\frac{1}{x} + \frac{1}{1-x}\right)\left\{\zeta_3\left(\frac{x}{x-1}\right) + 2\zeta_3(x) - \frac{1}{6}\log^3(1-x) - \zeta_2(x)\log(1-x)\right\} \\ &= \left\{\zeta_4\left(\frac{x}{x-1}\right) + 2\zeta_4(x) + \frac{1}{24}\log^4(1-x)\right. \\ &\quad \left. + \frac{1}{2}\zeta_2(x)\log^2(1-x) - 2\zeta_3(x)\log(1-x) - \frac{1}{2}\zeta_2^2(x)\right\}' \\ &\quad - 2\frac{1}{6}\frac{\zeta_1^3(x)}{x} - 2\frac{1}{6}\frac{\zeta_1^3(x)}{x} \overset{(6.134)}{=} -\frac{2}{x}\cdot\{G_{(1,1,1)}(x) - G_{(1,2)}(x) - G_{(2,1)}(x) + \zeta_3(x)\} \\ &= \{-2G_{(2,1,1)}(x) + 2G_{(2,2)}(x) + 2G_{(3,1)}(x) - 2\zeta_4(x)\}' \\ &\overset{(6.95)}{=} \left\{-2G_{(2,1,1)}(x) + G_{(2,2)}(x) + \zeta_4(x) + \frac{1}{2}\zeta_2^2(x)\right\}' \\ &\overset{(6.138)}{=} \left\{2\zeta_4\left(\frac{x}{x-1}\right) + 2\zeta_3\left(\frac{x}{x-1}\right)\log(1-x) - \frac{1}{2}\zeta_2^2(x)\right. \\ &\quad \left. - \zeta_2(x)\log^2(1-x) - \frac{1}{4}\log^4(1-x) + \zeta_4(x) + G_{(2,2)}(x)\right\}'. \end{aligned} \tag{6.143}$$

It follows that $G_{(1,2,1)}(x) = B(x) + C$, where $B(x)$ is the expression on the right-hand side of formula (6.141). For $x = 0$, we have $C = 0$, and so (6.141) is proved. □

Now by (6.141), (6.116), (6.99), and (6.107) we obtain

$$G_{(1,2,1)}(-1) = -G_{(2,2)}\left(\frac{1}{2}\right). \tag{6.144}$$

From the same set of formulas it follows that

$$G_{(1,2,1)}\left(\frac{1}{2}\right) = -G_{(2,2)}(-1). \tag{6.145}$$

Similarly to the proof of (6.141), we arrive at

$$\begin{aligned} G_{(1,1,2)}(x) = &-3\zeta_4\left(\frac{x}{x-1}\right) - \zeta_3\left(\frac{x}{x-1}\right)\log(1-x) - \frac{1}{2}G_{(2,2)}(x) - \frac{3}{2}\zeta_4(x) + \frac{1}{4}\zeta_2^2(x) \\ &+ \zeta_3(x)\log(1-x) + \frac{1}{24}\log^4(1-x) \quad (-1 \leqslant x \leqslant \tfrac{1}{2}). \end{aligned} \tag{6.146}$$

Formulas (6.146), (6.116), (6.99), and (6.108) together give

$$G_{(1,1,2)}(-1) = -G_{(3,1)}\left(\frac{1}{2}\right), \tag{6.147}$$

whereas (6.146), (6.116), (6.107), (6.99), (6.95), and (6.108) lead to

$$G_{(1,1,2)}\left(\frac{1}{2}\right) = -G_{(3,1)}(-1). \tag{6.148}$$

Further, by (6.72) and (6.76) we have

$$G_{(1,1)}(x) = -\zeta_2\left(\frac{x}{x-1}\right), \tag{6.149}$$

whereas (6.138), (6.94), and (6.72) imply the equality

$$G_{(2,1,1)}(x) = -G_{(1,3)}\left(\frac{x}{x-1}\right). \tag{6.150}$$

Starting from (6.90), we obtain

$$\begin{aligned} &\frac{1}{6}\int_0^{\frac{x}{x-1}} \frac{\log^3(1-t)}{t}\,\partial t \\ &= \frac{1}{2}G_{(2,2)}\left(\frac{x}{x-1}\right) + \zeta_4(x) - \zeta_3(x)\log(1-x) + \frac{1}{2}\zeta_4\left(\frac{x}{x-1}\right) - \frac{1}{2}\zeta_2\left(\frac{x}{x-1}\right)\log^2(1-x) \end{aligned}$$

$$-\frac{1}{4}\zeta_2^2\left(\frac{x}{x-1}\right)-\frac{1}{8}\log^4(1-x)\quad(-1\leqslant x\leqslant\tfrac{1}{2}),\tag{6.151}$$

$$\frac{1}{6}\int_0^{\frac{x}{x-1}}\frac{\log^3(1-t)}{t}\,\partial t=-\frac{1}{6}\int_0^x\frac{\log^3(1-s)}{s}\,\partial s+\frac{1}{24}\log^4(1-x)\quad(-1\leqslant x\leqslant\tfrac{1}{2}).\tag{6.152}$$

By (6.152), (6.90), (6.151), and (6.72) we get

$$\begin{aligned}G_{(2,2)}\left(\frac{x}{x-1}\right)&=-3\zeta_4\left(\frac{x}{x-1}\right)-2\zeta_3\left(\frac{x}{x-1}\right)\log(1-x)-G_{(2,2)}(x)-3\zeta_4(x)+\zeta_2^2(x)\\&\quad+\frac{1}{2}\zeta_2(x)\log^2(1-x)+2\zeta_3(x)\log(1-x)+\frac{5}{24}\log^4(1-x)\quad(-1\leqslant x\leqslant\tfrac{1}{2}).\end{aligned}\tag{6.153}$$

Combining (6.141) and (6.153), we arrive at

$$G_{(1,2,1)}(x)=-G_{(2,2)}\left(\frac{x}{x-1}\right).\tag{6.154}$$

Making the replacement $x\to\frac{x}{x-1}$ in (6.95), from (6.153) and (6.72) we obtain

$$\begin{aligned}G_{(3,1)}\left(\frac{x}{x-1}\right)&=3\zeta_4\left(\frac{x}{x-1}\right)+3\zeta_3\left(\frac{x}{x-1}\right)\log(1-x)+\frac{1}{2}G_{(2,2)}(x)+\frac{3}{2}\zeta_4(x)\\&\quad-\frac{1}{4}\zeta_2^2(x)-\zeta_3(x)\log(1-x)-\frac{1}{24}\log^4(1-x)\quad(-1\leqslant x\leqslant\tfrac{1}{2}).\end{aligned}\tag{6.155}$$

From (6.146) and the last relation it follows that

$$G_{(1,1,2)}(x)=-G_{(3,1)}\left(\frac{x}{x-1}\right)\quad(-1\leqslant x\leqslant\tfrac{1}{2}),\tag{6.156}$$

and if in (6.78) we replace $x\to\frac{x}{x-1}$, then with the help of (6.80) we arrive at

$$G_{(1,2)}(x)=-G_{(2,1)}\left(\frac{x}{x-1}\right)\quad(-1\leqslant x\leqslant\tfrac{1}{2}).\tag{6.157}$$

Similarly, by (6.157), replacing $x\to\frac{x}{x-1}$, we get

$$G_{(2,1)}(x)=-G_{(1,2)}\left(\frac{x}{x-1}\right)\quad(-1\leqslant x\leqslant\tfrac{1}{2}).\tag{6.158}$$

We will further prove the formula

$$G_{(1,1,\dots,1)_r}(x)=-\zeta_r\left(\frac{x}{x-1}\right)\quad(r\geqslant 2,\ -1\leqslant x\leqslant\tfrac{1}{2})\tag{6.159}$$

(r in the index indicating the number of 1s in the expression).

Proof. We will use mathematical induction. The formula is true for $r = 2$ and $r = 3$, as follows from (6.149) and (6.135). We have

$$\begin{aligned}(G_{(1,1,\dots,1)_{r+1}}(x))' &= \left(\sum_{k_1=1}^{\infty}\frac{x^{k_1}}{k_1}\sum_{k_2=1}^{k_1}\frac{1}{k_2}\cdots\sum_{k_{r+1}=1}^{k_r}\frac{1}{k_{r+1}}\right)' \\ &= \frac{1}{x}\sum_{k_1=1}^{\infty}x^{k_1}\sum_{k_2=1}^{k_1}\frac{1}{k_2}f(k_2) = \frac{1}{x}\sum_{k_2=1}^{\infty}\frac{1}{k_2}f(k_2)\sum_{k_1=k_2}^{k_1}x^{k_1} \\ &= \frac{1}{x(1-x)}\sum_{k_2=1}^{\infty}\frac{x^{k_2}}{k_2}f(k_2) = \frac{1}{x(1-x)}G_{(1,1,\dots,1)_r}(x) \\ &= -\frac{1}{x(1-x)}\zeta_r\left(\frac{x}{x-1}\right) = \left(-\zeta_{r+1}\left(\frac{x}{x-1}\right)\right)'. \end{aligned} \tag{6.160}$$

□

We will also prove the following formula:

$$\zeta_{r+1}\left(\frac{x}{x-1}\right) = \sum_{n=1}^{\infty}\frac{x^n}{n}\sum_{k=1}^{n}(-1)^k\binom{n}{k}\frac{1}{k^r} \quad (-1 \leqslant x \leqslant \tfrac{1}{2}). \tag{6.161}$$

Proof. We will use the following equalities:

$$\sum_{k=m}^{n}\frac{1}{k}\binom{k}{m} = \frac{1}{m}\binom{m}{n}, \tag{6.162}$$

$$b_{r,n} = \sum_{k_1=1}^{n}\frac{1}{k_1}\sum_{k_2=1}^{k_1}\frac{1}{k_2}\cdots\sum_{k_r=1}^{k_{r-1}}\frac{1}{k_r} = \sum_{k=1}^{n}(-1)^{k-1}\binom{n}{k}\frac{1}{k^r}. \tag{6.163}$$

Formula (6.162) is simply proved by induction on n. Formula (6.163) for $r = 1$ is in fact the well-known formula

$$\sum_{k=1}^{n}\frac{1}{k} = \sum_{k=1}^{n}(-1)^{k-1}\binom{n}{k}\frac{1}{k}. \tag{6.164}$$

From (6.164) it follows that

$$\begin{aligned} b_{r+1,n} &\overset{(5.94)}{=} \sum_{k_1=1}^{n}\frac{1}{k_1}\sum_{k_2=1}^{k_1}(-1)^{k_2-1}\binom{k_1}{k_2}\frac{1}{k_2^r} = \sum_{k_2=1}^{n}\frac{1}{k_r}(-1)^{k_2-1}\frac{1}{k^r}\sum_{k_1=k_2}^{n}\frac{1}{k_1}\binom{k_1}{k_2} \\ &\overset{(5.93)}{=} \sum_{k_2=1}^{n}(-1)^{k_2-1}\binom{n}{k_2}\frac{1}{k_2^{r+1}}. \end{aligned} \tag{6.165}$$

So formula (6.163) is proved. Formula (6.161) follows from (6.159) and (6.163). □

Let us also prove that the series of the form

$$\sum_{k=1}^{n}\frac{z^{k-1}}{k}\sum_{k_1=1}^{k}\frac{z^{k_1-1}}{k_1}\sum_{k_2=1}^{k_1}\frac{z^{k_2-1}}{k_2}\cdots\sum_{k_r=1}^{k_{r-1}}\frac{z^{k_r-1}}{k_r}=P_r(z) \tag{6.166}$$

absolutely converges for $|z|<1$ and converges for $z=-1$.

Proof. Let $a_k^{(r)}(z)$ be the members of the series $P_r(z)$. Then for $|z|<1$,

$$\left|a_k^{(r)}(z)\right|\leqslant a_k^{(r)}(|z|)\leqslant\frac{|z|^{k-1}}{k}\sum_{k_1=1}^{k}\frac{1}{k_1}\sum_{k_2=1}^{k_1}\frac{1}{k_2}\cdots\sum_{k_r=1}^{k_{r-1}}\frac{1}{k_r}. \tag{6.167}$$

Then for $|z|\leqslant|z_0|<1$, by (6.121) (property 1) and by *Weiestrass*' criterion it follows that the series $P_r(z)$ for $|z|\leqslant|z_0|<1$ absolutely converges, because of the arbitrariness of z_0, in the region $|z|<1$. It remains to prove the convergence of the series $P_r(z)$ for $z=-1$. Based on the results obtained by *Lyashko* [193] (p. 34. problem 65), we have that the series $\sum_{k=1}^{\infty}b_k$ converges if the following conditions are satisfied:

1. $\lim_{k\to\infty}b_k=0$.
2. The series $\sum_{n=1}^{\infty}B_n$, where $B_n=\sum_{j=q_n}^{q_{n+1}-1}$, $1=q_1<q_2<\cdots$, converges if the number of members in each B_n is finite and the order of the members of b_j does not change.

In our case,

$$b_k=a_k^{(r)}(-1)\leqslant a_k^{(r)}=\frac{(-1)^{k-1}}{k}\sum_{k_1=1}^{k}\frac{(-1)^{k_1-1}}{k_1}\sum_{k_2=1}^{k_1}\frac{(-1)^{k_2-1}}{k_2}\cdots\sum_{k_r=1}^{k_{r-1}}\frac{(-1)^{k_r-1}}{k_r}=\frac{(-1)^{k-1}}{k}\nu(r,k), \tag{6.168}$$

$$q_l=2l-1\quad(l=1,2,\dots),\quad s_k=\sum_{n=1}^{k}B_n,\quad B_n=a_{2n-1}^{(r)}+a_{2n}^{(r)}. \tag{6.169}$$

1.

$$\begin{aligned}0\leqslant\left|\lim_{k\to\infty}b_k\right|&=\left|\lim_{k\to\infty}\frac{(-1)^{k-1}}{k}\nu(r,k)\right|\leqslant\lim_{k\to\infty}\frac{|\nu(r,k)|}{k}\\&\leqslant\lim_{k\to\infty}\frac{\mu(r,k)}{k}\overset{(6.121)}{=}0\implies\lim_{k\to\infty}b_k=0.\end{aligned} \tag{6.170}$$

2.1.

$$d_t^{(s)}=\frac{1}{2t-1}\nu(s,2t-1)-\frac{1}{2t}\nu(s,2t)>0. \tag{6.171}$$

2.2.

$$\nu(s,t)>0\quad(s,t\in\mathbb{N}). \tag{6.172}$$

2.3.

$$\nu(s,t)<c_s=\frac{1}{2\sqrt{2}}\{(1+\sqrt{2})^{s+1}-(1-\sqrt{2})^{s+1}\}. \tag{6.173}$$

2.4.

$$0<s_k<c_r\log 2\quad(\forall_k),\ s_{k+1}>s_k. \tag{6.174}$$

2.1. Induction by s:

$$d_t^{(1)} = \frac{1}{(2t-1)2t}\left\{\sum_{l=1}^{2t}\frac{(-1)^{l-1}}{l}+1\right\} > 0 \quad (\forall t), \tag{6.175}$$

$$d_t^{(s+1)} = \frac{1}{(2t-1)2t}\{\nu(s+1,2t)+\nu(s,2t-1)\}, \tag{6.176}$$

$$d_t^{(2)} = \frac{1}{(2t-1)2t}\{\nu(2,2t)+\nu(1,2t-1)\}, \tag{6.177}$$

$$\nu(2,2t) = \sum_{l=1}^{t} d_l^{(1)} > 0, \quad \nu(1,2t-1) > 0 \implies d_t^{(2)} > 0 \quad (\forall t),$$

$$d_t^{(3)} = \frac{1}{(2t-1)2t}\{\nu(3,2t)+\nu(2,2t-1)\}, \quad \nu(3,2t) = \sum_{l=1}^{t} d_l^{(2)} > 0, \tag{6.178}$$

$$\nu(2,2t-1) = \nu(2,2t) + \frac{1}{2t}\nu(1,2t) > 0 \implies d_t^{(3)} > 0 \quad (\forall t). \tag{6.179}$$

It follows that

$$d_t^{(s)} > 0 \quad (s = 1,2,3;\ \forall t), \tag{6.180}$$

$$\nu(s+1,2t) = \sum_{l=1}^{t} d_l^{(s)}, \quad \nu(s,2t-1) = \nu(s,2t) + \frac{1}{2t}\nu(s-1,2t), \tag{6.181}$$

$$d_t^{(s+1)} = \frac{1}{(2t-1)2t}\left\{\sum_{l=1}^{t} d_l^{(s)} + \sum_{l=1}^{t} d_l^{(s-1)} + \frac{1}{2t}\sum_{l=1}^{t} d_l^{(s-2)}\right\} \quad (s \geqslant 3,\ \forall t). \tag{6.182}$$

By the induction assumption we conclude that $d_t^{(s+1)} > 0$, which leads to the proof of Proposition 2.1.

2.2. Induction by s:

$$\nu(1,t) > 0 \quad (\forall t), \tag{6.183}$$

$$\nu(2,t) > 0 \quad (\forall t), \quad \text{(according to 2.1.)}, \tag{6.184}$$

$$\nu(s+1,2t) = \sum_{l=1}^{t} d_l^{(s)} = \sum_{l=1}^{t}\frac{1}{(2l-1)2l}\{\nu(s,2l)+\nu(s-1,2l-1)\} \quad (s \geqslant 2, \forall t), \tag{6.185}$$

$$\nu(s+1,2t-1) = \nu(s+1,2t) + \frac{1}{2t}\nu(s,2t) \quad (s \geqslant 2,\ \forall t), \tag{6.186}$$

and the required inequality follows.

2.3. Induction by s:

$$\left.\begin{aligned} \nu(1,2t) &< 1 \\ \nu(1,2t-1) = \nu(1,2t) + \tfrac{1}{2t} &< 2 \end{aligned}\right\} \implies \nu(1,t) < 2 = c_1, \tag{6.187}$$

$$\left.\begin{aligned} \nu(2,2t) = \textstyle\sum_{l=1}^{t} \frac{1}{(2l-1)2l}\{\nu(1,2l)+1\} < 3\sum_{l=1}^{t}\frac{1}{(2l-1)2l} < 3 \\ \nu(2,2t-1) = \nu(2,2t) + \tfrac{1}{2t}\nu(1,2t) < 5 \end{aligned}\right\} \implies \nu(2,t) < 5 = c_2. \quad (6.188)$$

From the relation given in 2.2 it follows that

$$\left.\begin{aligned} \nu(s+1,2t) < (c_s + c_{s-1})\cdot \textstyle\sum_{l=1}^{t}\frac{1}{(2l-1)2l} < c_s + c_{s-1} \\ \nu(s+1,2t-1) < 2c_s + c_{s-1} = c_{s+1} \end{aligned}\right\} \implies \nu(s+1,t) < c_{s+1}, \quad (6.189)$$

and the required inequality follows.

2.4.

$$B_n = d_n^{(r)} > 0 \implies s_k = \sum_{n=1}^{k} d_n^{(r)} > 0; \quad s_{k+1} - s_k = d_{k+1}^{(r)} > 0, \quad (6.190)$$

$$\begin{aligned} s_k &= \sum_{t=1}^{k} \frac{1}{(2t-1)2t}\{\nu(r-1,2t) + \nu(r-2,2t-1)\} \\ &< (c_{r-1} + c_{r-2}) \sum_{t=1}^{k} \frac{1}{(2t-1)2t} < c_r \sum_{t=1}^{k} \frac{1}{(2t-1)2t} \\ &< c_r \sum_{t=1}^{\infty} \frac{1}{(2t-1)2t} = c_r \log 2 \quad (\forall k). \end{aligned} \quad (6.191)$$

The sequence of the partial sums of $\sum_{n=1}^{\infty} B_n$ is increasing and bounded from above, so it has a finite limit, i. e., the series $\sum_{n=1}^{\infty} B_n$ converges. In addition to the other evidence above, this proves the convergence of the series $P_r(z)$ for $z = -1$. □

Let us introduce the following notation:

$$\begin{aligned} F_r(x) &= \sum_{k_1=1}^{\infty} \frac{x^{k_1}}{k_1} \sum_{k_2=1}^{k_1} \frac{x^{k_2}}{k_2} \cdots \sum_{k_r=1}^{k_{r-1}} \frac{x^{k_r}}{k_r} \quad (r \geqslant 2), \\ F_0(x) &= 1, \quad F_1(x) = \zeta_1(x) \quad (-1 \leqslant x < 1). \end{aligned} \quad (6.192)$$

By the established relation it is clear that $F_r(x) = x^r P_{r-1}(x)$, so that $F_r(x)$ absolutely converges for $|x| < 1$ and converges for $x = -1$.

Let us prove the following formulas:

$$F_r'(x) = \frac{1}{x} \sum_{j=0}^{r-1} \zeta_j(x^{j+1}) F_{r-1-j}(x) \quad (-1 \leqslant x < 1), \quad (6.193)$$

$$(r+1)F_{r+1}(x) = \sum_{j=0}^{r} \zeta_{j+1}(x^{j+1}) F_{r-j}(x) \quad (-1 \leqslant x < 1). \quad (6.194)$$

Proof. We have

$$F_r'(x) = \sum_{k_1=1}^{\infty} x^{k_1-1} \sum_{k_2=1}^{k_1} \frac{x^{k_2}}{k_2} \cdots \sum_{k_r=1}^{k_{r-1}} \frac{x^{k_r}}{k_r} + \sum_{p=2}^{r-1} \sum_{k_1=1}^{\infty} \frac{x^{k_1}}{k_1} \cdots \sum_{k_p=1}^{k_{p-1}} x^{k_p-1} \cdots \sum_{k_r=1}^{k_{r-1}} \frac{x^{k_r}}{k_r}$$
$$+ \sum_{k_1=1}^{\infty} \frac{x^{k_1}}{k_1} \sum_{k_2=1}^{k_1} \frac{x^{k_2}}{k_2} \cdots \sum_{k_{r-1}=1}^{k_{r-2}} \frac{x^{k_{r-1}}}{k_{r-1}} \sum_{k_r=1}^{k_{r-1}} x^{k_r-1} = S_1 + \sum_{p=2}^{r-1} S_p + S_r, \tag{6.195}$$

$$S_1 = \sum_{k_2=1}^{\infty} \frac{x^{k_2}}{k_2} \cdots \sum_{k_r=1}^{k_{r-1}} \frac{x^{k_r}}{k_r} \sum_{k_1=k_2}^{\infty} x^{k_1-1} = \frac{1}{x(1-x)} \sum_{k_2=1}^{\infty} \frac{x^{2k_2}}{k_2} \sum_{k_3=1}^{k_2} \frac{x^{k_3}}{k_3} \cdots \sum_{k_r=1}^{k_{r-1}} \frac{x^{k_r}}{k_r}. \tag{6.196}$$

Let T_p be the expression obtained from S_p by the replacement $x^{k_p-1} \to x^{2k_p}$. Then

$$S_1 = \frac{1}{x(1-x)} T_1, \tag{6.197}$$

$$S_p = \sum_{k_1=1}^{\infty} \frac{x^{k_1}}{k_1} \cdots \sum_{k_{p-1}=1}^{k_{p-2}} \frac{x^{k_{p-1}}}{k_{p-1}} \sum_{k_{p+1}=1}^{k_{p-1}} \frac{x^{k_{p+1}}}{k_{p+1}} \cdots \sum_{k_r=1}^{k_{r-1}} \frac{x^{k_r}}{k_r} \sum_{k_p=k_{p+1}}^{k_{p-1}} x^{k_p-1} \quad (2 \leqslant p \leqslant r-1), \tag{6.198}$$

whereas for $p = 2$, there is no expression $\sum_{k_{p-1}=1}^{k_{p-2}} \frac{x^{k_{p-1}}}{k_{p-1}}$, and for $p = r-1$ in the penultimate sum, the upper bound is k_{r-2}. Using the formula

$$\sum_{l=m}^{n} x^{l-1} = \frac{x^{m-1} - x^n}{1-x}, \tag{6.199}$$

we get

$$S_p = \frac{1}{x(1-x)} T_p - \frac{1}{1-x} T_{p-1}, \tag{6.200}$$
$$S_r = \frac{1}{1-x} F_{r-1}(x) - \frac{1}{1-x} T_{r-1}. \tag{6.201}$$

From the relations derived above it follows that

$$F_r'(x) = \frac{1}{x(1-x)} T_1 + \sum_{p=2}^{r-1} \left\{ \frac{1}{x(1-x)} T_p - \frac{1}{1-x} T_{p-1} \right\} + \frac{1}{1-x} F_{r-1}(x) - \frac{1}{1-x} T_{r-1}, \tag{6.202}$$
$$F_r'(x) = \frac{1}{x} \sum_{p=1}^{r-1} T_p + \frac{1}{1-x} F_{r-1}(x). \tag{6.203}$$

Next,

$$T_{r-1} = \sum_{k_1=1}^{\infty} \frac{x^{k_1}}{k_1} \sum_{k_2=1}^{k_1} \frac{x^{k_2}}{k_2} \cdots \sum_{k_{r-2}=1}^{k_{r-3}} \frac{x^{k_{r-2}}}{k_{r-2}} \sum_{k_{r-1}=1}^{k_{r-2}} \frac{x^{2k_{r-1}}}{k_{r-1}}$$

$$= \sum_{k_1=1}^{\infty}\frac{x^{k_1}}{k_1}\sum_{k_2=1}^{k_1}\frac{x^{k_2}}{k_2}\cdots\sum_{k_{r-1}=1}^{k_{r-3}}\frac{x^{2k_{r-1}}}{k_{r-1}}\sum_{k_{r-2}=k_{r-1}}^{k_{r-3}}\frac{x^{k_{r-2}}}{k_{r-2}}$$

$$= \sum_{k_1=1}^{\infty}\frac{x^{k_1}}{k_1}\sum_{k_2=1}^{k_1}\frac{x^{k_2}}{k_2}\cdots\sum_{k_{r-2}=1}^{k_{r-3}}\frac{x^{3k_{r-2}}}{k_{r-2}^2}+\sum_{k_1=1}^{\infty}\frac{x^{k_1}}{k_1}\sum_{k_2=1}^{k_1}\frac{x^{k_2}}{k_2}\cdots\sum_{k_{r-2}=1}^{k_{r-3}}\frac{x^{2k_{r-2}}}{k_{r-2}}\sum_{k_{r-1}=1}^{k_{r-3}}\frac{x^{k_{r-1}}}{k_{r-1}}-T_{r-2}$$

$$= \sum_{k_1=1}^{\infty}\frac{x^{k_1}}{k_1}\sum_{k_2=1}^{k_1}\frac{x^{k_2}}{k_2}\cdots\sum_{k_{r-2}=1}^{k_{r-4}}\frac{x^{3k_{r-2}}}{k_{r-2}^2}\sum_{k_{r-3}=k_{r-2}}^{k_{r-4}}\frac{x^{k_{r-3}}}{k_{r-3}}$$

$$+ \sum_{k_1=1}^{\infty}\frac{x^{k_1}}{k_1}\sum_{k_2=1}^{k_1}\frac{x^{k_2}}{k_2}\cdots\sum_{k_{r-2}=1}^{k_{r-4}}\frac{x^{2k_{r-2}}}{k_{r-2}}\sum_{k_{r-3}=k_{r-2}}^{k_{r-4}}\frac{x^{k_{r-3}}}{k_{r-3}}\sum_{k_{r-1}=1}^{k_{r-3}}\frac{x^{k_{r-1}}}{k_{r-1}}-T_{r-2}$$

$$= \sum_{k_1=1}^{\infty}\frac{x^{k_1}}{k_1}\sum_{k_2=1}^{k_1}\frac{x^{k_2}}{k_2}\cdots\sum_{k_{r-3}=1}^{k_{r-4}}\frac{x^{4k_{r-3}}}{k_{r-3}^3}++\sum_{k_1=1}^{\infty}\frac{x^{k_1}}{k_1}\sum_{k_2=1}^{k_1}\frac{x^{k_2}}{k_2}\cdots\sum_{k_{r-3}=1}^{k_{r-4}}\frac{x^{3k_{r-3}}}{k_{r-3}^2}\sum_{k_{r-2}=1}^{k_{r-4}}\frac{x^{k_{r-2}}}{k_{r-2}}$$

$$+ \sum_{k_1=1}^{\infty}\frac{x^{k_1}}{k_1}\sum_{k_2=1}^{k_1}\frac{x^{k_2}}{k_2}\cdots\sum_{k_{r-3}=1}^{k_{r-4}}\frac{x^{2k_{r-3}}}{k_{r-3}}\sum_{k_{r-2}=1}^{k_{r-4}}\frac{x^{k_{r-2}}}{k_{r-2}}\sum_{k_{r-1}=1}^{k_{r-2}}\frac{x^{k_{r-1}}}{k_{r-1}}-T_{r-3}-T_{r-2}$$

$$= \sum_{k_1=1}^{\infty}\frac{x^{k_1}}{k_1}\sum_{k_2=1}^{k_1}\frac{x^{k_2}}{k_2}\cdots\sum_{k_{r-4}=1}^{k_{r-5}}\frac{x^{5k_{r-4}}}{k_{r-4}^4}+\sum_{k_1=1}^{\infty}\frac{x^{k_1}}{k_1}\sum_{k_2=1}^{k_1}\frac{x^{k_2}}{k_2}\cdots\sum_{k_{r-4}=1}^{k_{r-5}}\frac{x^{4k_{r-4}}}{k_{r-4}^3}\sum_{k_{r-3}=1}^{k_{r-5}}\frac{x^{k_{r-3}}}{k_{r-3}}$$

$$+ \sum_{k_1=1}^{\infty}\frac{x^{k_1}}{k_1}\sum_{k_2=1}^{k_1}\frac{x^{k_2}}{k_2}\cdots\sum_{k_{r-4}=1}^{k_{r-5}}\frac{x^{3k_{r-4}}}{k_{r-4}^2}\sum_{k_{r-3}=1}^{k_{r-5}}\frac{x^{k_{r-3}}}{k_{r-3}}\sum_{k_{r-2}=1}^{k_{r-3}}\frac{x^{k_{r-2}}}{k_{r-2}}$$

$$+ \sum_{k_1=1}^{\infty}\frac{x^{k_1}}{k_1}\sum_{k_2=1}^{k_1}\frac{x^{k_2}}{k_2}\cdots\sum_{k_{r-4}=1}^{k_{r-5}}\frac{x^{2k_{r-4}}}{k_{r-4}}\sum_{k_{r-3}=1}^{k_{r-5}}\frac{x^{k_{r-3}}}{k_{r-3}}\sum_{k_{r-2}=1}^{k_{r-3}}\frac{x^{k_{r-2}}}{k_{r-2}}\sum_{k_{r-1}=1}^{k_{r-2}}\frac{x^{k_{r-1}}}{k_{r-1}}$$

$$- T_{r-4}-T_{r-3}-T_{r-2}=\cdots$$

$$= -\sum_{j=2}^{r-2}T_j+\sum_{k_1=1}^{\infty}\frac{x^{k_1}}{k_1}\sum_{k_2=1}^{k_1}\frac{x^{2k_2}}{k_2}\sum_{k_3=1}^{k_1}\frac{x^{k_3}}{k_3}\cdots\sum_{k_{r-1}=1}^{k_{r-2}}\frac{x^{k_{r-1}}}{k_{r-1}}$$

$$+ \sum_{k_1=1}^{\infty}\frac{x^{k_1}}{k_1}\sum_{k_2=1}^{k_1}\frac{x^{3k_2}}{k_2^2}\sum_{k_3=1}^{k_1}\frac{x^{k_3}}{k_3}\cdots\sum_{k_{r-2}=1}^{k_{r-3}}\frac{x^{k_{r-2}}}{k_{r-2}}$$

$$+ \sum_{k_1=1}^{\infty}\frac{x^{k_1}}{k_1}\sum_{k_2=1}^{k_1}\frac{x^{4k_2}}{k_2^3}\sum_{k_3=1}^{k_1}\frac{x^{k_3}}{k_3}\cdots\sum_{k_{r-3}=1}^{k_{r-4}}\frac{x^{k_{r-3}}}{k_{r-3}}+\cdots+\sum_{k_1=1}^{\infty}\frac{x^{k_1}}{k_1}\sum_{k_2=1}^{k_1}\frac{x^{(r-1)k_2}}{k_2^{r-2}}$$

$$= -\sum_{j=2}^{r-2}T_j+\left(\sum_{k_2=1}^{k_1}\frac{x^{3k_2}}{k_2^2}\sum_{k_3=1}^{k_2}\frac{x^{k_3}}{k_3}\cdots\sum_{k_{r-1}=1}^{k_{r-2}}\frac{x^{k_{r-1}}}{k_{r-1}}+\zeta_1(x^2)F_{r-2}(x)-T_1\right)$$

$$+ \left(\sum_{k_2=1}^{\infty}\frac{x^{4k_2}}{k_2^3}\sum_{k_3=1}^{k_2}\frac{x^{k_3}}{k_3}\cdots\sum_{k_{r-2}=1}^{k_{r-3}}\frac{x^{k_{r-2}}}{k_{r-2}}+\zeta_2(x^3)F_{r-3}(x)-\sum_{k_2=1}^{\infty}\frac{x^{3k_2}}{k_2^2}\sum_{k_3=1}^{k_2}\frac{x^{k_3}}{k_3}\cdots\sum_{k_{r-1}=1}^{k_{r-2}}\frac{x^{k_{r-1}}}{k_{r-1}}\right)$$

$$+ \left(\sum_{k_2=1}^{\infty}\frac{x^{5k_2}}{k_2^4}\sum_{k_3=1}^{k_2}\frac{x^{k_3}}{k_3}\cdots\sum_{k_{r-3}=1}^{k_{r-4}}\frac{x^{k_{r-3}}}{k_{r-3}}+\zeta_3(x^4)F_{r-4}(x)\right.$$

$$- \sum_{k_2=1}^{\infty} \frac{x^{4k_2}}{k_2^3} \sum_{k_3=1}^{k_2} \frac{x^{k_3}}{k_3} \cdots \sum_{k_{r-3}=1}^{k_{r-3}} \frac{x^{k_{r-2}}}{k_{r-2}} \Bigg) + \cdots$$

$$+ \left(\zeta_{r-1}(x^r) + \zeta_{r-2}(x^{r-1}) F_1(x) - \sum_{k_2=1}^{\infty} \frac{x^{(r-1)k_2}}{k_2^{r-2}} \sum_{k_3=1}^{k_2} \frac{x^{k_3}}{k_3} \right)$$

$$= - \sum_{j=1}^{r-2} T_j + \sum_{j=1}^{r-1} \zeta_j(x^{j+1}) F_{r-1-j}(x). \tag{6.204}$$

From this we obtain

$$T_{r-1} = - \sum_{j=1}^{r-2} T_j + \sum_{j=1}^{r-1} \zeta_j(x^{j+1}) F_{r-1-j}(x). \tag{6.205}$$

Formula (6.193) follows from the last two relations, provided that $\zeta_0(x) = \frac{x}{1-x}$. Relation (6.194) will be proved by induction on r. Formula (6.194) for $r = 0$ follows from $F_0(x) = 1$ and $F_1(x) = \zeta_1(x)$. The proof presented below will make it clear that in the induction assumption, it should be taken that formula (6.194) is true for all $l = 0, 1, \ldots, r$. We have

$$\left(\sum_{j=0}^{r+1} \zeta_{j+1}(x^{j+1}) F_{r+1-j}(x) \right)'$$

$$\overset{(5.124)}{=} \sum_{j=1}^{r+1} (j+1) \frac{1}{x} \zeta_j(x^{j+1}) F_{r+1-j}(x) + \sum_{j=0}^{r} \zeta_{j+1}(x^{j+1}) \frac{1}{x} \sum_{s=0}^{r-j} \zeta_s(x^{s+1}) F_{r-j-s}(x)$$

$$= \frac{1}{x} \sum_{s=0}^{r+1} (s+1) \zeta_s(x^{s+1}) F_{r+1-s}(x) + \frac{1}{x} \sum_{s=0}^{r} \zeta_s(x^{s+1}) \cdot \sum_{j=0}^{r-s} \zeta_{j+1}(x^{j+1}) F_{r-s-j}(x) = \left| \begin{matrix} (5.125) \\ r \to r-s \end{matrix} \right|$$

$$= \frac{1}{x}(r+2) \zeta_{r+1}(x^{r+2}) + \frac{1}{x} \sum_{s=0}^{r} (s+1) \zeta_s(x^{s+1}) F_{r+1-s}(x) + \frac{1}{x} \sum_{s=0}^{r} (r-s+1) \zeta_s(x^{s+1}) F_{r-s+1}(x)$$

$$= (r+2) \frac{1}{x} \sum_{s=0}^{r+1} \zeta_s(x^{s+1}) F_{r+1-s}(x) \overset{(5.124)}{=} ((r+2) F_{r+2}(x))'. \tag{6.206}$$

From this equality it follows that

$$(r+2) F_{r+2}(x) = \sum_{j=0}^{r+1} \zeta_{j+1}(x^{j+1}) F_{r+1-j}(x) + C. \tag{6.207}$$

Taking $x = 0$, we get that $C = 0$. So formula (5.125) is proved for $r = 1$. From the relations above, by the principle of mathematical induction it follows that it is also true for every nonnegative r. □

By (6.194) we obtain the following:

$$F_0(x) = 1, \tag{6.208}$$

$$F_1(x) = \zeta_1(x), \tag{6.209}$$

$$F_2(x) = \frac{1}{2}\zeta_2(x^2) + \frac{1}{2}\zeta_1^2(x), \tag{6.210}$$

$$F_3(x) = \frac{1}{3}\zeta_3(x^3) + \frac{1}{2}\zeta_2(x^2)\zeta_1(x) + \frac{1}{6}\zeta_1^3(x), \tag{6.211}$$

$$F_4(x) = \frac{1}{4}\zeta_4(x^4) + \frac{1}{3}\zeta_3(x^3)\zeta_1(x) + \frac{1}{8}\zeta_2^2(x^2) + \frac{1}{4}\zeta_2(x^2)\zeta_1^2(x) + \frac{1}{24}\zeta_1^4(x), \tag{6.212}$$

$$\begin{aligned} F_5(x) = {} & \frac{1}{5}\zeta_5(x^5) + \frac{1}{4}\zeta_4(x^4)\zeta_1(x) + \frac{1}{6}\zeta_3(x^3)\zeta_1^2(x) + \frac{1}{8}\zeta_2^2(x^2)\zeta_1(x) \\ & + \frac{1}{12}\zeta_2(x^2)\zeta_1^3(x) + \frac{1}{6}\zeta_3(x^3)\zeta_2(x^2) + \frac{1}{5!}\zeta_1^5(x), \end{aligned} \tag{6.213}$$

$$\begin{aligned} F_6(x) = {} & \frac{1}{6}\zeta_6(x^6) + \frac{1}{5}\zeta_5(x^5)\zeta_1(x) + \frac{1}{8}\zeta_4(x^4)\zeta_1^2(x) + \frac{1}{8}\zeta_4(x^4)\zeta_2(x^2) \\ & + \frac{1}{18}\zeta_3^2(x^3) + \frac{1}{6}\zeta_3(x^3)\zeta_2(x^2)\zeta_1(x) + \frac{1}{18}\zeta_3(x^3)\zeta_1^3(x) + \frac{1}{48}\zeta_2^3(x^2) \\ & + \frac{1}{16}\zeta_2^2(x^2)\zeta_1^2(x) + \frac{1}{48}\zeta_2(x^2)\zeta_1^4(x) + \frac{1}{6!}\zeta_1^6(x). \end{aligned} \tag{6.214}$$

By relation (6.211), for $x = -1$, we have

$$F_0(-1) = 1, \tag{6.215}$$

$$F_1(-1) = -\log 2, \tag{6.216}$$

$$F_2(-1) = \frac{1}{2}\zeta_2(1) + \frac{1}{2}\log^2 2,$$

$$F_3(-1) = -\frac{1}{4}\zeta_3(1) - \frac{1}{2}\zeta_2(1)\log 2 - \frac{1}{6}\log^2 2, \tag{6.217}$$

$$F_4(-1) = \frac{1}{24}\log^4 2 + \frac{1}{4}\zeta_2(1)\log^2 2 + \frac{1}{4}\zeta_3(1)\log 2 + \frac{9}{16}\zeta_4(1), \tag{6.218}$$

$$\begin{aligned} F_5(-1) = {} & -\frac{1}{5!}\log^5 2 - \frac{1}{12}\zeta_2(1)\log^3 2 - \frac{1}{8}\zeta_3(1)\log^2 2 \\ & - \frac{1}{8}\zeta_3(1)\zeta_2(1) - \frac{9}{16}\zeta_4(1)\log 2 - \frac{3}{16}\zeta_5(1), \end{aligned} \tag{6.219}$$

$$\begin{aligned} F_6(-1) = {} & \frac{1}{6!}\log^6 2 + \frac{1}{48}\zeta_2(1)\log^4 2 + \frac{1}{24}\zeta_3(1)\log^3 2 + \frac{1}{32}\zeta_3^2(1) \\ & + \frac{9}{32}\zeta_4(1)\log^2 2 + \frac{1}{8}\zeta_3(1)\zeta_2(1)\log 2 + \frac{3}{16}\zeta_5(1)\log 2 + \frac{61}{128}\zeta_6(1). \end{aligned} \tag{6.220}$$

7 The sums in which the *Riemann* zeta function appears

7.1 About the sums in which the *Riemann* zeta function appears

Srivastava [197] provided several different classes of the summation formulas with the series in which the *Riemann* zeta function appears. Some of these formulas are also given in *Euler*'s papers [3, 5] and *Legendre*'s textbook [6]. In [197] the author provides an overview of the results regarding this topic, starting from *Euler* and up to the 1990s. In this chapter, we present completely new results regarding general formulas for the corresponding series, which have only been calculated for the initial values $n = 0, 1, 2$. So, for example, *Johnson* [23] presented the formula

$$\sum_{k=1}^{\infty} \frac{\zeta(2k) - 1}{k} = \log 2, \tag{7.1}$$

whereas in this chapter, we will calculate the formula for the series

$$L_n = \sum_{k=1}^{\infty} \frac{\zeta(2k) - 1}{n + k} \quad (n \in \mathbb{N}). \tag{7.2}$$

Wilton [16], from 1922–1923, introduced the formula ([13], 2.35)

$$\sum_{k=1}^{\infty} \frac{\zeta(2k)}{k(2k+1)} = \log(2\pi) - 1, \tag{7.3}$$

whereas in this chapter, we obtain a formula for the following series:

$$\sum_{k=1}^{\infty} \frac{\zeta(2k)}{k(2k+n)} \quad (n \in \mathbb{N}). \tag{7.4}$$

Euler [3] obtained the formula

$$\sum_{k=2}^{\infty} (-1)^k \frac{\zeta(k) - 1}{k} = \gamma - 1 + \log 2, \tag{7.5}$$

and *Suryanarayana* [118] gave the formula

$$\sum_{k=2}^{\infty} (-1)^k \frac{\zeta(k) - 1}{k+1} = 1 + \frac{\gamma}{2} - \frac{1}{2}\log(2\pi), \tag{7.6}$$

from which it follows that

$$\sum_{k=2}^{\infty} (-1)^k \frac{\zeta(k) - 1}{k+1} = \frac{3}{2} + \frac{\gamma}{2} - \frac{1}{2}\log(8\pi). \tag{7.7}$$

https://doi.org/10.1515/9783112233276-007

In this chapter, we also give a formula for the following series:

$$K_n = \sum_{k=2}^{\infty} (-1)^k \frac{\zeta(k) - 1}{n + 1} \quad (n \in \mathbb{N}). \tag{7.8}$$

In addition to the result given in [197],

$$\sum_{k=2}^{\infty} (-1)^k \frac{\zeta(2k) - 1}{2k + 1} = \frac{3}{2} - \frac{1}{2} \log(4\pi), \tag{7.9}$$

we will derive the formula for the series

$$P_n = \sum_{k=1}^{\infty} \frac{\zeta(2k) - 1}{2n + 1 + 2k} \quad (n \in \mathbb{N}). \tag{7.10}$$

Also, in addition to the already known result [197]

$$\sum_{k=1}^{\infty} \frac{\zeta(2k+1) - 1}{k + 1} = \log 2 - \gamma, \tag{7.11}$$

in this chapter, we also provide a formula for the series

$$M_n = \sum_{k=1}^{\infty} \frac{\zeta(2k) - 1}{n + k} \quad (n \in \mathbb{N}) \tag{7.12}$$

and

$$Q_n = \sum_{k=2}^{\infty} (-1)^k \frac{\zeta(k)}{2^k (k + n)} \quad (n \in \mathbb{N}). \tag{7.13}$$

First of all, we will start from the well-known formula for *Raabe* integral:

$$I_0 = \int_0^1 \log \Gamma(x)\, \partial x = \frac{1}{2} \log(2\pi). \tag{7.14}$$

One of the natural generalizations of this integral is of the following form:

$$I_n = \int_0^1 x^n \log \Gamma(x)\, \partial x \quad (n \text{ is a nonnegative integer}). \tag{7.15}$$

Since the integral I_0 exists, it follows that so do the integrals I_n, $n \geqslant 1$. In relation to the integral I_0 is the integral $J_0 = \int_0^1 \log \sin \pi x\, \partial x$, from which follows that the latter exists:

$$J_0 < 0, \quad J_0 = \frac{2}{\pi} \int_0^{\frac{\pi}{2}} \log \sin x\, \partial x > \frac{2}{\pi} \int_0^{\frac{\pi}{2}} \log \frac{2}{\pi} x\, \partial x = -1. \tag{7.16}$$

For J_0, we introduce the following generalization:

$$J_n = \int_0^1 x^n \log \sin \pi x \, \partial x \quad (n \text{ is a nonnegative integer}). \tag{7.17}$$

We will prove the following formula:

$$\begin{aligned} I_n = \int_0^1 x^n \log \Gamma(x) \, \partial x &= \frac{\log(2\pi)}{2(n+1)} - \frac{n(\gamma + \log(2\pi))}{2(n+1)(n+2)} \\ &+ \frac{1}{2} \sum_{2 \leqslant 2k \leqslant n} (-1)^{k+1} \frac{n!}{(n-2k+1)!} \frac{\zeta(2k+1)}{(2\pi)^{2k}} \\ &+ \frac{1}{\pi} \sum_{2 \leqslant 2k \leqslant n+1} (-1)^{k+1} \frac{n!}{(n-2k+2)!} \frac{\zeta'(2k+1)}{(2\pi)^{2k-1}} \quad (n \geqslant 0, \gamma \text{ is } \textit{Euler's} \text{ constant}). \end{aligned} \tag{7.18}$$

Proof. The following formula is known as the formula for *Kummer*'s series of the function $\log \Gamma(x)$, where $\Gamma(x)$ is *Euler*'s Γ-function:

$$\log \Gamma(x) = \frac{1}{2} \log(2\pi) + \sum_{m=1}^{\infty} \left\{ \frac{1}{2m} \cos 2m\pi x + (\gamma + \log 2m\pi) \frac{\sin 2m\pi x}{m\pi} \right\} \quad (0 < x < 1). \tag{7.19}$$

This formula was given in [7] and can also be found in the textbook by *Erdélyi et al.* [54]. The series in (7.19) uniformly converges in each closed subinterval $(0, 1)$. This follows from the following *Abel*'s result: Let $\{a_m\}$ be a decreasing sequence tending to zero. Then the series $\sum_{m=1}^{\infty} a_m \cos mx$ and $\sum_{m=1}^{\infty} a_m \sin mx$ uniformly converge in each closed subset of the interval $(0, 2\pi)$. This result is a particular case of *the Dirichlet* criterion. All that remains is to check if the sequence $a_m = \frac{\log m}{m}$ $(m \geqslant 3)$ is decreasing. However, since:

$$\begin{aligned} a_m > a_{m+1} \iff \frac{\log m}{m} > \frac{\log(m+1)}{m+1} &\iff \log\left(1 + \frac{1}{m}\right) < \frac{1}{m} \log m, \\ \log\left(1 + \frac{1}{m}\right) < \frac{1}{m} < \frac{1}{m} \log m \quad (m \geqslant 3), \end{aligned} \tag{7.20}$$

and since

$$\lim_{x \downarrow 0} x^n \log \Gamma(x) = \lim_{x \downarrow 0} x^n \{\log \Gamma(x+1) - \log x\} = 0 = \lim_{x \downarrow 0} x^n \log \Gamma(x) \quad (n \geqslant 1), \tag{7.21}$$

we can integrate term-by-term the series (7.19) multiplied by x^n in the interval $[\varepsilon_1, 1 - \varepsilon_2]$ and then pass to the limit as $\varepsilon_1, \varepsilon_2 \to 0$. From

$$\int_0^1 x^n \cos 2m\pi x \, \partial x = \sum_{2 \leqslant 2k \leqslant n} (-1)^{k+1} \frac{n!}{(n-2k+1)!} \frac{1}{(2m\pi)^{2k}}, \tag{7.22}$$

$$\int_0^1 x^n \sin 2m\pi x \,\partial x = \sum_{2\leqslant 2k\leqslant n+1} (-1)^k \frac{n!}{(n-2k+2)!} \frac{1}{(2m\pi)^{2k-1}}, \tag{7.23}$$

$$\zeta'(s) = -\sum_{m=1}^{\infty} \frac{\log m}{m^s} \quad (\Re\mathfrak{e}(s) > 1) \tag{7.24}$$

we get

$$\begin{aligned} I_n = {} & \frac{\log(2\pi)}{2(n+1)} + 2(\gamma + \log(2\pi)) \sum_{2\leqslant 2k\leqslant n+1} (-1)^k \frac{n!}{(n-2k+2)!} \frac{\zeta(2k)}{(2\pi)^{2k}} \\ & + \frac{1}{2} \sum_{2\leqslant 2k\leqslant n} (-1)^{k+1} \frac{n!}{(n-2k+1)!} \frac{\zeta(2k+1)}{(2\pi)^{2k}} \\ & + \frac{1}{\pi} \sum_{2\leqslant 2k\leqslant n+1} (-1)^{k+1} \frac{n!}{(n-2k+2)!} \frac{\zeta'(2k)}{(2\pi)^{2k-1}}. \end{aligned} \tag{7.25}$$

Using the formula

$$\zeta(2k) = (-1)^{k-1} \frac{2^{2k-1} B_{2k}}{(2k)!} \pi^{2k}, \tag{7.26}$$

we get that the first sum on the right-hand side of relation (7.26) can be presented as

$$-\frac{1}{2(n+1)(n+2)} \sum_{2\leqslant 2k\leqslant n+1} \binom{n+2}{2k} B_{2k}. \tag{7.27}$$

By the relation

$$\sum_{2\leqslant 2k\leqslant n+1} \binom{n+2}{2k} B_{2k} = \frac{n}{2}, \tag{7.28}$$

and the resulting expression for the first sum on the right-hand side of relation (7.26) we obtain (7.19). □

From (7.19) it follows that

$$I_1 = \int_0^1 x \log\Gamma(x)\,\partial x = \frac{\zeta'(2)}{2\pi^2} - \frac{\gamma}{12} + \frac{\log(2\pi)}{6}, \tag{7.29}$$

$$I_2 = \int_0^1 x^2 \log\Gamma(x)\,\partial x = \frac{1}{12}(\log(2\pi) - \gamma) + \frac{1}{4\pi^2}(\zeta(3) + 2\zeta'(2)). \tag{7.30}$$

It further follows that for I_n, we have the formula

$$I_n = \int_0^1 x^n \log\Gamma(x)\,\partial x = \frac{\log \pi}{(n+1)(n+2)} - \frac{n(\gamma + \log 2)}{2(n+1)(n+2)} - \frac{1}{2} J_n$$

$$+\frac{1}{\pi}\sum_{2\leqslant 2k\leqslant n+1}(-1)^{k+1}\frac{n!}{(n-2k+2)!}\frac{\zeta'(2k)}{(2\pi)^{2k-1}}\quad (n\geqslant 0). \tag{7.31}$$

Formula (7.32) follows from (7.24), (7.25), and another form of *Kummer*'s series for $\log\Gamma(x)$ (this formula can be found in the same sources as formula (7.20):

$$\log\Gamma(x)=\left(\frac{1}{2}-x\right)(\gamma+\log 2)+(1-x)\log\pi \\ -\frac{1}{2}\log\sin\pi x+\sum_{m=1}^{\infty}\frac{\log m}{m\pi}\sin 2m\pi x\quad (0<x<1). \tag{7.32}$$

Starting from (7.19) and (7.32), we can also conclude that

$$J_n=\int_0^1 x^n\log\sin\pi x\,\partial x=-\frac{\log 2}{n+1}+\sum_{2\leqslant 2k\leqslant n}(-1)^{k+1}\frac{n!}{(n-2k+1)!}\frac{\zeta(2k+1)}{(2\pi)^{2k}}\quad (n\geqslant 0). \tag{7.33}$$

For $n=0,1$, we arrive at the well-known results

$$J_0=-\log 2, J_1=-\frac{1}{2}\log 2, \tag{7.34}$$

whereas for $n=2,3$, we have

$$J_2=-\frac{1}{3}\log 2-\frac{\zeta(3)}{2\pi^2}, \tag{7.35}$$

$$J_3=-\frac{1}{4}\log 2-\frac{3\zeta(3)}{4\pi^2}. \tag{7.36}$$

Let us prove the following formula:

$$I_m=\int_0^1 x^n\log\Gamma(x)\,\partial x=\frac{\gamma}{(n+1)(n+2)}+\frac{1}{(n+1)^2}+\frac{1}{n+1}\sum_{m=2}^{\infty}(-1)^{m+1}\frac{\zeta(m)}{n+1+m}\quad (n\geqslant 0). \tag{7.37}$$

Proof. Since

$$\frac{\Gamma'(x)}{\Gamma(x)}=-\gamma-\frac{1}{x}-\sum_{m=1}^{\infty}\left(\frac{1}{x+m}-\frac{1}{m}\right)\quad (x>0), \tag{7.38}$$

where the series on the rights uniformly converges, integrating by parts, we obtain

$$I_m = -\frac{1}{n+1}\int_0^1 x^{n+1}\frac{\Gamma'(x)}{\Gamma(x)}\,\partial x$$
$$= \frac{1}{n+1}\int_0^1 x^{n+1}\left\{\gamma + \frac{1}{x} + \sum_{m=1}^{\infty}\left(\frac{1}{x+m} - \frac{1}{m}\right)\right\}\partial x$$
$$= \frac{\gamma}{(n+1)(n+2)} + \frac{1}{(n+1)^2} + \frac{1}{n+1}\sum_{m=1}^{\infty}\int_0^1 x^{n+1}\left(\frac{1}{x+m} - \frac{1}{m}\right)\partial x. \tag{7.39}$$

For $m = 1$, we have

$$\int_0^1 x^{n+1}\left(\frac{1}{x+m} - \frac{1}{m}\right)_{m=1}\partial x$$
$$= -\int_0^1 \frac{x^{n+2}}{1+x}\,\partial x = -\lim_{\varepsilon\downarrow 0}\int_0^{1-\varepsilon} x^{n+2}\sum_{p=1}^{\infty}(-1)^{p-1}x^{p-1}\,\partial x$$
$$= -\lim_{\varepsilon\downarrow 0}\sum_{p=1}^{\infty}(-1)^{p-1}\int_0^{1-\varepsilon} x^{n+p+1}\,\partial x = \lim_{\varepsilon\downarrow 0}\sum_{p=1}^{\infty}(-1)^p\frac{(1-\varepsilon)^{n+p+2}}{n+p+2} = \sum_{p=1}^{\infty}\frac{(-1)^p}{n+p+2}. \tag{7.40}$$

The above statement is based on the application of *Abel*'s theorem at the boundary of the convergence circle, where the convergence of a number series is established by the *Leibniz* criterion. In particular, we obtain that the series is conditionally convergent. For $m \geqslant 2$, we have

$$x^{n+1}\left(\frac{1}{x+m} - \frac{1}{m}\right) = -x^{n+2}\frac{1}{1+\frac{x}{m}}\frac{1}{m^2} = -\frac{1}{m^2}\sum_{p=1}^{\infty}(-1)^{p-1}\frac{x^{p+n+1}}{m^{p-1}}, \tag{7.41}$$

so that

$$\sum_{m=2}^{\infty}\int_0^1 x^{n+1}\left(\frac{1}{x+m} - \frac{1}{m}\right)\partial x \overset{(A)}{=} \sum_{m=2}^{\infty}\frac{1}{m^2}\sum_{p=1}^{\infty}\frac{(-1)^p}{m^{p-1}(n+p+2)} \overset{(B)}{=} \sum_{p=1}^{\infty}(-1)^{p-1}\frac{\zeta(p+1)-1}{n+p+2}. \tag{7.42}$$

Equality (A) in this equation follows from the uniform convergence of the series

$$\sum_{p=1}^{\infty}(-1)^p\frac{x^{p+n+1}}{m^{p-1}} \quad (0 \leqslant x \leqslant 1,\ m \geqslant 2). \tag{7.43}$$

From this it follows that

$$\left|\sum_{p=q}^{\infty}(-1)^p\frac{x^{p+n+1}}{m^{p-1}}\right| \leqslant \sum_{p=q}^{\infty}\frac{1}{m^{p-1}} = \frac{1}{m^{q-1}}\frac{1}{1-\frac{1}{m}} \to 0 \quad (q \to \infty) \tag{7.44}$$

uniformly in x. Equality (B) (in relation (7.43)) follows after changing the order of summation, which is allowed for an absolutely convergent double series [182]:

$$\sum_{m=2}^{\infty}\frac{1}{m^2}\sum_{p=1}^{\infty}\frac{1}{m^{r-1}(n+p+2)} < \frac{1}{n+3}\sum_{m=2}^{\infty}\frac{1}{m^2}\sum_{p=1}^{\infty}\left(\frac{1}{m}\right)^{p-1} = \frac{1}{n+3}\sum_{m=2}^{\infty}\frac{1}{m^2}\frac{1}{1-\frac{1}{m}} = \frac{1}{n+3}. \tag{7.45}$$

It follows from (7.46) that the series behind (B) absolutely converges. This fact can be directly proved. Since

$$t_q = \frac{1}{(2^q+2)^n}+\dots+\frac{1}{(2^{q+1})^n} < \frac{2^q}{2^{qn}} = \left(\frac{1}{2^{n-1}}\right)^q, \tag{7.46}$$

$$\zeta(n) = 1+\frac{1}{2^n}+\sum_{q=1}^{\infty}t_q < 1+\frac{1}{2^n}+\frac{1}{2^{n-1}-1} < \frac{1}{2^{n-2}}+1 \quad (n \geqslant 3), \tag{7.47}$$

$$\left|(-1)^p\frac{\zeta(p+1)-1}{n+p+2}\right| < \frac{2}{2^p(n+p+2)} \overset{(p\geqslant 5)}{<} \frac{2}{p^2(n+p+2)} < \frac{2}{p^3}, \tag{7.48}$$

the following series absolutely converges:

$$\sum_{q=1}^{\infty}(-1)^p\frac{\zeta(p+1)-1}{n+p+2}. \tag{7.49}$$

From this we can conclude that

$$\begin{aligned}\sum_{m=1}^{\infty}\int_0^1 x^{n+1}\left(\frac{1}{x+m}-\frac{1}{m}\right)\partial x &= \sum_{p=1}^{\infty}\frac{(-1)^p}{n+p+2}+\sum_{p=1}^{\infty}(-1)^p\frac{\zeta(p+1)-1}{n+p+2}\\ &= \sum_{p=1}^{\infty}(-1)^p\frac{\zeta(p+1)}{n+p+2} = \sum_{p=2}^{\infty}(-1)^{p-1}\frac{\zeta(p)}{n+1+p},\end{aligned} \tag{7.50}$$

and (7.38) follows. □

By (7.19) and (7.38) we obtain

$$\begin{aligned}K_n' = \sum_{k=2}^{\infty}(-1)^k\frac{\zeta(k)}{n+k} &= \frac{1}{n}+\frac{\gamma}{2}-\frac{\log(2\pi)}{n+1}\\ &+\frac{1}{2}\sum_{2\leqslant 2k\leqslant n-1}(-1)^k\frac{n!}{(n-2k)!}\frac{\zeta(2k+1)}{(2\pi)^{2k}}\\ &+\frac{1}{\pi}\sum_{2\leqslant 2k\leqslant n}(-1)^k\frac{n!}{(n-2k+1)!}\frac{\zeta'(2k)}{(2\pi)^{2k-1}} \quad (n\geqslant 1).\end{aligned} \tag{7.51}$$

In particular, for $n=1$, we obtain the known result

$$K_1' = 1+\frac{\gamma}{2}-\frac{1}{2}\log(2\pi), \tag{7.52}$$

and for $n = 2, 3$, we get

$$K_2' = \frac{1+\gamma}{2} - \frac{1}{3}\log(2\pi) - \frac{\zeta'(2)}{\pi^2}, \tag{7.53}$$

$$K_3' = \frac{2+3\gamma}{6} - \frac{1}{4}\log(2\pi) - \frac{3\zeta(3)}{4\pi^2} - \frac{3\zeta'(2)}{2\pi^2}. \tag{7.54}$$

Further, we can conclude that

$$J_n = \int_0^1 x^n \log\sin\pi x\,\partial x = \frac{\log\pi}{n+1} - \frac{1}{(n+1)^2} - \sum_{k=1}^{\infty}\frac{\zeta(2k)}{k(n+2k+1)} \quad (n \geqslant 0). \tag{7.55}$$

This formula is an immediate consequence of the formulas

$$\log\frac{\sin t}{t} = \sum_{k=1}^{\infty}(-1)^k\frac{2^{2k-1}}{k(2k)!}B_{2k}t^{2k} \quad (|t| < \pi) \tag{7.56}$$

and

$$J_n = \frac{1}{\pi^{n+1}}\int_0^{\pi} t^n\left(\log\frac{\sin t}{t} + \log t\right)\partial t \tag{7.57}$$

after integrating and applying formula (7.27). From (7.56) and (7.34) it follows that

$$\sum_{k=1}^{\infty}\frac{\zeta(2k)}{k(n+2k)} = \frac{1}{n}\log(2\pi) - \frac{1}{n^2} + \sum_{2\leqslant 2k\leqslant n-1}(-1)^{k+1}\frac{(n-1)!}{(n-2k)!}\frac{\zeta(2k+1)}{(2\pi)^{2k}} \quad (n \geqslant 1). \tag{7.58}$$

Let $L_n = \sum_{k=1}^{\infty}\frac{\zeta(2k)-1}{n+k}$. It is known that $L_0 = \log 2$, and therefore, starting from (7.59) in the case $n = 2$, we obtain

$$\sum_{k=1}^{\infty}\frac{\zeta(2k)-1}{k+1} = \frac{3}{2} - \log\pi. \tag{7.59}$$

Furthermore, from the equalities

$$\Psi(x) = \frac{\Gamma'(x)}{\Gamma(x)}, \quad \Psi(1+x) = -\gamma + x\sum_{k=1}^{\infty}\frac{1}{k(k+1)} \quad (x > 0) \tag{7.60}$$

it follows that

$$\sum_{k=1}^{\infty}\frac{1}{k(n+2k)} = \frac{1}{n}\left\{\Psi\left(1+\frac{n}{2}\right) + \gamma\right\}. \tag{7.61}$$

Starting from (7.59) and (7.62) with the replacement $n \to 2n$ and using the formulas

$$\Psi(1+n)+\gamma=\sum_{p=1}^{\infty}\frac{1}{p},\quad \sum_{k=1}^{\infty}\frac{\zeta(2k)-1}{k}=\log 2, \tag{7.62}$$

we are now able to obtain that

$$L_n=\sum_{k=1}^{\infty}\frac{\zeta(2k)-1}{n+k}=1+\frac{1}{2}+\frac{1}{3}+\cdots+\frac{1}{n}-\log\pi+\frac{1}{2n}+\sum_{k=1}^{n-1}(-1)^k\frac{(2n)!}{(2n-2k)!}\frac{\zeta(2k+1)}{(2\pi)^{2k}}. \tag{7.63}$$

In particular, for $n=2$, we have

$$L_2=\frac{7}{4}-\log\pi-\frac{3\zeta(3)}{\pi^2}. \tag{7.64}$$

From (7.59), after replacement $n \to 2n+1$, from the derived formula for L_0 and from

$$\Psi\left(\frac{1}{2}\right)=-\gamma-2\log 2,\quad \Psi(1+x)=\frac{1}{x}+\Psi(x),\quad \Psi\left(\frac{2n+1}{2}\right)=2\sum_{p=1}^{n}\frac{1}{2p-1}-\gamma-\log 2 \tag{7.65}$$

we have

$$\begin{aligned}P_n=\sum_{k=1}^{\infty}\frac{\zeta(2k)-1}{2n+1+2k}&=-\frac{1}{2}\log(4\pi)+1+\frac{1}{3}+\cdots+\frac{1}{2n+1}\\&+\frac{1}{2}\frac{1}{2n+1}+\frac{1}{2}\sum_{k=1}^{n}(-1)^k\frac{(2n+1)!}{(2n+1-2k)!}\frac{\zeta(2k+1)}{(2\pi)^{2k}}\quad(n\geqslant 0).\end{aligned} \tag{7.66}$$

In particular, for $n=0$, we have the result already stated. In addition,

$$P_1=\sum_{k=1}^{\infty}\frac{\zeta(2k)-1}{2k+3}=\frac{3}{2}-\log 2-\frac{1}{2}\log\pi-\frac{3\zeta(3)}{4\pi^2}. \tag{7.67}$$

By (7.52) it follows that

$$\begin{aligned}K_n=\sum_{k=2}^{\infty}(-1)^k\frac{\zeta(k)-1}{n+k}&=\frac{1}{n}+\frac{\gamma}{2}-\frac{\log(2\pi)}{n+1}\\&+(-1)^n\left\{\log 2-\left(1-\frac{1}{2}+\frac{1}{3}-\cdots+(-1)^n\frac{1}{n+1}\right)\right\}\\&+\frac{1}{2}\sum_{2\leqslant 2k\leqslant n-1}(-1)^k\frac{n!}{(n-2k)!}\frac{\zeta(2k+1)}{(2\pi)^{2k}}\\&+\frac{1}{\pi}\sum_{2\leqslant 2k\leqslant n}(-1)^k\frac{n!}{(n-2k+1)!}\frac{\zeta'(2k)}{(2\pi)^{2k-1}}\quad(n\geqslant 1).\end{aligned} \tag{7.68}$$

By (7.67) and (7.69), since the corresponding series absolutely converge, we obtain

$$M_n = \sum_{k=2}^{\infty} \frac{\zeta(2k+1)-1}{n+k} = -\gamma - \log\pi + \frac{\log(2\pi)}{n} + 1 + \frac{1}{2} + \cdots + \frac{1}{n}$$
$$- \frac{1}{2n-1} + \frac{2}{\pi}\sum_{k=1}^{n-1}(-1)^{k+1}\frac{(2n-1)!}{(2n-2k)!}\frac{\zeta'(2k)}{(2\pi)^{2k-1}} \quad (n \geqslant 1). \tag{7.69}$$

By the above, for $n = 1$, we have the result already stated, whereas for $n = 2, 3$,

$$M_2 = \frac{7}{6} - \gamma - \frac{1}{2}\log\frac{\pi}{2} + \frac{\zeta'(2)}{\pi^2}, \tag{7.70}$$
$$M_3 = \frac{49}{30} - \gamma + \frac{\log 2}{3} - \frac{2}{3}\log\pi + 5\frac{\zeta'(2)}{\pi^2} - 15\frac{\zeta'(4)}{\pi^4}, \tag{7.71}$$

and by (7.59) and (7.69) it follows that

$$\sum_{k=2}^{\infty}\frac{\zeta(2k+1)-1}{2k+3} = \frac{13}{12} - \frac{2}{3}\log 2 - \frac{1}{6}\log\pi - \frac{\gamma}{2} + \frac{\zeta'(2)}{\pi^2}. \tag{7.72}$$

Let us introduce the function

$$Q_n(x) = \int_0^x t^n \log\Gamma(t)\, \partial t \quad (0 \leqslant x \leqslant 1). \tag{7.73}$$

Applying the same method as in the case $x = \frac{1}{2}$ for the series

$$Q_n = \sum_{k=2}^{\infty}(-1)^k \frac{\zeta(k)}{2^k(k+n)} \quad (n \geqslant 1), \tag{7.74}$$

we arrive at the following expressions:

$$Q_1 = 1 - \frac{2}{3}\log 2 - \frac{1}{4}\log\pi + \frac{3}{2}\frac{\zeta'(2)}{\pi^2}, \tag{7.75}$$
$$Q_2 = \frac{1}{2} - \frac{1}{2}\log 2 - \frac{1}{6}\log\pi + \frac{7}{4}\frac{\zeta(3)}{\pi^2} + \frac{\zeta'(2)}{\pi^2}, \tag{7.76}$$
$$Q_n = \frac{\gamma}{2n(n+1)} + \frac{1}{n^2} - \frac{\log 2}{2n} - \frac{(n-1)!\cos\frac{n\pi}{2}}{2\pi^n}\zeta(n+1)$$
$$- \frac{(n-1)!\sin\frac{n\pi}{2}}{\pi^{n+1}}(\gamma + \log(2\pi))\zeta(n+1) + \frac{(n-1)!\sin\frac{n\pi}{2}}{\pi^{n+1}}\zeta'(n+1)$$
$$+ \frac{\gamma + \log(4\pi)}{n(n+1)}\left(B_{n+1} - \frac{n-1}{2}\right) - \frac{\gamma + \log(2\pi)}{n(n+1)}\sum_{2\leqslant 2k\leqslant n+1}\binom{n+1}{2k}2^{2k-1}B_{2k}$$
$$+ \frac{1}{2}\sum_{2\leqslant 2k\leqslant n}(-1)^{k+1}\frac{(n-1)!}{(n-2k)!}\frac{1-2^{-2k}}{\pi^{2k}}\zeta(2k+1)$$

$$+\sum_{2\leqslant 2k\leqslant n+1}(-1)^{k+1}\frac{(n-1)!}{(n-2k+1)!}\frac{1-2^{-2k+1}}{\pi^{2k}}\zeta'(2k)\quad (n\geqslant 3). \tag{7.77}$$

In the even case, by (5.47), which we will write in the form

$$\sum_{k=1}^{n}\binom{2n+1}{2k}2^{2k-1}B_{2k}=n, \tag{7.78}$$

we get

$$\begin{aligned}Q_{2n}=\sum_{k=2}^{\infty}(-1)^k\frac{\zeta(k)}{2^k(k+2n)}&=\frac{1-2n}{2n(n+1)}\gamma+\frac{1}{4n^2}+\frac{1-8n}{4n(2n+1)}\log 2\\&+\frac{1-4n}{4n(2n+1)}\log\pi+(-1)^{n-1}\frac{(2n-1)!}{2\pi^{2n}}\zeta(2n+1)\\&+\frac{1}{2}\sum_{k=1}^{n}(-1)^{k+1}\frac{(2n-1)!}{(2n-2k)!}\frac{1-2^{-2k}}{\pi^{2k}}\zeta(2k+1)\\&+\sum_{k=1}^{n}(-1)^{k+1}\frac{(2n-1)}{(2n-2k+1)!}\frac{1-2^{-2k+1}}{\pi^{2k}}\zeta'(2k+1).\end{aligned} \tag{7.79}$$

In the odd case, the following functional relation can be established on the basis of (5.40):

$$\sum_{k=1}^{n-1}\binom{2n+2}{2k}2^{2k-1}B_{2k}=\frac{2n+1}{2}-(2^{2n+1}-1)B_{2n+2}. \tag{7.80}$$

Therefore it follows that

$$\begin{aligned}Q_{2n+1}=\sum_{k=2}^{\infty}(-1)^k\frac{\zeta(k)}{2^k(k+2n+1)}&=-\frac{n}{(n+1)(2n+1)}\gamma+\frac{1}{(2n+1)^2}\\&-\frac{4n+1}{4(n+1)(2n+1)}\log\pi+\frac{2B_{2n+2}-(8n+3)}{4(n+1)(2n+1)}\log 2\\&+(-1)^{n-1}\frac{(2n)!}{\pi^{2n+2}}\zeta'(2n+2)+\frac{1}{2}\sum_{k=1}^{n}(-1)^k\frac{(2n)!}{(2n-2k+1)!}\frac{1-2^{-2k}}{\pi^{2k}}\zeta(2k+1)\\&+\sum_{k=1}^{n+1}(-1)^{k+1}\frac{(2n)!}{(2n-2k+2)!}\frac{1-2^{-2k+1}}{\pi^{2k}}\zeta'(2k).\end{aligned} \tag{7.81}$$

7.2 *Ramanujan*'s formula and some other formulas for $\zeta(2n+1)$

There is a formula for $\zeta(2n+1)$ $(n\in N)$, which is a natural complement to the formula for $\zeta(2n)$. It was discovered by *Ramanujan* [65, 106, 163, 168], whereas *Malurkar* [37] was the first to prove it. The formula is given in the following theorem.

Theorem 7.1 (*Ramanujan*'s formula for $\zeta(2n+1)$). *For positive numbers α and β such that $\alpha\beta = \pi^2$ and natural number n, we have*

$$\alpha^{-n}\left(\frac{1}{2}\zeta(2n+1) + \sum_{k=1}^{\infty}\frac{k^{-2n-1}}{e^{2\alpha k}-1}\right)$$
$$= (-\beta)^{-n}\left(\frac{1}{2}\zeta(2n+1) + \sum_{k=1}^{\infty}\frac{k^{-2n-1}}{e^{2\alpha k}-1}\right) - 2^{2n}\sum_{k=0}^{n+1}\frac{B_{2k}}{(2k)!}\frac{B_{2n+2-2k}}{(2n+2-2k)!}\alpha^{n+1-k}\beta^{k}. \quad (7.82)$$

In particular, for $\alpha = \beta = \pi$ and odd n,

$$\zeta(2n+1) = \pi^{2n+1}\cdot 2^{2n}\sum_{k=0}^{n+1}(-1)^{k+1}\frac{B_{2k}}{(2k)!}\frac{B_{2n+2-2k}}{(2n+2-2k)!} - 2\sum_{k=1}^{\infty}\frac{k^{-2n-1}}{e^{2\pi k}-1}. \quad (7.83)$$

Berndt [130] also states the following formula:

$$(1-\rho^{2n})\zeta(2n+1) = (-1^{n})(2\pi)^{2n+1}\mathrm{i}\sum_{k=0}^{n+1}\frac{B_{2k}}{(2k)!}\frac{B_{2n+2-2k}}{(2n+2-2k)!}\rho^{2k-1}$$
$$+ 2(\rho^{2n}-1)\sum_{k=1}^{\infty}\frac{k^{-2n-1}}{(-1)^{k}e^{k\pi\sqrt{3}}-1} \qquad (\rho = e^{\frac{2\pi}{3}\mathrm{i}}). \quad (7.84)$$

Formula (7.84) was first given and proved by *Smart* [109]. Assuming that

$$\sigma_b(n) = \sum_{d|n} d^{b}, \quad (7.85)$$

$$\sum_{k=1}^{\infty}\sigma_v(k)e^{-2\pi ky} = \sum_{k=1}^{\infty}\sum_{d=1}^{\infty} d^{v}e^{-2\pi rdy} = \sum_{d=1}^{\infty}\frac{d^{v}}{e^{-2\pi dy}-1}, \quad (7.86)$$

for $v = -2n-1$ and $y = \frac{\alpha}{\pi}$, formula (7.82) can also be written in the following form.

Theorem 7.2. *We have the following equality:*

$$\alpha^{-n}\left(\frac{1}{2}\zeta(2n+1) + \sum_{k=1}^{\infty}\sigma_{-(2n+1)}(k)e^{-2k\alpha}\right)$$
$$= (-\beta)^{-n}\left(\frac{1}{2}\zeta(2n+1) + \sum_{k=1}^{\infty}\sigma_{-(2n+1)}(k)e^{-2k\beta}\right) - 2^{2n}\sum_{k=0}^{n+1}(-1)^{k}\frac{B_{2k}}{(2k)!}\frac{B_{2n+2-2k}}{(2n+2-2k)!}\alpha^{n+1-k}\beta^{k}, \quad (7.87)$$

where $\alpha,\beta > 0$, $\alpha\beta = \pi^2$, and n is a natural number.

In relation to the sums similar to those appearing in (7.82), there are some results in the professional literature. We list some formulas below, whereas their proof can be found in [130].

Theorem 7.3. (a) *Let $\alpha, \beta > 0$ be such that $\alpha\beta = \pi^2$, and let $n > 1$ be a natural number. Then*

$$\alpha^n \sum_{k=1}^{\infty} \frac{k^{2n-1}}{e^{2\alpha k} - 1} - (-\beta)^n \sum_{k=1}^{\infty} \frac{k^{2n-1}}{e^{2\beta k} - 1} = (\alpha^n - (-\beta)^n) \frac{B_{2n}}{4n}. \tag{7.88}$$

(b) *If in (a), n is odd, then*

$$\sum_{k=1}^{\infty} \frac{k^{2n-1}}{e^{2\pi k} - 1} = \frac{B_{2n}}{4n}. \tag{7.89}$$

Formulas (7.88) and (7.89) are given in [65], the proof of (7.89) is given in [14, 41], whereas the proof of (7.88) is given by *Rao* and *Aiyar* [36] and *Grosswald* [97].

Theorem 7.4. *For $\alpha, \beta > 0$ such that $\alpha\beta = \pi^2$, we have the following formulas:*

$$\alpha \sum_{k=1}^{\infty} \frac{k}{e^{2\alpha k} - 1} + \beta \sum_{k=1}^{\infty} \frac{k}{e^{2\beta k} - 1} = \frac{\alpha + \beta}{24} - \frac{1}{4}, \tag{7.90}$$

$$\sum_{k=1}^{\infty} \frac{k}{e^{2\pi k} - 1} = \frac{1}{24} - \frac{1}{8\pi}. \tag{7.91}$$

The last two formulas are given in [65]. It appears that they were first proved by *Schlömilch* [12]. The proof can also be found in *Grosswald* [97] (1972).

Theorem 7.5. *For $\alpha, \beta > 0$ such that $\alpha\beta = \pi^2$, we have*

$$\sum_{k=1}^{\infty} \frac{1}{k(e^{2\alpha k} - 1)} - \frac{1}{4} \log \alpha + \frac{\alpha}{12} = \sum_{k=1}^{\infty} \frac{1}{k(e^{2\beta k} - 1)} - \frac{1}{4} \log \beta + \frac{\beta}{12}. \tag{7.92}$$

Formula (7.92) is also mentioned in [65], and the proof is given in [97]. Formulas (7.82)–(7.92) were proved by *Berndt* [105] using modular transformations of the form

$$V(s) = \frac{as + b}{cs + d}, \quad a, b, c, d \in Q, \ c > 0, \ ad - bc = 1. \tag{7.93}$$

The following theorem was proved by *Grosswald* [87] in 1970.

Theorem 7.6. *We have*

$$\zeta(2n+1) = \pi^{2n+1} \cdot r_{2n+1} - S_n, \tag{7.94}$$

where $r_{2n+1} \in Q$, and

$$S_n = 2 \sum_{m=1}^{\infty} \frac{1}{m^{2n+1}(e^{2\pi m - 1})} + \frac{2\pi}{n} [1 + (-1)^n] \sum_{m=1}^{\infty} \frac{e^{2\pi m}}{m^{2n}(e^{2\pi m} - 1)^2}. \tag{7.95}$$

We also have the following formula for $\zeta(2n+1)$.

Theorem 7.7. *We have*

$$\zeta(2n+1) = \left(\frac{\pi}{2}\right)^{2n+1} \lim_{m\to\infty} \frac{1}{m^{2n+1}} \sum_{k=1}^{\infty} \operatorname{ctan}^{2n+1} \frac{k\pi}{2m+1}. \tag{7.96}$$

This formula was derived by *Apostol* [104], using the idea of *Holme* [88] and *Papadimitriou* [108], to calculate the following sum:

$$\sum_{k=1}^{\infty} \operatorname{ctan}^2 \frac{k\pi}{2m+1} = \frac{m(2m-1)}{3}. \tag{7.97}$$

The following formulas for $\zeta(2n+1)$ are derived on the basis of (1.148), (1.161), and (1.163) for $s = 2n+1$:

$$\zeta(2n+1) = \frac{2^{2n}}{2^{2n}-1} \frac{1}{(2n)!} \int_0^{\infty} \frac{x^{2n}}{e^x+1}\, \partial x, \tag{7.98}$$

$$\zeta(2n+1) = \frac{1}{(2n)!} \int_0^{\infty} \frac{x^{2n}}{e^x-1}\, \partial x, \tag{7.99}$$

$$\zeta(2n+1) = \frac{2^{2n+1}}{2^{2n+1}-1} \frac{1}{(2n)!} \int_0^{\infty} \frac{e^x x^{2n}}{e^x-1}\, \partial x. \tag{7.100}$$

After appropriate replacements, they become

$$\zeta(2n+1) = \frac{2^{2n}}{2^{2n}-1} \frac{1}{(2n)!} \int_0^{\infty} \frac{\log^{2n} x}{1+x}\, \partial x, \tag{7.101}$$

$$\zeta(2n+1) = \frac{1}{(2n)!} \int_0^{\infty} \frac{\log^{2n} x}{1-x}\, \partial x, \tag{7.102}$$

$$\zeta(2n+1) = \frac{1}{(2n-1)!} \int_0^{\infty} \frac{\log^{2n-1} x \log(1-x)}{x}\, \partial x, \tag{7.103}$$

$$\zeta(2n+1) = \frac{1}{(2n-1)!} \int_0^{\infty} \frac{\log^{2n-1}(1-x) \log x}{1-x}\, \partial x, \tag{7.104}$$

$$\zeta(2n+1) = \frac{\log^{2n+1} 2}{(2n)!} \int_0^{\infty} \frac{x^{2n}}{2^x-1}\, \partial x. \tag{7.105}$$

We will prove the following three formulas:

$$\frac{1}{(2n)!}\int_a^\infty \frac{x^{2n}}{e^x-1}\,\partial x = \sum_{k=0}^{2n}\frac{a^k}{k!}\sum_{r=1}^\infty \frac{e^{-ra}}{r^{2n+1-k}} \quad (a>0), \tag{7.106}$$

$$\frac{\log^{2n+1} m}{(2n)!}\int_1^\infty \frac{x^{2n}}{m^x-1}\,\partial x = \sum_{k=0}^{2n}\frac{\log^k m}{k!}\zeta_{2n+1-k}\left(\frac{1}{m}\right) \quad (m\geqslant 2), \tag{7.107}$$

$$\zeta(2n+1) = \sum_{k=0}^{2n}\frac{\log^k m}{k!}\zeta_{2n+1-k}\left(\frac{1}{m}\right) + \frac{1}{(2n)!}\int_0^{m-1}\frac{\log^{2n}(1+x)}{x}\,\partial x - \frac{\log^{2n+1} m}{(2n+1)!}. \tag{7.108}$$

Proof. Let

$$I(a) = \frac{1}{(2n)!}\int_0^a \frac{x^{2n}}{e^x-1}\,\partial x \quad (a>0). \tag{7.109}$$

Then

$$I(a) = |e^{-x}=y| = \frac{1}{(2n)!}\int_{e^{-a}}^1 \frac{\log^{2n} x}{1-x}\,\partial x = \frac{1}{(2n)!}\sum_{k=0}^\infty \int_{e^{-a}}^1 x^k \log^{2n} x\,\partial x, \tag{7.110}$$

$$\int x^n \log^m x\,\partial x = \frac{x^{n+1}}{m+1}\sum_{k=0}^m (-1)^k (m+1)m(m-1)\cdots(m-k+1)\frac{\log^{m-k} x}{(n+1)^{k+1}}, \tag{7.111}$$

and accordingly it follows that

$$\begin{aligned} I(a) &= \frac{1}{(2n)!}\sum_{k=0}^\infty\left((2n)!\frac{1}{(k+1)^{2n+1}} - \frac{e^{-(k+1)a}}{2n+1}\sum_{r=0}^{2n}\frac{(2n+1)!}{(2n-r)!}\frac{a^{2n-r}}{(k+1)^{r+1}}\right) \\ &= \zeta(2n+1) - \sum_{k=0}^\infty e^{-(k+1)a}\sum_{r=0}^\infty \frac{a^{2n-r}}{(2n-r)!(k+1)^{r+1}}. \end{aligned} \tag{7.112}$$

It follows that

$$\frac{1}{(2n)!}\int_a^\infty \frac{x^{2n}}{e^x-1}\,\partial x = \sum_{k=0}^\infty e^{-(k+1)a}\sum_{r=0}^\infty \frac{a^{2n-r}}{(2n-r)!(k+1)^{r+1}}, \tag{7.113}$$

and (7.106) directly follows. Further,

$$\sum_{k=1}^\infty e^{-ka}\sum_{r=0}^{2n}\frac{a^r}{r!k^{2n+1-r}} = \sum_{k=1}^\infty \frac{e^{-ka}}{k^{2n+1}}\sum_{r=0}^{2n}\frac{(ak)^r}{r!} = \sum_{k=1}^\infty \frac{e^{-ka}}{k^{2n+1}}\left(e^{ka} - \sum_{r=2n+1}^\infty \frac{(ak)^r}{r!}\right)$$

$$= \zeta(2n+1) - \sum_{k=1}^{\infty} \frac{e^{-ak}}{k^{2n+1}} \sum_{r=2n+1}^{\infty} \frac{(ak)^r}{r!}. \tag{7.114}$$

By the last formula we have

$$\frac{1}{(2n)!} \int_0^a \frac{x^{2n}}{e^x - 1}\, \partial x = \sum_{k=1}^{\infty} \frac{e^{-ak}}{k^{2n+1}} \sum_{r=2n+1}^{\infty} \frac{(ak)^r}{r!} \tag{7.115}$$

and

$$\zeta(2n+1) = \lim_{a\to\infty} \sum_{k=1}^{\infty} \frac{e^{-ak}}{k^{2n+1}} \sum_{r=2n+1}^{\infty} \frac{(ak)^r}{r!}. \tag{7.116}$$

Taking $a = \log m$ in (7.106) and replacing $e^x = m^y$, by the definition of the function $\zeta_k(x)$ we obtain relation (7.107). Finally,

$$\begin{aligned} \int_0^1 \frac{x^{2n}}{m^x - 1}\, \partial x &= |m^x - 1 = y| = \frac{1}{\log^{2n+1} m} \int_0^{m-1} \frac{\log^{2n}(1+x)}{x(1+x)}\, \partial x \\ &= \frac{1}{\log^{2n+1} m} \left(\int_0^{m-1} \frac{\log^{2n}(1+x)}{x(1+x)}\, \partial x - \frac{\log^{2n+1} m}{2n+1} \right). \end{aligned} \tag{7.117}$$

Since

$$\zeta(2n+1) = \frac{\log^{2n+1} m}{(2n)!} \int_0^{\infty} \frac{x^{2n}}{m^x - 1}\, \partial x, \tag{7.118}$$

relation (7.108) follows from (7.107), (7.117), and (7.118). □

7.3 Formulas for $\zeta(2n+1)$

The importance of determining this formula is evidenced by the extensive literature and numerous papers in the previous period [203, 204, 206, 207, 209–211, 213, 215, 217, 219, 228]. These papers offer a very large collection of the closed forms for the sums of the series that include the zeta function. We present a new formula for $\zeta(2n+1)$, which has a more compact form and faster convergence than the relations described in the aforementioned papers.

For $\zeta(2n+1)$ $(n \in \mathbb{N})$, we have [232]

$$\zeta(2n+1) = (-1)^{n-1} \frac{2(2\pi)^{2n}}{(2n+2)!} \left(\sum_{j=1}^{\infty} \left(\frac{z_{0,n}}{j+1} + \frac{z_{1,n}}{j+2} + \cdots + \frac{z_{n,n}}{j+n+1} \right) (\zeta(2j) - \tau_n) \right), \tag{7.119}$$

where

$$
\begin{aligned}
z_{k,n} &= (2n-2k-1)\binom{2n+2}{2k+2}B_{2(n-k)} \quad (k = 0,1,\dots,n),\\
\tau_n &= -\frac{1}{2}\frac{\sum_{k=0}^{n}\frac{z_{k,n}}{k+1}}{\sum_{k=0}^{n} z_{k,n}H_{k+1}},\\
H_k &= 1+\frac{1}{2}+\cdots+\frac{1}{k} \quad (k \in \mathbb{N}),
\end{aligned} \tag{7.120}
$$

and B_m are *Bernoulli's* numbers. Moreover, for $\zeta(2n+1)$ $(n \in \mathbb{N})$, we have

$$
\zeta(2n+1) = (-1)^{n-1}\frac{(2\pi)^{2n}}{(2n)!}\left(\sum_{k=1}^{\infty}\frac{P_{n-1}(k)}{(k+1)(k+2)\cdots(k+n+1)}[\zeta(2k)-1]+q_n\right), \tag{7.121}
$$

where $P_{n-1}(k) \in Q[k]$, $\deg(P_{n-1}(k)) = n-1$, $q_n \in \mathbb{Q}$. For $\zeta(2n+1)$ $(n \in \mathbb{N})$, we can write

$$
\zeta(2n+1) = (-1)^{n-1}\frac{(2\pi)^{2n}}{(2n)!}\left(\sum_{k=1}^{\infty}\frac{P_{n-1}(k)}{(k+1)(k+2)\cdots(k+n+1)}[\zeta(2k)-r_n]\right), \tag{7.122}
$$

for $r_n \in \mathbb{Q}$ or a formula similar to (7.123), in which π^{2n} or $\zeta(2n)$ should be written instead of $(-1)^{n-1}\frac{(2\pi)^{2n}}{(2n)!}$.

Proof. By replacing n in (7.64) with $1,2,\dots,n+1$, we get a linear system with respect to $\zeta(3),\zeta(5),\dots,\zeta(2n+1)$. Now making the replacement

$$
\begin{aligned}
\omega_j &= (-1)^j\frac{\zeta(2j+1)}{(2\pi)^{2j}},\\
N_j &= \frac{1}{(2k+2)!}\left(L_{k+1}-L_1\left[H_{k+1}+\frac{1}{2(k+1)}-\frac{3}{2}\right]\right),
\end{aligned} \tag{7.123}
$$

we obtain the system

$$
\frac{1}{(2j)!}\omega_1+\frac{1}{(2j-2)!}\omega_2+\cdots+\frac{1}{2!}\omega_j = N_j \quad (j = 1,2,\dots,n), \tag{7.124}
$$

which can be expressed in matrix form

$$
A[\omega_1\,\omega_2\cdots\omega_n]^T = [N_1N_2\cdots N_n]^T, \tag{7.125}
$$

where the matrix A is given by its rows

$$
\begin{array}{llllll}
[\frac{1}{2!} & 0 & 0 & 0 & \cdots & 0], \\
[\frac{1}{4!} & \frac{1}{2!} & 0 & 0 & \cdots & 0], \\
[\frac{1}{6!} & \frac{1}{4!} & \frac{1}{2!} & 0 & \cdots & 0], \\
\vdots & & & & & \\
[\frac{1}{(2n)!} & \frac{1}{(2n-2)!} & \frac{1}{(2n-4)!} & \frac{1}{(2n-6)!} & \cdots & \frac{1}{2!}].
\end{array}
\tag{7.126}
$$

The matrix A^{-1} has the following columns:

$$
\begin{array}{lllll}
[a_0 & a_2 & a_4 & \cdots & a_{2(n-1)}]^T, \\
[0 & a_0 & a_2 & \cdots & a_{2(n-2)}]^T, \\
[0 & 0 & a_0 & \cdots & a_{2(n-3)}]^T, \\
\vdots & & & & \\
[0 & 0 & 0 & \cdots & a_0]^T,
\end{array}
$$

with elements a_{2j} satisfying the relation

$$\sum_{m=0}^{k} \frac{a_{2m}}{(2k+2-2m)!} = 0 \tag{7.127}$$

for every $k = 1, 2, \ldots, n-1$ with the initial value $a_0 = 2$. Based on these relations, we can easily obtain

$$\sum_{k=0}^{\infty} \frac{x^{2k}}{(2k+2)!} \sum_{m=0}^{\infty} a_{2m} x^{2m} = 2 \Longrightarrow \sum_{m=0}^{\infty} a_{2m} x^{2m} = \frac{2x^2}{e^x + e^{-x} - 2} = \frac{2x^2 e^x}{(e^x - 1)^2}. \tag{7.128}$$

Differentiating the equality

$$\frac{x}{e^x - 1} = \sum_{m=0}^{\infty} \frac{B_m}{m!} x^m, \tag{7.129}$$

we get

$$\frac{2x^2 e^x}{(e^x - 1)^2} = 2 + \sum_{m=1}^{\infty} \frac{2(1-2m)}{(2m)!} B_{2m} \Longrightarrow a_{2m} = \frac{2(1-2m)}{(2m)!} B_{2m}, \quad m = 0, 1, 2, \ldots. \tag{7.130}$$

Based on our matrix equation, we have that $\omega_n = \sum_{k=1}^{n} a_{2(n-k)} N_k$, which leads us to the formula [232]

$$
\begin{aligned}
(-1)^n \frac{\zeta(2n+1)}{(2\pi)^{2n}} = 2 \sum_{k=1}^{n} \frac{2n-2k-1}{(2n-2k)!} \frac{B_{2(n-k)}}{(2k+2)!} \\
\cdot \left(k \sum_{j=1}^{\infty} \frac{\zeta(2j) - 1}{(j+1)(j+1+k)} + H_{k+1} + \frac{1}{2(k+1)} - \frac{3}{2} \right).
\end{aligned}
\tag{7.131}
$$

Taking

$$\lambda_n = (-1)^{n-1} 2 \frac{(2\pi)^{2n}}{(2n+2)!}, \tag{7.132}$$

we have

$$\zeta(2n+1) = -\lambda_n \sum_{k=1}^{n} z_{k,n} \left(k \sum_{j=1}^{\infty} \frac{\zeta(2j)-1}{(j+1)(j+1+k)} + H_{k+1} + \frac{1}{2(k+1)} - \frac{3}{2} \right). \tag{7.133}$$

Since

$$-k \sum_{j=1}^{\infty} \frac{\zeta(2j)-1}{(j+1)(j+1+k)} + H_{k+1} + \frac{1}{2(k+1)} - \frac{3}{2} = -\frac{k}{2(k+1)}, \tag{7.134}$$

we can conclude that

$$\zeta(2n+1) = -\lambda_n \sum_{k=1}^{n} z_{k,n} \left(k \sum_{j=1}^{\infty} \frac{\zeta(2j)}{(j+1)(j+1+k)} - \frac{k}{2(k+1)} \right). \tag{7.135}$$

We further have

$$\zeta(2n+1) = -\lambda_n \left(\sum_{j=1}^{\infty} \zeta(2j) \left(\frac{\sum_{k=1}^{n} z_{k,n}}{j+1} - \frac{z_{1,n}}{j+2} - \cdots - \frac{z_{n,n}}{j+n+1} \right) - \sum_{k=1}^{n} \frac{k}{2(k+1)} z_{k,n} \right). \tag{7.136}$$

It follows that

$$\begin{aligned} \sum_{k=1}^{n} z_{k,n} &= \sum_{k=1}^{n} (2n-2k-1) \binom{2n+2}{2k+2} B_{2(n-k)} = \sum_{m=0}^{n-1} (2m-1) \binom{2n+2}{2m} B_{2m} \\ &= \sum_{m=0}^{n-1} \left((2n+1) \binom{2n+2}{2m} - (2n+2) \binom{2n+1}{2m} \right) B_{2m} = -z_{0,n}, \end{aligned} \tag{7.137}$$

because of

$$\sum_{m=0}^{n} \binom{2n+2}{2m} B_{2m} = n+1, \quad \sum_{m=0}^{n} \binom{2n+1}{2m} B_{2m} = n + \frac{1}{2}. \tag{7.138}$$

Based on the above equations, we can conclude that

$$\zeta(2n+1) = \lambda_n \left(\sum_{j=1}^{\infty} \zeta(2j) \left(\frac{z_{0,n}}{j+1} + \frac{z_{1,n}}{j+2} + \cdots + \frac{z_{n,n}}{j+n+1} \right) + \sum_{k=1}^{n} \frac{k}{2(k+1)} z_{k,n} \right). \tag{7.139}$$

For the denominator t_n of the quotient $\tau_n = s_n/t_n$, we can write

$$\begin{aligned}
-t_n &= \sum_{j=1}^{\infty}\left(\frac{z_{0,n}}{j+1}+\frac{z_{1,n}}{j+2}+\cdots+\frac{z_{n,n}}{j+n+1}\right)\\
&= z_{0,n}\sum_{j=1}^{\infty}\left(\frac{1}{j+1}-\frac{1}{j+2}\right)+(z_{0,n}+z_{1,n})\sum_{j=1}^{\infty}\left(\frac{1}{j+2}-\frac{1}{j+3}\right)+\cdots\\
&\quad+(z_{0,n}+z_{1,n}+\cdots+z_{n-1,n})\sum_{j=1}^{\infty}\left(\frac{1}{j+n}-\frac{1}{j+n+1}\right)\\
&= \sum_{r=2}^{n+1}\frac{1}{r}\sum_{k=0}^{r-2}z_{k,n}=\sum_{k=0}^{n-1}z_{k,n}\sum_{r=k+2}^{n+1}\frac{1}{r}=\sum_{k=0}^{n-1}z_{k,n}(H_{n+1}-H_{k+1})\\
&= -z_{n,n}H_{n+1}-\sum_{k=0}^{n-1}z_{k,n}H_{k+1}=-\sum_{k=0}^{n}z_{k,n}H_{k+1}.
\end{aligned}\tag{7.140}$$

For the numerator s_n, we have

$$\sum_{k=1}^{n}\frac{k}{2(k+1)}z_{k,n}=\frac{1}{2}\sum_{k=1}^{n}\left(z_{k,n}-\frac{z_{k,n}}{k+1}\right)=\frac{1}{2}\left(-z_{0,n}-\sum_{k=1}^{n}\frac{z_{k,n}}{k+1}\right)=-\frac{1}{2}\sum_{k=0}^{n}\frac{z_{k,n}}{k+1},\tag{7.141}$$

from which all the other relations lead us to the end of the proof of (7.120). □

In [208], another serial representation of $\zeta(3)$ was given, which played a key role in the glorious proof of the irrationality of $\zeta(3)$ carried out by *Roger Aperi*. Determining the values of $\zeta(3)$, $\zeta(5)$, and so on [215, 216] is of great importance for many fields of engineering and practical applications, such as electrostatics, theory of elastic deformations, and quantum field theory. Symbolic and numerical calculations performed in the Mathematica software package (version 4.0) for Linux showed, among other things, that only 50 elements of one of these series can offer the accuracy of seven decimal places.

After recalculating based on formula (7.120), we have [232]

$$\zeta(3)=2\zeta(2)\sum_{j=1}^{\infty}\left(\frac{1}{j+1}-\frac{1}{j+2}\right)\left(\zeta(2j)-\frac{1}{2}\right),\tag{7.142}$$

$$\zeta(5)=2\zeta(4)\sum_{j=1}^{\infty}\left(\frac{3}{j+1}-\frac{5}{j+2}+\frac{2}{j+3}\right)\left(\zeta(2j)-\frac{7}{10}\right),\tag{7.143}$$

$$\zeta(7)=\zeta(6)\sum_{j=1}^{\infty}\left(\frac{10}{j+1}-\frac{21}{j+2}+\frac{14}{j+3}-\frac{3}{j+4}\right)\left(\zeta(2j)-\frac{41}{50}\right),\tag{7.144}$$

$$\zeta(9)=\frac{2}{3}\zeta(8)\sum_{j=1}^{\infty}\left(\frac{21}{j+1}-\frac{50}{j+2}+\frac{42}{j+3}-\frac{15}{j+4}+\frac{2}{j+5}\right)\left(\zeta(2j)-\frac{399}{442}\right),\tag{7.145}$$

$$\zeta(11)=\frac{2}{5}\zeta(8)\sum_{j=1}^{\infty}\left(\frac{90}{j+1}-\frac{231}{j+2}+\frac{220}{j+3}-\frac{99}{j+4}+\frac{22}{j+5}-\frac{2}{j+6}\right)\left(\zeta(2j)-\frac{1629}{1690}\right).\tag{7.146}$$

Further, for every natural n, we have

$$\zeta(2n+1) = (-1)^{n-1} 2 \frac{(2\pi)^{2n}}{(2n+1)!} \left(\sum_{j=1}^{\infty} \left(\frac{v_{0,n}}{2j+1} + \frac{v_{1,n}}{2j+3} + \cdots + \frac{v_{n,n}}{2j+2n+1} \right) (\zeta(2j) - \mu_n) \right), \tag{7.147}$$

where

$$v_{k,n} = (2^{2(n-k)} - 2) \binom{2n+1}{2k+1} B_{2(n-k)} \quad (k = 0, 1, 2, \ldots, n), \tag{7.148}$$

$$\mu_n = -\frac{1}{2} \frac{\sum_{k=0}^{n} \frac{v_{k,n}}{2k+1}}{\sum_{k=0}^{n} v_{k,n} H^*_{2k+1}},$$

$$H^*_{2k+1} = 1 + \frac{1}{3} + \cdots + \frac{1}{2k+1} \quad (k = 0, 1, 2, \ldots).$$

The proof of this formula is essentially the same as that of (7.120). Here we start from formula (7.67), and to find the inverse matrix, we use the function $2xe^x/(e^{2x}-1)$ and the equality for *Bernoulli*'s numbers given in *Nielsen*'s book and in [212]:

$$\sum_{j=1}^{2k} 2^j \binom{2k}{j} B_j = (2 - 2^{2k}) B_{2k}. \tag{7.149}$$

In particular, we have

$$\zeta(3) = 8\zeta(2) \sum_{j=1}^{\infty} \left(\frac{1}{2j+1} - \frac{1}{2j+3} \right) (\zeta(2j) - 1), \tag{7.150}$$

$$\zeta(5) = 8\zeta(4) \sum_{j=1}^{\infty} \left(\frac{7}{2j+1} - \frac{10}{2j+3} + \frac{3}{2j+5} \right) \left(\zeta(2j) - \frac{16}{13} \right), \tag{7.151}$$

$$\zeta(7) = 8\zeta(6) \sum_{j=1}^{\infty} \left(\frac{31}{2j+1} - \frac{49}{2j+3} + \frac{21}{2j+5} - \frac{3}{2j+7} \right) \left(\zeta(2j) - \frac{121}{94} \right), \tag{7.152}$$

$$\zeta(9) = \frac{8}{3}\zeta(8) \sum_{j=1}^{\infty} \left(\frac{381}{2j+1} - \frac{620}{2j+3} + \frac{294}{2j+5} - \frac{60}{2j+7} + \frac{5}{2j+9} \right) \left(\zeta(2j) - \frac{2216}{1703} \right). \tag{7.153}$$

Bibliography

[1] Euler–Goldbach. Correspondance. 1742–1743. Bd. I., 163–200.

[2] Euler, L.: *Corresp. Math. Phys.* I, 191 (1743); *Novi commentarii Academiae Petropolitanae*, Bd. 20 (1775/1776).

[3] Euler, L.: *Novi Comm. Acad. Petropolitenae* 14, 158–159 (1769/1770).

[4] Euler, L.: Meditationes circa singulare serierum genus. *Novi Comm. Acad. Sci. Petropolitanae* 20, 140–186 (1775).

[5] Euler, L.: *Acta Acad. Petrop.* 1751/II; *Nova Acta Acad. Petrop.*, 4 (1786) 1789.

[6] Legendre A., M.: *Traité des Fonctions Elliptiques et des Intégrales Eulériennes*, vol. 2. Huzard-Courcier, Paris (1826).

[7] Kummer, E. J.: Beitrag zur Theorie der Funktion $\Gamma(x)$. *J. Reine Angew. Math.* 35, 1–4 (1847).

[8] Dirichlet, L.: Über die Bestimmung der mittleren Werte in der Yahlentheorie. *Abh. Akad. Berlin Math. Abh.* 2, 49–66 (1849). Werke.

[9] Riemann, B.: Über die Anyahl der Primyahlen unter einer gegebenen Grösse. *Berliner Monatsberichte*, 1859 (reprint 1892), S. 671–680; Werke, vol. 2, 145–155.

[10] Boole, G. A.: *Treatise on the Calculus of Finite Differences*, 2nd edn. Macmillan and Company, London (1872) (Reprinted by Dover Publications, New York, 1960).

[11] Glaisher, J. W. L.: On the history of Euler's constant. *Messenger Math.* 1, 25–30 (1872).

[12] Schlömilch, O.: Über einige unendliche Reihen. *Ber. Verh. K. Sächs. Ges. Wiss. Leipz.* 29, 101–105 (1877).

[13] Stieltjes, T. J.: Table des valeurs des sommes $S_k = \sum_1^\infty n^{-k}$. *Acta Math.* 10, 299–302 (1887).

[14] Glaisher, J. W.: On the series which represent the twelve elliptic and the four zeta functions. *Messenger Math.* 18, 1–84 (1889).

[15] Hadamard, J.: Étude sur les proprietes des fonctions entieres et en particulier d'une fonction consideree par Riemann. *J. Math. Pures Appl., Ser. 4* 9, 171–215 (1893).

[16] Von Mangoldt, H.: Auszug aus einer Arbeit unter dem Titel: Zu Riemann's Abhandlung "Über die Anyalh der Primyahlen unter einer gegebenen Grõsse", 337–350, 883–896. Berlin (1894).

[17] Mangoldt H. Zu, v.: Riemann's Abhandlung "Über die Anzahl der Primzahlen unter einer gegebenen Grösse". *J. Reine Angew. Math.* 114, 255–305 (1895).

[18] Hadamard, J.: Sur la distribution des zéros de la fonction cet ses conséquences arithmétiques. *Bull. Soc. Math. Fr.* 24, 199–220 (1896).

[19] De la Vallée Poussin, Ch.: Recherches analytiques sur la théorie des nombres premiers. Premiére partie: La fonction $\zeta(s)$ de Riemann et les nombres premiers en général. *Ann. Soc. Sci. Brux.* 20, 183–256 (1896).

[20] De la Vallée Poussin, Ch.: Sur la fonction $\zeta(s)$ de Riemann et le nombre des nombres premiers inférieurs á une limite donnée. *Mém. Couronnés Autres Mém. Publiés Acad. R. Sci. Lett. Beaux-Arts Belg.* 59(1), 1–74 (1899–1900).

[21] Voronoi, G.: Sur une problemé du calcul des fonctions asymptotiques. *J. Math.* 126, 241–282 (1903).

[22] Von Mangoldt, H.: Zur Verteilung der Nullstellen der Riemannschen Funktion $\zeta(s)$. *Math. Ann.* 60, 1–19 (1905).

[23] Johnson, W. W.: Note on the numerical transcendents S_n and $S_n = S_n - 1$. *Bull. Am. Math. Soc.* 12, 477–482 (1905–1906).

[24] Nielsen, N.: *Handbuch der theorie der Gammafunktion*. Teubner, Leipzig (1906). Reprint: Chelsea, New York (1965), p. 47.

[25] Landau, E.: *Handbuch der Lehre von der Verteilung der Primzahlen*. Teubner, Leipzig (1909). Reprint: Chesea, New York (1953).

[26] Landau, E.: Über die Anzahl der Gitterpunkte in gewissen Bereichen. *Nachr. Königl. Ges. Wiss. Gött.* 6, 687–771 (1912).

[27] Hardy, G. H.: Sur les zéros de la fonction $\zeta(s)$ de Riemann. *C. R. Acad. Sci.* 158, 1012–1014 (1914).

https://doi.org/10.1515/9783112233276-008

[28] Landau, E.: Über die Hardysche Eutdeckung unendlich vieler Nullstellen der Zeta-funktion mit reelen Teil 1/2. *Math. Ann.* 76, 212–243 (1915).

[29] Ramanujan, S.: A series for Euler's constant γ. *Messenger Math.* 46, 73–80 (1916–1917).

[30] Backlung, R. J.: Über die Nullstellen der Riemannschen Zeta-funktion. *Acta Math.* 41, 345–375 (1918).

[31] Hardy, G. H., Littlwood, J. E.: Contributions to the theory of the Riemann zeta-function and the theory of the distribution of primes. *Acta Math.* 41, 119–196 (1918).

[32] Hardy, G. H., Littlwood, J. E.: The zeros of Riemann's zeta-function on the critical line. *Math. Z.* 10, 283–317 (1921).

[33] Hardy, G. H., Littlwood, J. E.: The approximate functional equation in the theory of the zeta-function, with applicationss to the divisor problems of Dirichlet and Piltz. *Proc. Lond. Math. Soc.* 21(2), 39–74 (1922).

[34] Wilton, J. R.: A proof of Burnside's formula for $\log \Gamma(x+1)$ and certain allied properties of Riemann's ζ-funcition. *Messenger Math.* 52, 90–93 (1922–1923).

[35] Nielsen, N.: *Elémentaire des Nombres de Bernoulli*. Gauthier-Villars et Cie, Paris (1923).

[36] Rao Bhimasena, M., Aiyar Venkatarama, M.: On some infinite series and products, Part I. *J. Indian Math. Soc.* 15, 150–162 (1923–1924).

[37] Malurkar, S. L.: On the application of Herr Mellin's integral to some series. *J. Indian Math. Soc.* 16, 130–138 (1925–1926).

[38] Bromwich, T. J. I. A.: *An Introduction to the Theory of Infinite Series*, 2nd edn. Macmillan and Company, London (1926).

[39] Kluyver, J. C.: On certain series of Mr. Hardy. *Q. J. Pure Appl. Math.* 50, 185–192 (1927).

[40] Whittaker, E. T., Watson, G. N.: *A Course of Modern Analysis*, 4th edn. Cambridge University Press, Cambridge (1927).

[41] Watson, G. N.: Theorems stated by Ramanujan (II): theorems on summation of series. *J. Lond. Math. Soc.* 3, 216–225 (1928).

[42] Ramaswami, V.: Notes on Riemann's ζ-function. *J. Lond. Math. Soc.* 9, 165–169 (1934).

[43] Rutledge, G., Douglas, R. D.: Evaluation of $\int_0^1 (\log u/u) \log^2(1+\mu) du$ and related integrals. *Am. Math. Mon.* 41, 29–36 (1934).

[44] Selberg, A.: On the zeros of Riemann's zeta-function. *Skr. Nor. Vidensk.-Akad. Oslo* 10, 1–59 (1942).

[45] Selberg, A.: On the zeros of Riemann's zeta-function on the critical line. *Arch. Math. Naturvidensk.* 45, 101–114 (1942).

[46] Bromwich, T. J.: *An Introduction to the Theory of Infinite Series*. Macmillan and Co., London (1949).

[47] Hardy, G. H.: *Divergent Series*. Oxford University Press, Oxford (1949).

[48] Левин, В.И.: По поводу одноj задачи С. Рамануджана. *Усп. Мат. Наук* V(3), 161–166 (1950).

[49] Torheim, L.: Harmonic double series. *Am. J. Math.* 72, 303–314 (1950).

[50] Titchmarsh, E. C.: *The Theory of the Riemann Zeta-function*. Clarendon Press, Oxford (1951).

[51] Knopp, K.: *Theory and Application of Infinite Series*, second English edn. Hafner Publishing Company, New York (1951) (Translated from the Second German ed. and Revised in accordance with the Fourth German ed. by Young R. C. H).

[52] Klamkin, M. S.: Problem 4431. Soln. by R. Steinberg. *Am. Math. Mon.* 59, 471–472 (1952).

[53] Chandrasekharan, K.: *Lectures on the Riemann Zeta-function*. Tata Institute of Fundamental Research, Bombay (1953).

[54] Erdelyi, A., et al.: *Bateman Manuscript Project. Higher Transcendental Functions, vol. I*. McGraw-Hill Book Co., New York (1953).

[55] Landau, E.: *Handbuch der Lehre von der Verteilung der Primzahlen*, 2nd edn. Chelsea Publishing Company, New York (1953).

[56] Williams, G. T.: A new method of evaluating $\zeta(2n)$. *Am. Math. Mon.* 60, 19–25 (1953).

[57] Hjortnaes, M. M.: Overforing av rekken $\sum_{k=1}^{\infty} \frac{1}{k^3}$ til et bestemt integral. In: *Proc. 12th Cong. Scand. Maths*, Lund 10–15 Aug. 1953. Lund (1954).

[58] Riemann, B.: *Gesammelte Werke*, 2nd edn. Teubner, Leipzig (1892). Reprint: Dover Books, New York (1953).

[59] Apostol, T. M.: Some series involving the Riemann zeta function. *Proc. Am. Math. Soc.* 5, 239–243 (1954).

[60] Briggs, W. E., Chowla, S.: The power series coefficients of $\zeta(s)$. *Am. Math. Mon.* 62, 323–325 (1955).

[61] Robbins, H.: A remark on Stirling's formula. *Am. Math. Mon.* 62, 26–29 (1955).

[62] Klamkin, M. S.: Problem 4564. Solns. by J. V. Whittaker and the proposer. *Am. Math. Mon.* 62, 129–130 (1955).

[63] Lehmer, D. H.: Extented computation of the Riemann Zeta-function. *Mathematika* 3, 102–108 (1956).

[64] Lehmer, D. H.: On the roots of the Riemann Zeta-function. *Acta Math.* 95, 291–298 (1956).

[65] Ramanujan, S.: *Notebooks of Srinivasa Ramanujan (2 volumes)*. Tata Institute of Fundamental Research, Bombay (1957).

[66] Mordell, L. J.: On the evaluation of some multiple series. *J. Lond. Math. Soc.* 33, 368–371 (1958).

[67] Виноградов, И.М.: Новая оценка функции $\zeta(1+it)$. *Изв. Акад. Наук СССР* 22, 161–164 (1958).

[68] Carlitz, L.: Note on the integral of the product of several Bernoulli polynomials. *J. Lond. Math. Soc.* 34, 361–363 (1959).

[69] Menon, P. K.: Summation of certain series. *J. Indian Math. Soc.* XXV, 121–128 (1961).

[70] Menon, P. K.: Some series involving the zeta function. *Math. Stud.* 29, 77–80 (1961).

[71] Fuchs, W. H. J.: Problem 4917, Proposed by Lewin L. *Am. Math. Mon.*, June–July 580–581 (1961).

[72] Verma, D. P.: A note on Euler's constant. *Math. Stud.* 29, 140–141 (1961).

[73] Chowla, S., Hawkins, D.: Asymptotic expansions of some series involving the Riemann Zeta-function. *J. Indian Math. Soc.* XXV, 115–124 (1962).

[74] Уиттекер, Э. Т., Ватсон, Дж. Н.: *Курс Современного Анализа*. Госуд. Издат-во Физ.-Мат. Литературы, Москва (1963).

[75] Abramowitz, M., Stegun, I. A.: *Handbook of Mathematical Functions with Formulas, Graphs, and Mathematical Tables*. National Bureau of Standards, Washington, D. C. (1964).

[76] Chrystal, G.: *Algebra: An Elementary Text-Book (Part II)*, 7th edn. Chelsea Publishing Company, New York (1964).

[77] Barnes, E. R., Kaufman, W. E.: The Euler–Mascheroni constant. *Am. Math. Mon.* 72, 1023 (1965).

[78] Jordan, C.: *Calculus of Finite Differences*, 3rd edn. Chelsea Publishing Company, New York (1965).

[79] Gupta, H.: An identity. *Res. Bull. Panjab Univ. (N. S.)* 15, 347–349 (1964/1965).

[80] Jordan, C.: *Calculus of Finite Differences*. Chelsea, New York (1965).

[81] Grosswald, E.: *Topics from the Theory of Numbers*. Macmillan Co., New York (1966).

[82] Magnus, W., Oberhettinger, F., Soni, R. P.: *Formulas and Theorems for the Special Functions of Mathematical Physics*, 3rd enlarged edn. Springer-Verlag, New York (1966).

[83] Greenberg, R., Marsh, D. C. B., Danese, A. E.: A zeta-function summation. *Am. Math. Mon.* 74, 80–81 (1967).

[84] Прахар, К.: *Распределение Простых Чисел*. Мир, Москва (1967).

[85] Lehner, J., Newman, M.: Sums involving Farey fractions. *Acta Arith.* 15, 181–187 (1968/1969).

[86] Adamović, D. D., Tasović, M. R.: Monotony and the best possible bounds of some sequences of sums. *Univ. Beogr. Publ. Elektroteh. Fak. Ser. Mat. Fiz.* 247(273), 41–50 (1969).

[87] Grosswald, E.: Die Werte der Riemannschen Zeta-function an ungeraden Argumentstellen. *Nachr. Akad. Wiss. Götingen* 9–13 (1970).

[88] Holme, F.: En enkel beregnig av $\sum_{k=1}^{\infty}\frac{1}{k^2}$. *Nord. Mat. Tidskr.* 91–92 (1970).

[89] Sherman Lehman, R.: on the distribution of zeros of Riemann Zeta-function. *Proc. Lond. Math. Soc.* 20, 303–320 (1970).

[90] Монтгомери, Х. Л.: Нули L-функции. *Математика* 14(5), 133–140 (1970).

[91] Стечкин, С.Б.: О некоторых экстремальных свойствах положительных тригонометрических полиномов. *Мат. Заметки* 7(4), 411–422 (1970).

[92] Стечкин, С.Б.: О нулях дзета функции Римана. *Мат. Заметки* 8(4), 419–429 (1970).

[93] Девенпорт, Г.: *Мультипликативная Теория Чисел*. Наука, Москва (1971).
[94] Градштейн, М.С., Рыжик, И.М.: *Таблицы Интегралов, Сумм, Рядов и Произведений*. Наука, Москва (1971).
[95] Berndt, B. C.: On the Hurwity zeta-function. *Rocky Mt. J. Math.* 2, 151–157 (1972).
[96] Гелјфонд, А.О.: *Исчисление Конечних Разностеj*. Наука, Москва (1971).
[97] Grosswald, E.: Comments on some formulae of Ramanujan. *Acta Arith.* XXI, 25–34 (1972).
[98] Grosswald, E.: Remarks concerning the values of the Riemann zeta function at integral, odd arguments. *J. Number Theory* 4, 225–235 (1972).
[99] Huxley, M. N.: On the difference between consecutive primes. *Invent. Math.* 15, 164–170 (1972).
[100] Карацуба, А.А.: Равномернаjа оценка остаточного члена в проблеме делителеj Дирихле. *Изв. Акад. Наук СССР* 36, 475–483 (1972).
[101] Montgomery, H. L.: The pair correlation of zeros of the zeta-function. In: *Analytic Number Theory (Proc. Sympos. Pure Math.)*, vol. XXIV, pp. 181–193. St. Louis Univ., St. Louis (1972).
[102] Серр, Ж.П.: *Курс Арифметики*. Мир, Москва (1972).
[103] Виноградов, И.М.: *Основи Теории Чисел*. Наука, Москва (1972).
[104] Apostol, T. M.: Another elementary proof of Euler's formula for $\zeta(2n)$. *Am. Math. Mon.* 80(4), 425–431 (1973).
[105] Berndt, C. B.: Generalized Dedekind eta-functions and generalized Dedekind sums. *Trans. Am. Math. Soc.* 178, 495–508 (1973).
[106] Katayama, K.: On Ramanujan's formula values of Riemann zeta-function at positive old integers. *Acta Arith.* XXII, 149–155 (1973).
[107] Jordan, P. F.: Infinite sums of psi functions. *Bull. Am. Math. Soc.* 79, 681–683 (1973).
[108] Papadimitrou, I.: A simple proof of the formula $\sum_{k=1}^{\infty} k^{-2} = \frac{\pi^2}{6}$. *Am. Math. Mon.* 80(4), 424–425 (1973).
[109] Smart, J. R.: On the values of the Epstein zeta function. *Glasg. Math. J.* 14, 1–12 (1973).
[110] Melzak, Z. A.: *Companion to Concrete Mathematics, vol. I: Mathematical Techiques and Various Applications*. John Wiley and Sons, New York (1973).
[111] Ayoub, R.: Euler and the Zeta function. *Am. Math. Mon.* 81(10), 1067–1086 (1974).
[112] Чандрасекхаран, К.: *Введение в Аналитическую Теорию Чисел*. Мир, Москва (1974).
[113] Levinson, N.: At least one-third of zeros of Rieman's zeta-function are on $\sigma = (1/2)$. *Proc. Natl. Acad. Sci. USA* 71(4), 1013–1015 (April 1974).
[114] Levinson, N.: Generalization of recent method giving lower bound for $N_0(T)$ of Rieman's zeta-function. *Proc. Natl. Acad. Sci. USA* 71(10), 3984–3987 (1974).
[115] Levinson, N.: Zeros of derivative of Rieman's ξ-function. *Bull. Am. Math. Soc.* 80(5), 951–954 (1974).
[116] Levinson, N.: More than one third of zeros of Rieman's zeta-function are on $\sigma = (1/2)$. *Adv. Math.* 13, 383–436 (1974).
[117] Gunter, B.: Problem 5885, Proposed by Haring F. and Nelson G. T. *Am. Math. Mon.* 2, 180–181 (February 1974).
[118] Suryanarayana, D.: Sums of the Riemann zeta function. *Math. Stud.* XLII(2), 141–143 (1974).
[119] Чандрасекхаран, К.: *Арифметические Функции*. Наука, Москва (1975).
[120] Levinson, N.: Deduction of semi-optimal mollifier obtaining lower bound for $N_0(T)$ of Rieman's zeta-function. *Proc. Natl. Acad. Sci. USA* 72(1), 294–297 (January 1975).
[121] Rosser, J. B., Schoenfeld, L.: Sharper bounds for the Chebyshev functions $\theta(x)$ and $\psi(x)$. *Math. Comput.* 29, 243–269 (1975).
[122] Туран, П.: *О Некоторых Новых Результатах в Аналитической Теории Чисел*. Сб. Математика. Мир, Москва (1975).
[123] Hansen, E. R.: *A Table of Series and Products*. Prentice-Hall, Englewood Cliffs, New Jersey (1975).
[124] Klambauer, G.: *Mathematical Analysis*. Marcel Dekker, New York (1975).
[125] Лаврик, А.Ф.: О главном члене проблемы делителей и степенном ряде дзета-функции Римана в окрестности ее полюса. *Тр. Мат. Инст. Стеклова АН СССР* 142, 165–173 (1976).
[126] Apostol, T. M.: *Modular Functions and Dirichlet Series in Number Theory*. Springer-Verlag, Berlin (1976).

[127] Apostol, T. M.: *Introduction to Analytic Number Theory*. Springer-Verlag, New York (1976).

[128] Adamović, D. D.: An identity and asymptotic behavior of some sequences. *Univ. Beogr. Publ. Elektroteh. Fak. Ser. Mat. Fiz.* 544–576, 149–154 (1976).

[129] Ashcroft, N. W., Mermin, N. D.: *Solid State Physics*. Saunders College, Philadelphia (1976).

[130] Berndt, C. B.: Modular transformations and generalizations of several formulae of Ramanujan. *Rocky Mt. J. Math.* 7(1), 147–189 (Winter 1977).

[131] *Изобранние Задачи (The Otto Dunkel Memorial Problem Book), Problem №275*. Мир, Москва (1977).

[132] Jutila, M.: Zero density estimates for L functions. *Acta Arith.* 32, 55–62 (1977).

[133] Klusch, D.: On certain infinite series. *Arch. Math.* 29(3), 280–283 (1977).

[134] Balasubramanian, R.: An improvement of a theorem of Titchmarsh on mean square of $|\zeta((1/2)+it)|$. *Proc. Lond. Math. Soc.* 36, 540–576 (1978).

[135] Полиа, Г., Сеге, Г.: *Задачи и Теоремы из Анализа, Ч. II*. Наука, Москва (1978).

[136] Beukers, F.: A note on the irritionality of $\zeta(2)$ and $\zeta(3)$. *Bull. Lond. Math. Soc.* 11, 268–272 (1979).

[137] Heath-Brown, D. R.: Simple zeros of the Riemann zeta-function on the critical line. *Bull. Lond. Math. Soc.* 11, 17–18 (1979).

[138] Matsuoka, Y.: On the values of the Riemann Zeta Function at half integers. *Tokyo J. Math.* 2(2), 371–377 (1979).

[139] Мозер, Я.: Об одноj теореме Харди–Литтлвуда в теории дзета-функции Римана. *Acta Arith.* 31, 45–51 (1976) Добавление: Acta Arith 35, 403–404 (1979).

[140] Kanemitsu, S.: On some sums involving Farey fractions. *Math. J. Okayama Univ.* 20, 101–113 (1978). Corrigendum. ibid., 22, 223–224 (1980).

[141] Van der Poorten, A.: A proof that Euler missed. *Math. Intell.* 1, 195–203 (1979).

[142] Sita Rama Chandra Rao, R., Siva Rama Sarma, A.: Some identities involving the Riemann Zeta Function. *Indian J. Pure Appl. Math.* 10, 602–607 (May 1979).

[143] Sitaramachandrarao, R., Sivaramasarma, A.: Some identities involving the Riemann zeta-function. *Indian J. Pure Appl. Math.* 10, 602–607 (1979).

[144] Ivić, A.: Exponent pairs and zeta-function of Riemann. *Studia Sci. Math. Hung.* 15, 157–181 (1980).

[145] Der van, P. A. J.: Some wonderful formulas … an introduction to polylogarithms. In: *Proceedings of the Queen's Number Theory Conference*. Queen's Papers Pure Appl., vol. 54, pp. 269–286 (1980).

[146] Лаврик, А.Ф., Исраилов, М.И., Едгоров, Ж.: Об интегралах, содержащих остаточный член проблемы делителей. *Acta Arith.* 37, 381–389 (1980).

[147] Sitaramachandrarao, R., Sivaramasarma, A.: Two identities due to Ramanujan. *Indian J. Pure Appl. Math.* 11, 1139–1140 (1980).

[148] Sivaramasarma, A.: *Some problems in the theory of Farey series and the Euler totient function*. Doctoral dissertation, Andhra University, India, Waltair (1980).

[149] Титчмарш, Е.: *Теория Функций*. Наука, Москва (1980).

[150] Toyoizumi, M.: Formulae for the Riemann Zeta Function at half integers. *Tokyo J. Math.* 3(1), 177–186 (1980).

[151] Виноградов, И.М.: *Метод Тригонометрических Сумм в Теории Чисел*. Наука, Москва (1980).

[152] Gradshteyn, I. S., Ryzhik, I. M.: *Table of Integrals, Series, and Products*, corrected and enlarged edn. Academic Press, New York (1980).

[153] Cohen, H.: Generalisation d'une construction de R. Apery. *Bull. Soc. Math. Fr.* 109, 269–281 (1981).

[154] Heath-Brown, D. R.: The mean values of the Riemann zeta function. In: *Proc. Conf. on the Progress in Analytic Number Theory, vol. 1*, Durham, pp. 115–119 (1979–1981).

[155] Исраилов, М.И.: О разложении Лорана дзета-функции Римана. *Тр. Мат. Инст. Стеклова АН СССР* 158 98–104 (1981).

[156] Карацуба, А.А.: О расстоянии между соседними нулями дзета-функции Римана лежащими на критической прямой. *Тр. Мат. Инст. Стеклова АН СССР* 157, 49–64 (1981).

[157] Leshchiner, D.: Some new identities for $\zeta(k)$. *J. Number Theory* 13(3), 355–362 (1981).

[158] Richmond, B., Szekeres, G.: Some formulas related to dilogarithms, the zeta function and the Andrews–Gordon identities. *J. Aust. Math. Soc. A* 31(3), 362–373 (1981).

[159] Lewin, L.: *Polylogarithms and Associated Functions*. North-Holland, New York (1981).

[160] Kolesnik, G.: On the order of $\zeta(\frac{1}{2} + it)$ and $\triangle(R)$. *Pac. J. Math.* 98(1), 107–122 (1982).

[161] Bruckman, P. S.: Problem H-320. *Fibonacci Q.* 20, 186–187 (1982).

[162] Lewin, L.: The dilogarithm in algebraic fields. *J. Aust. Math. Soc. A* 33(3), 302–330 (1982).

[163] Matsuoka, Y.: Generalizations of Ramanujan's formulae. *Acta Arith.* 41, 19–26 (1982).

[164] Matsuoka, Y.: One the values of a certain Dirichlet series at rational integers. *Tokyo J. Math.* 5(2), 399–403 (1982).

[165] Anderson, R. J.: Simple zeros of the Riemann zeta-function. *J. Number Theory* 17(2), 176–182 (1983).

[166] Bernd, V. S., Jochi, P. T.: *Infinite series identities, transformations and evaluations*. Ramanujan's Second Notebook. Contemporary Mathematics, vol. 23. Amer. Math. Soc., Providence R. I. (1983). Chap. 9.

[167] Georghiou, S., Philipou, A. N.: Harmonic sums and the zeta-function. *Fibonacci Q.* 21(1), 29–36 (1983).

[168] Berndt, B. C.: Chapter 8 of Ramanujan's second notebook. *J. Reine Angew. Math.* 338, 1–55 (1983).

[169] Jutila, M.: Zeros of the zeta function near the critical line. In: *Stud. Pure Math. Mem. Paul Turán*, Budapest (1983).

[170] Карацуба, A.A.: *Основы Аналитической Теории Чисел*. Наука, Москва (1983).

[171] Мозер, Я.: Улучшение теоремы Харди–Литтлвуда о плотности нулей функции $\zeta(\frac{1}{2} + it)$. *Acta Math. Univ. Comen. Bratislava* 42–43, 41–50 (1983).

[172] Singh, R. J., Verma, D. P.: Some series involving Riemann zeta function. *Yokohama Math. J.* 31, 1–4 (1983).

[173] Verma, D. P., Kaur, A.: Summation of some series involving Riemann zeta function. *Indian J. Math.* 25(2), 181–184 (May 1983).

[174] Галочкин, А.И., Нестеренко, Ю.В., Шидловский, А.Б.: *Введение в Терию Чисел*. Изд МГУ, Москва (1984).

[175] Apostol, T. M., Vu, T. H.: Dirichlet series related to the Riemann zeta-function. *J. Number Theory* 19(1), 85–102 (1984).

[176] Sitaramachandrarao, R., Subbarao, M. V.: Transformation formulae for multiple series. *Pac. J. Math.* 113(2), 471–479 (1984).

[177] Ivić, A.: A zero-density theorem for the Riemann zeta-function. *Тр. Мат. Инст. Стеклова АН СССР* 163, 85–89 (1984).

[178] Карацуба, А.А.: О нулях функции $\zeta(s)$ на коротких промежутках критической прямой. *Изв. Акад. Наук СССР* 48, 569–584 (1984).

[179] Карацуба, А.А.: Метод тригонометрических сумм И. М. Виноградова. *Тр. Мат. Инст. Стеклова АН СССР* 163, 97–103 (1984).

[180] Loxton, J. H.: Special values of the dilogarithm function. *Acta Arith.* 43, 155–166 (1984).

[181] Матиясевич, Ю.В.: Одно аналитическое представление для суммы величин, обратных к нетривиальным нулям дзета-функции Римана. *Тр. Мат. Инст. Стеклова АН СССР* 163, 181–182 (1984).

[182] Привалов, И.И.: *Введение в Теорию Функции Комплексного Переменного*, Издание Тринадцатое. Наука, Москва (1984).

[183] Мозер, Я.: Некоторые следствия из формулы Римана–Зигеля. *Тр. Мат. Инст. Стеклова АН СССР* 163, 183–185 (1984).

[184] Conrey, J. B., Ghosh, A.: A simpler proof of Levinson's theorem. *Math. Proc. Camb. Philos. Soc.* 97(3), 385–395 (1985).

[185] Карацуба, А.А.: Дзета-функция Римана и ее нули. *Усп. Мат. Наук* 40(5), 19–70 (1985).

[186] Kolesnik, G.: On the method of exponent pairs. *Acta Arith.* 45(2), 115–143 (1985).

[187] Lehmer, D. H.: Interesting series involving the central binomial coefficient. *Am. Math. Mon.* 92(7), 449–457 (Aug.–Sept. 1985).

[188] Van de Lune, J., Te Riele, H. J. J., Winter, D. T.: On the zeros of the Riemann zeta function in the critical strip. *Cent. Wiskd. Inform.* 46(174), 667–681 (June 1985).

[189] Subbarao, M. V., Sitaramachandrarao, R.: On some infinite series of J. L. Mordell and their analoques. *Pac. J. Math.* 119(1), 245–255 (1985).

[190] Zucker, I. J.: On the series $\sum_{k=1}^{\infty}\sum_{k=1}^{\infty}\binom{2k}{k}^{-1}k^{-n}$ and related sums. *J. Number Theory* 20(1), 92–102 (1985).

[191] Ivić, A.: *The Riemann Zeta-Function.* John Wiley and Sons, New York (1985).

[192] Nan-Yue, Z. H., Shun-Yan, Z. H.: Riemann zeta-function. *Contemp. Math.* 48, 235–241 (1985).

[193] Ляшко, И.И., Борячук, А.К., Гай, Г., Головчак, Г.Н.: *Справочное Пособие по Математическому Анализу, Ряды, Функции Векторного Аргумента, Кратные и Криволинейные Интегралы.* Виша Школа, Киев (1986).

[194] Christophe, L. M. Jr., Finn, M. V., Crow, J. A.: Two series involving zete function values. *Math. Mag.* 59, 176–178 (1986).

[195] Shallit, J. D., Zikan, K.: A theorem of Goldbach. *Am. Math. Mon.* 93, 402–403 (1986).

[196] Sitaramachandrarao, R.: A formula of S. Ramanujan. *J. Number Theory* 25(1), 1–19 (1987).

[197] Srivastava, H. M.: A unified presentation of certain classes of series of the Riemann zeta function. *Riv. Mat. Univ. Parma* 14(4), 1–23 (1988).

[198] Stevanović, M.: *Riemann-ova funkcija i njena primena na izračunavanje suma i integrala.* Magistarski rad, PMF Sarajevo (1988).

[199] Исраилов, М.И.: О сумме рядов, содержащих значения дзета-функций Риманова и Гурвица в целых точках. *Вопр. Вычисл. Прикл. Мат. Ташк.* 86, 64–69 (1989).

[200] Nan-Yao, Z. H.: Euler constant and some sums associated with the Riemann zeta function. *Math. Pract. Theory* 4, 62–70 (1990).

[201] Dušan, A.: Corrections and supplements of some details in two former papers. *Univ. Beogr. Publ. Elektroteh. Fak. Ser. Mat.* 2, 42–48 (1991).

[202] Stevanović, M.: *Višestruko sumiranje, Rimanova zeta funkcija i primene.* Doktorska disertacija, PMF Beograd (1993).

[203] Ewell, J. A.: On the Zeta function values $\zeta(2k+1)$; $k = 1, 2, \ldots$. *Rocky Mt. J. Math.* 25(3), 1003–1012 (1995).

[204] Zhang, N.-Y., Williams, K. S.: Values of the Riemann Zeta function and integrals involving $\log(2\sinh\theta/2)$ and $\log(2\sin\theta/2)$. *Pac. J. Math.* 168(2), 271–289 (1995).

[205] Dabrowski, A.: A note on the values of the Riemann Zeta function at positive odd integers. *Nieuw Arch. Wiskd. (Ser. 4)* 14(2), 199–207 (1996).

[206] Cvijovic, D., Klinowski, J.: New rapidly convergent series representations for $\zeta(2n+1)$. *Proc. Am. Math. Soc.* 125(5), 1263–1271 (1997).

[207] Srivastava, H. M.: Certain families of rapidly convergent series representations for $\zeta(2n+1)$. *Math. Sci. Res. Hot-Line* 1(6), 1–6 (1997) (Research Announcement).

[208] Chen, M.-P., Srivastava, H. M.: Some families of series representations for the Riemann $\zeta(3)$. *Result. Math.* 33, 179–197 (1998).

[209] Srivastava, H. M.: Further series representations for $\zeta(2n+1)$. *Appl. Math. Comput.* 97(1), 1–15 (1998).

[210] Srivastava, H. M.: Some rapidly converging series for $\zeta(2n+1)$. *Proc. Am. Math. Soc.* 127(2), 385–396 (1999).

[211] Srivastava, H. M.: Some simple algorithms for the evaluations and representations of the Riemann Zeta function at positive integer arguments. *J. Math. Anal. Appl.* 246(2), 331–351 (2000).

[212] Srivastava, H. M.: Some formulas for the Bernoulli and Euler polynomials at rational arguments. *Math. Proc. Camb. Philos. Soc.* 129(1), 77–84 (2000).

[213] Srivastava, H. M., Glasser, M. L., Adamchik, V. S.: Some definite integrals associated with the Riemann Zeta function. *Z. Anal. Anwend.* 19(3), 831–846 (2000).

[214] Borwein, J. M., Bradley, D. M., Crandall, R. E.: Computational strategies for the Riemann Zeta function. *J. Comput. Appl. Math.* 121(1), 247–296 (2000).

[215] Srivastava, M., Tsumura, H.: A certain class of rapidly convergent series representations for $\zeta(2n+1)$. *J. Comput. Appl. Math.* 118(1–2), 323–335 (2000).

[216] Srivastava, H. M., Tsumura, H.: New rapidly convergent series representations for $\zeta(2n+1)$, $L(2n;\chi)$ and $L(2n+1;\chi)$. *Math. Sci. Res. Hot-Line* 4(7), 17–24 (2000) (Research Announcement).

[217] Srivastava, H. M., Choi, J.: *Series Associated with the Zeta and Related Functions*. Kluwer Academic Publishers, Dordrecht, Boston and London (2001).

[218] Ivić, A.: *The Riemann Zeta-Function. Theory and Applications*. Dover Publications, Inc., Mineola, NY (2003). Reprint of the 1985 original, Wiley, New York.

[219] Srivastava, H. M., Tsumura, H.: Inductive construction of rapidly convergent series representations for $\zeta(2n+1)$. *Int. J. Comput. Math.* 80(9), 1161–1173 (2003).

[220] Choi, J., Cho, Y. J., Srivastava, H. M.: Series involving the Zeta and multiple Gamma functions. *Appl. Math. Comput.* 159(2), 509–537 (2004).

[221] Shiroishi, M., Takahashi, M.: Exact calculation of correlation functions for spin-1/2 Heisenberg chain. *J. Phys. Soc. Jpn.* 74, 47–52 (2005).

[222] Choi, J., Srivastava, H. M.: Explicit evaluation of Euler and related sums. *Ramanujan J.* 10, 51–70 (2005).

[223] Alzer, H., Karayannakis, D., Srivastava, H. M.: Series representations for some mathematical constants. *J. Math. Anal. Appl.* 320(1), 145–162 (2006).

[224] Segarra, E.: *An exploration of the Riemann zeta function and its application to the theory of prime number distribution*. PhD Dissertation, Harvey Mudd College, USA (2006).

[225] Babolian, E., Hajikandi, A. A.: Numerical computation of the Riemann zeta function and prime counting function by using Gauss–Hermite and Gauss–Laguerre quadratures. *Int. J. Comput. Math.* 87(15), 3420–3429 (2010).

[226] Connon, D. F.: Some infinite series involving the Riemann zeta function. *Int. J. Math. Comput. Sci.* 7(1), 11–83 (2012).

[227] Lima, F. M. S.: A simpler proof of a Katsurada's theorem and rapidly converging series for $\zeta(2n+1)$ and $\beta(2n)$. arXiv:1203.5660v2 (2012).

[228] Srivastava, H. M., Choi, J.: *Zeta and q-Zeta Functions and Associated Series and Integrals*. Elsevier Science Publishers, Amsterdam, London and New York (2012).

[229] Choi, J.: Rapidly converging series for $\zeta(2n+1)$ from Fourier series. *Abstr. Appl. Anal.* 2014, 457620 (2014), 9 pages. https://doi.org/10.1155/2014/457620.

[230] Choi, J., Srivastava, H. M.: Series involving the Zeta functions and a family of generalized Goldbach–Euler series. *Am. Math. Mon.* 121(3), 229–236 (2014).

[231] Qiang, L., Wang, Z.: Numerical calculation of the Riemann zeta function at odd-integer arguments: a direct formula method. *Math. Sci.* 9, 39–45 (2015).

[232] Stevanović, M., Petrović, P.: The Euler–Riemann zeta function in some series formulae and its values at odd integer points. *J. Number Theory* 180, 769–786 (2017).

[233] Ivić, A.: A note on the zeros of the zeta-function of Riemann. arXiv:1801.04711 (2018).

[234] Bondarenko, A., Ivić, A., Saksman, E., Seip, K.: On certain sums over ordinates of zeta-zeros II. arXiv: 1808.01763 (2018).

Index

https://doi.org/10.1515/9783112233276-009

www.ingramcontent.com/pod-product-compliance
Lightning Source LLC
LaVergne TN
LVHW081315110826
845149LV00006B/1509

9783112233269